Aamir Raina
Samiullah Khan
Editors

Faba Beans

Breeding, Production, Cultivation and Uses

nova
science publishers
www.novapublishers.com

DOI: https://doi.org/10.52305/XIAI4185

NOTICE TO THE READER

The Publisher has taken reasonable care in the preparation of this book, but makes no expressed or implied warranty of any kind and assumes no responsibility for any errors or omissions. No liability is assumed for incidental or consequential damages in connection with or arising out of information contained in this book. The Publisher shall not be liable for any special, consequential, or exemplary damages resulting, in whole or in part, from the readers' use of, or reliance upon, this material. Any parts of this book based on government reports are so indicated and copyright is claimed for those parts to the extent applicable to compilations of such works.

Independent verification should be sought for any data, advice or recommendations contained in this book. In addition, no responsibility is assumed by the Publisher for any injury and/or damage to persons or property arising from any methods, products, instructions, ideas or otherwise contained in this publication.

The Publisher assumes no responsibility for any statements of fact or opinion expressed in the published contents.

This publication is designed to provide accurate and authoritative information with regard to the subject matter covered herein. It is sold with the clear understanding that the Publisher is not engaged in rendering legal or any other professional services. If legal or any other expert assistance is required, the services of a competent person should be sought. FROM A DECLARATION OF PARTICIPANTS JOINTLY ADOPTED BY A COMMITTEE OF THE AMERICAN BAR ASSOCIATION AND A COMMITTEE OF PUBLISHERS.

Additional color graphics may be available in the e-book version of this book.

Library of Congress Cataloging-in-Publication Data

ISBN: 979-8-89530-447-1 (Hardcover)
ISBN: 979-8-89530-554-6 (eBook)

Published by Nova Science Publishers, Inc. † New York

Contents

Chapter 1

Faba Beans: Origin, History, Taxonomy and Breeding

Nedaravalli Manjappa Kavyashree[1,*]
Vishal Reddy[2]
and Halehundi Bellegowda Sharada[2]
[1]PhD Scholar, Department of Genetics and Plant Breeding, College of Agriculture, University of Agricultural Sciences Dharwad, Karnataka, India
[2]PhD Scholar, Department of Genetics and Plant Breeding, College of Agriculture, Keladi Shivappa Nayaka University of Agricultural and Horticultural Sciences Shivamogga, Karnataka, India

Abstract

Faba bean (*Vicia faba* L.), one of the world's oldest domesticated crops, has played a crucial role in agricultural systems and human nutrition for millennia. This chapter presents a comprehensive overview of faba bean's history, origin, and taxonomic classification, synthesizing current knowledge and ongoing debates in the field. The crop's domestication history spans approximately 14,000 years, with archaeological evidence placing its earliest cultivation in the southern Levant. While the precise origin and wild progenitor remain subjects of scholarly debate, the Near East is widely recognized as a primary center of origin. Afghanistan and Ethiopia serve as secondary centers of diversity. Taxonomically, *V. faba* belongs to section Faba within the subgenus *Vicia*, though its unique characteristics have led some taxonomists to propose its classification as a monotypic genus. The chapter also explores the crop's global distribution patterns, germplasm diversity, and evolution of breeding

* Corresponding Author's Email: kavyamnaik@gmail.com

In: Faba Beans
Editors: Aamir Raina and Samiullah Khan
ISBN: 979-8-89530-447-1

methods - from traditional approaches to modern techniques like genomic selection and speed breeding. This review highlights faba bean's significance in sustainable agriculture and its potential role in addressing future food security challenges.

Keywords: faba bean, *Vicia faba*, faba bean taxonomy, faba bean origin, faba bean history

Introduction

Faba bean, recognized as broad bean or *Vicia faba*, is a legume crop adapted to cooler seasons and has been under cultivation for many years for its edible seeds. It is indigenous to the Mediterranean area and southern Asia and belongs to the Fabaceae family (Smykal et al., 2015). Faba beans are cultivated worldwide due to their nutritional value, versatility, and ability to fix atmospheric N2 in the soil (Singh et al., 2013; Karkanis et al., 2018). It is a diploid outcrossing species (2n = 12) with a genome size of approximately 13 Gb distributed on six chromosomes (Carrillo-Perdomo et al., 2020). Biologically, faba beans are annual plants that can reach a height of 1 to 1.5 meters and produce clusters of white flowers that produce long pods. Each pod typically contains 3 to 8 seeds, which are usually large and flattened, with a green or tan color. The plants possess a tap root system (Link et al., 2008). They are adaptable to a diverse range of soil types but thrive better under well-drained soils with a pH between 6 and 7. Cool climate favours crop growth, and thus, they are typically sown in early spring or autumn, depending on the region. Owing to the excellent N2 fixing ability (130-160 kg N/ha) and foliage cover, faba beans are often used as a cover crop to improve soil fertility, control weeds, and prevent erosion (Singh et al., 2013). Faba beans are rich in protein content (20-41%), making them an important source of plant-based protein (Dangi et al., 2022). They are also rich in dietary fiber, vitamins (such as vitamin K, vitamin B6, and folate), and minerals (including iron, manganese, and potassium). They are relatively low in fat and contain no cholesterol (Dhull et al., 2022). However, antinutritional substances such as phytic acid, lectins, trypsin inhibitors, saponins, condensed tannins, and favism-inducing compounds, reduce their nutritional value (Karkanis et al., 2018; Rahate et al., 2021). Faba beans have diverse culinary uses. They can be eaten fresh when young and tender, similar to green beans, or harvested when mature and dried for storage. Dried faba beans are commonly used in

soups, stews, and curries, and they are ground into flour for use in several cuisines (Allaire and Brady, 2010). Despite faba beans being consumed by humans for a long time, they are primarily used as animal feed in many developed nations (Labba et al., 2021). In addition, the economic, environmental, and health implications associated with the production and consumption of animal products have generated increased research and public interest in sustainable diets, which involve exploring alternatives to animal-based diets. This shift in dietary preferences has also influenced the utilization of faba beans, as research and practical initiatives promote a transition from resource-intensive animal proteins to sustainable diets that emphasize higher consumption of plant-based proteins (Augustin and Cole, 2022).

In addition, faba beans have the potential to address the needs of the gluten-free industry. They can serve as vegetable-based ingredients, offering sustainable, nutritional, economic, and health benefits in the form of flour, starch, fibre, and protein (Ofoedu et al., 2022). Although faba bean-based products are currently limited and not widely popular, it is popularly recognized that processing and refining pulses into isolates and individual components can enhance the value of the raw product (Tsolakis et al., 2019). This chapter aims to present an overview of faba beans with an emphasis on their history, origin, and taxonomical hierarchy, considering numerous benefits of faba beans and growing research interest.

History

Faba bean is one of the oldest crops in the world (Singh et al., 2013) and probably introduced into agriculture in the late Neolithic period (Link et al., 2008). The cultivation of faba bean dates back a long time, making it one of the earliest food legumes to be domesticated. The oldest faba bean discovered so far was found approximately 14,000 years ago in the southern Levant (Avramidou et al., 2023). Subsequently, its cultivation has spread across the world (Fouad et al., 2013). Cubero, in 1973 and 1974, documented that it extended its reach to central Europe and the Eurasiatic plains, following pathways through Anatolia, the Danube Valley, and the Caucasus. It also expanded to North Africa and the Iberian Peninsula *via* the Mediterranean coast and islands. From Mesopotamia, it made its way to India, and it is believed that it may have reached Ethiopia through the Sabean Route, with likely activity in the second millennium BC. The Near East saw its spread to China, probably during the first millennium AD, where only landraces of the

major type were known before the 20th century. In the 16th century, European colonizers introduced the crop to America (Torres et al., 2011). Around 5,000 years ago, the Chinese utilized these crops as a food source, while the Egyptians began cultivating them about 3,000 years ago. The Hebrews also cultivated them during biblical times, followed by the Greeks and Romans at a slightly later period. It is probable that Europeans introduced these crops to India as a garden crop during the Sultanic period (1206-1555), as historical references mention their cultivation during this era (Singh et al., 2013). The term "faba" derived from a variation of the Greek verb φάγέώ, meaning "to eat," emphasizing its historical association with food and animal feed in ancient Greek and Roman cultures (Duc et al., 2010).

Currently, faba beans play a significant role as a primary crop in numerous nations, notably China, Ethiopia, and Egypt. Additionally, they are extensively cultivated for human consumption across the Mediterranean region and certain areas of Latin America. The beans serve as a staple food source and a valuable cash crop in various regions of the world (Baye 2021).

Origin

Information regarding the wild progenitor and the specific region where *Vicia faba* originated from and the subsequent stage of its domestication is limited and is an ongoing debate (Karkanis et al., 2018). Duc et al. (2010) reviewed about the origin of faba beans - Near East served as the central point of origin for the genus *Vicia*, with four distinct routes originating from this center: (1) spreading to Europe, (2) following the north African coast towards Spain, (3) extending along the Nile to Ethiopia, and (4) from Mesopotamia reaching India (Cubero, 1973 and 1974). The secondary centers of diversity of faba beans are speculated as Afghanistan and Ethiopia (Fouad et al., 2013). In contrast, Ladizinsky (1975) advocated central Asia as its origin. Hajjar and Hodgkin (2007) also supported Ladizinsky (1975). However, Muratova (1931) and Maxted (1995) reported that southeastern Europe and southwestern Asia were the center of origin for the genus *Vicia*.

The precise origin and wild ancestor of the faba bean are still uncertain. Several wild species, such as *Vicia narbonensis, Vicia johannis* and *Vicia galilaea* that are closely related to the cultivated faba bean have a chromosome count of 2n = 14, whereas the cultivated faba bean has a chromosome count of 2n = 12 (Singh et al., 2013). Also, their DNA content is half of the faba bean, and even their protein content is different (Adhikari et al., 2021). *V.*

johannis and *V. narbonensis* have valuable agricultural traits, including the ability to withstand winter conditions, resistance to black aphids (*Aphis fabae*), and protection against chocolate spot disease (*Botrytis fabae*). However, interspecific crosses of *V. faba* with these useful *Vicia* species have not been successful (Maalouf et al., 2018). This might be due to post-zygotic barriers, which stop embryo development, and hence no viable seed could be formed (Adhikari et al., 2021).

The existing crop gene pool of *V. faba* exhibits significant genetic variations. Over 38,000 collections of faba bean genetic resources are preserved worldwide in multiple gene banks, including the International Center for Agricultural Research in Dry Areas (ICARDA) (Karkanis et al., 2018). Although there is a notable genetic similarity among Vicia species within eco-geographical areas, some variations can be observed with respect to plant architecture, leaf size and shape, seed colour, seed shape and size. Based on seed size, the faba beans can be grouped into three classes viz., (a) faba bean small-seeded forms in south-western Asia, (b) large-seeded forms in the west, and (c) the third group, medium-seeded type. Medium seeded type is an ancient type, found in large areas and can be traced back to Neolithic times (Maalouf et al., 2018). Muratova (1931) called these groups based on seed size as subspecies of *Vicia faba viz., minor* (small seed), *major* (large seed) and *equine* (medium seed). However, these sub-groups were cross-compatible, and hence, they were considered to be one species but different botanical groups by Cubero (1973).

Taxonomy

There are approximately 150 species in the genus *Vicia*, which is classified into the subgenera *Vicia* and *Vicilla.* The subgenus *Vicia* is composed of five sections, one of which is section *Faba*. Whereas, *Vicilla* consists of 18 sections. Section *Faba* is unique among the other sections due to several characteristics: the stipules have a nectariferous spot, the calyx is sub-regular, the legume has parallel sutures, the leaflets are few in number (1-3 pairs) and usually longer than 2 cm, the lateral veins on the leaflets are not prominent and curve towards the apex. Section *Faba* encompasses a small number of species, such as *V. faba, V. bithynica, V. narbonensis* L., and four similar forms that are sometimes recognized as distinct species. Nevertheless, some taxonomists argue that *Vicia faba* should be considered a monotypic genus

because it is distinguished by possessing an inflated pod with spongy partitions between the seeds and the absence of tendrils (Birch et al., 1985).

Vicia faba is an annual herb that possesses coarse upright stems. The leaves contain 2 to 6 leaflets, each up to 8 cm long, and are alternate and pinnate. *V. faba* lacks tendrils or possesses only rudimentary ones. Despite its profuse flowering habit, it yields a small percentage of pods. The flowers appear white with dark purple markings and are larger in size. Around 1-5 flowers form clusters on each axillary raceme, typically between the 5th and 10th node. Each cluster produces 1-4 pods. The plants typically possess indeterminate growth habit, although some determinate mutants exist. Bumblebees serve as the primary pollinators for approximately 30% of cross-fertilized plants in a population. *Vicia faba* boasts a robust taproot system with extensively branched secondary roots (Singh et al., 2013).

Distribution of Faba Bean

It is believed to be one of the oldest cultivated plants, with evidence of its domestication dating back to around 10,000 years ago in the Mediterranean region. Today, faba beans are widely distributed across various continents and adapted to diverse climates and regions. The faba bean is a traditional crop in Europe that is still vital to the continent's agricultural systems in nations like Spain, France, United Kingdom and Italy (Lepse et al., 2017). It is particularly well-suited for cultivation in cool temperate climates. In these regions, faba beans are often grown as a winter-spring crop or as a cover crop to fix nitrogen in the soil. The faba bean has been grown for many years and is still a vital component of agriculture and food in North Africa, the Middle East, and the Mediterranean region. In this region, major faba bean producers include Egypt, Ethiopia, Morocco, Algeria, and Tunisia. The crop's great nutritional value, wide range of culinary applications, and capacity to boost soil fertility make it highly valued.

In Asia, faba bean is grown in countries like India, China, Pakistan, and Afghanistan. In China, it is widely cultivated in the northern provinces, serving as an important protein source for humans and livestock. Similarly, it is an essential food crop in India and Pakistan, especially in regions with cooler climates. In Oceania, faba bean is grown in Australia, New Zealand, and some Pacific islands (Pietrzak et al., 2016). Australia is a major producer of faba bean, and the crop has gained popularity due to its ability to fix nitrogen in the soil and its potential as a rotational crop with cereals. In some

parts of the Americas, faba beans are also cultivated, albeit to a lesser extent. Countries like Canada, the United States, Argentina, and Chile grow faba bean primarily for livestock feed and as a cover crop to improve soil health. The distribution of faba beans is not limited to large-scale commercial farming; it is also grown in smallholder and subsistence farming systems in various regions. Its adaptability to different agroecological conditions and its ability to fix nitrogen makes it a valuable crop for sustainable agricultural practices (Zhao et al., 2018). In summary, faba bean is a widely distributed crop, grown across multiple continents and adapted to various climates and agricultural systems. Its significance lies in its rich nutritional content, versatility in culinary uses, and its role in enhancing soil fertility through N2 fixation. As global interest in sustainable and nutritious food sources grows, faba bean continues to be a valuable and essential crop for food security and sustainable agriculture.

Diversity in Germplasm

The faba bean has a diverse and rich germplasm encompassing an extensive range of genetic variation and is of great significance to farmers, researchers, and breeders. This diversity is found in various aspects of faba bean germplasm, including genetic, geographic, morphological, phenotypic, and end-use traits. Faba bean germplasm is known for its high level of genetic diversity, with different landraces, wild relatives, and cultivated varieties displaying distinct genetic characteristics. For example, accessions from Ethiopia's highlands exhibit unique genetic traits that confer drought tolerance, while those from Spain's coastal regions may possess traits related to salt tolerance (Shiran et al., 2014).

Geographic Diversity: Faba bean is cultivated in diverse geographic regions, leading to a wide array of germplasm. For instance, varieties from the Mediterranean basin, such as those grown in Morocco and Tunisia, are adapted to the region's mild winters and hot summers. In contrast, accessions from the Himalayan region, like those in Nepal and India, have evolved to withstand cooler temperatures and higher altitudes.

Morphological Diversity: Faba bean germplasm shows considerable morphological variation. Different accessions may vary in plant height, leaf size, flower color, pod shape, and seed size. For example, some landraces from Ethiopia are known for their tall and vigorous growth, while certain Chinese varieties exhibit compact plant architecture suitable for high-density planting.

Phenotypic Diversity: Faba bean germplasm displays various phenotypic traits related to various physiological and biochemical attributes. This includes resistance to pests and diseases, tolerance to environmental stresses like drought or heat, and nutritional content. For example, some accessions may have enhanced resistance to rust or chocolate spot diseases, while others may have higher protein content in their seeds.

Climatic Adaptation: Faba bean germplasm includes varieties adapted to different climatic conditions. Varieties from Mediterranean regions like Italy and Greece are well-suited for temperate climates. At the same time, accessions from North Africa, such as those from Egypt or Algeria, are better adapted to warmer and drier conditions.

Diversity in End Uses: Faba bean germplasm encompasses varieties with diverse end-use characteristics. Some accessions are primarily used for human consumption, such as whole beans or processed into flour or protein concentrates. For example, the "Aquadulce" type is popular for human consumption due to its large seeds and mild flavor. Other varieties are primarily grown as livestock feed, providing valuable protein for animals. Faba bean germplasm exhibits variations in nutritional traits, such as protein content, amino acid composition, mineral content, and anti-nutritional factors. For instance, some accessions may have higher protein content and balanced amino acid profiles, making them valuable sources of nutrition for human diets.

Conserving and utilizing the diverse germplasm of faba bean is essential for maintaining genetic resilience, improving crop productivity, and addressing food security challenges. International gene banks and germplasm collections, like those managed by the International Center for Agricultural Research in the Dry Areas (ICARDA) and national agricultural research institutes, play a crucial role in preserving and sharing faba bean germplasm. Through collaborative efforts, researchers and breeders can access this diversity to develop improved cultivars that meet the evolving needs of farmers and consumers.

Faba Bean Breeding Methods

Faba bean, a partially allogamous species, can be bred using line breeding, population breeding, and hybrid breeding approaches. The extent of outcrossing varies widely among genotypes (10-70%) and is influenced by environmental factors (Bishop et al., 2016). Modern faba bean cultivars are

developed using methods like mass selection, recurrent selection, and conventional techniques such as pedigree selection, single-seed descent line breeding, or creating synthetic cultivars. Parental lines are selected based on their pedigree and phenotypic characteristics in these breeding methods (Brünjes and Link, 2021).

Mutagenesis

Mutagenesis in faba bean breeding offers a promising approach to enrich its genetic diversity and develop improved cultivars. Mutations induced by physical (e.g., gamma-ray) or chemical (e.g., ethyl methane sulphonate-EMS) mutagens can lead to novel traits that may be beneficial for faba bean production (Nurmansyah et al., 2020). The faba bean, a self-fertilized crop, has recently exhibited notable results through induced mutagenesis (Gulfishan et al., 2015; Khursheed et al., 2018). The biological system's sensitivity and mutagenic potential are often assessed using biological damage from an earlier generation. Faba beans are extremely responsive to physical and chemical mutagens (Khursheed et al., 2019). Given this information, Vandana et al. (1996), Kumar et al. (1993), and Laskar and Khan (2014) investigated the mutagenic responsiveness of faba bean. They found that the use of mutagens had a negative impact on germination, seedling growth, pollen fertility, maturation time, and survival rate. One of the key objectives of seeking mutants is to increase resistance against root parasitic weeds like broomrape and stem parasitic dodders, which have become problematic in the Mediterranean basin. Developing faba bean varieties with enhanced resistance to these weeds can significantly improve faba bean production in the region. Additionally, resistance against herbicides used to control these weeds is also sought as a valuable trait. Another target for mutation breeding is to develop mutants that exhibit decreased pod wall thickness, which could increase grain yield. Such traits are valuable for improving the economic and agronomic performance of faba bean crops. As of the time of writing, 20 faba bean mutant varieties have already been developed by various countries listed in the International Atomic Energy Agency (IAEA) mutant variety database (International Atomic Energy Agency, 2021). These mutants can serve as valuable genetic resources for breeding programs seeking to incorporate novel traits and improve faba bean germplasm. Furthermore, integrating mutagenesis with biotechnology tools can accelerate the release of novel faba bean germplasm and improved cultivars. Biotechnology techniques, such as

marker-assisted selection and gene editing, can help identify and manipulate specific genes of interest, complementing traditional breeding approaches (Mejri et al., 2018).

Line Breeding

Half of the genetic variance in line cultivar breeding lies between the resulting lines from potential crosses and the other half lies between these crosses. The choice of parents is crucial, and it is important to eliminate crosses that show lower performance before investing resources in testing their offspring. Single Seed Descent (SSD) involves selecting and releasing new cultivars based on the performance of individual lines, while pedigree-type selection judges the yield potential of entire crosses (families) at earlier generations (F3 and F4) and eliminates inferior crosses before investing further in testing their offspring. The low propagation coefficient of faba beans limits breeding and yield testing. Controlled self-fertilization or isolation is necessary to multiply lines. One faba bean plant produces a limited amount of seed, which affects the size of observation trials and testing plots (Moehring et al., 2013).

Hybrid Breeding

Development of hybrid cultivars requires the combination of two lines; faba bean hybrid breeding faces three major obstacles. Firstly, there is a need for a reliable and efficient cytoplasmic male sterility (CMS) system, as the existing ones show inadequate male sterility and frequent reversion to fertility. Secondly, appropriate pollinators are required, as some insects tend to steal nectar without pollinating flowers effectively, affecting seed set in hybrid production (Maalouf et al., 2016). Lastly, faba bean's low propagation coefficient necessitates multiple generations of seed increase, making it challenging to reach the desired quantity for certified seed production. To compensate for the lack of CMS and gametocides, breeders have utilized monogenic traits as markers; despite their creativity, a significant breakthrough in faba bean hybrid breeding is yet to be achieved (Marzinzig et al., 2018).

Synthetic Breeding

In this approach, cultivars are "resynthesized" from four or more founder lines, aiming to exploit heterosis for improved agronomic performance. The selection of inbred lines as components of a synthetic population is based on their varietal ability, which is linked with general combining ability and breeding value. Varietal ability is realized by the performance of the line's polycross-progeny or top-cross progeny. The discrepancy between the per se performance (performance of individual lines) and the performance of their open-pollinated progenies (polycross-progenies or top cross-progenies) is attributed to several reasons. One major source of discrepancy is the line-specific General Combining Ability (GCA), which influences the performance of the progenies. Additionally, the extent of self-fertilization of candidate lines and paternal outcrossing success also contribute to the variation in progeny performance, making it line-specific (Ghaouti et al., 2008; Ghaouti and Link, 2009). The process of seed production for inbred lines and open-pollinated progenies differs, as inbred lines' seed is produced in cages, and progenies' seed is produced in the open field, leading to differences in seed size, seed quality, and resulting crop stands for non-genetic reasons. Polycross and top cross are two options to assess the performance of candidate lines' progenies, with both serving similar purposes. A top cross might be easier to conduct, as the pollinator (tester) is sown in strips, mainly facilitating cross-pollination with the candidates according to their individual extent of outcrossing (Brünjes and Link, 2021).

Speed Breeding

Speed breeding is an essential tool in modern breeding programs, aiming to increase the genetic gains per year and accelerate the development of improved cultivars. By shortening the reproductive cycle, breeders can achieve multiple generations in a year, allowing for faster selection and breeding progress. This technique has significantly succeeded in pulse crops like chickpeas and faba beans (Cazzola et al., 2021). In faba bean, an *in vivo* speed breeding protocol has been developed, utilizing cytokinin or cold treatment to increase pollen viability and seed set. This leads to a 22-day reduction in the breeding cycle from seed to seed. Further advances in faba bean speed breeding have been achieved by regulating light and temperature

in Australia, enabling the production of five generations per year. However, there are some limitations to speed breeding protocols. The small amount of seed produced on speed-bred plants and the expenses associated with maintaining controlled environments make speed breeding more suitable for the primary cross/backcross and inbreeding generations (Bhatta et al., 2021). Additional research is needed to explore the possibility of extending speed breeding applications to other breeding steps.

Marker Assisted Selection

Marker-assisted selection (MAS) is a valuable tool in faba bean breeding, as it allows for indirect selection of traits of interest using genetic markers that are closely linked to those traits. This approach is particularly useful when phenotyping for certain traits is difficult, expensive, or time-consuming. One practical example of MAS is the case of vicine and convicine (v-c), which are antinutritional factors limiting the usage of faba bean seeds (Khazaei et al., 2019; Tacke et al., 2021). The v-c gene was mapped to a specific region on chromosome 1, and a diagnostic SNP assay was developed for use in selection. This marker allows for straightforward foreground selection of low v-c genotypes; this is especially important because low v-c faba bean cultivars are safe for individuals susceptible to favism and are preferred for feed conversion efficiency in animals. Similarly, other DNA markers have been developed for traits like low seed coat tannins, growth habit, rust resistance (Uvf1 and Uvf2 genes), and herbicide tolerance (imidazolinone). These markers have been integrated into faba bean breeding programs, enabling more efficient selection of desired traits. The discovery of the biosynthetic pathway of v-c and the availability of DNA markers for various traits in faba bean have enhanced the efficiency of breeding efforts (Zanotto et al., 2020). Nevertheless, there is still potential for further development of DNA markers for other important traits, such as resistance to chocolate spot and ascochyta blight. The continuous discovery and validation of new DNA markers for various biotic and abiotic stresses in faba beans will greatly benefit breeding programs, allowing for a more targeted and efficient selection of desirable traits.

Genomic Selection

In this approach, a genetically elite and diverse breeding population is created, where individuals are heterozygous due to the partial allogamy of the faba bean. The breeding population is genotyped with a high number of representative individuals. Genotypic data helps predict varietal ability, and selection is applied between chipped seeds (Gaynor et al., 2017). The selected chipped seed is then directly allowed to propagate further through open pollination without any inbreeding or multiplication phase. By conducting recurrent genomic selection, where multiple cycles of selection are performed within a year, it is possible to achieve rapid turnover and accelerate the breeding process in faba bean. Overall, genomic selection holds great potential for accelerating faba bean breeding, increasing genetic gains, and shortening breeding cycles. However, to fully realize its benefits, a highly cost-effective genome-wide genotyping platform is required, which would further optimize the selection process and enhance the development of improved cultivars in a shorter time span.

The integration of low-cost genome-wide genotyping and high-throughput phenotyping platforms has opened up new possibilities for accelerating the selection process for complex traits in faba bean breeding. Genomic selection, a black-box approach focusing on multigenic traits, has shown great promise in improving the efficiency of recurrent selection within base populations of allogamous crop species like faba bean (Müller, 2017). Genomic selection allows for more precise estimation of breeding values for individual plants compared to the heritability of their performance. This precision enables genomic selection based on single plants, significantly reducing breeding cycles. For faba bean breeding, implementing genomic selection may lead to a cross-to-cross cycle length of 6 months to 1 year, significantly increasing genetic gains per unit time.

Conclusion

The review of literature on the faba bean's history, origin, and taxonomy reveals a crop of remarkable antiquity and significance whose importance continues to grow in contemporary agriculture. The extensive history of faba bean cultivation, spanning over 14,000 years and multiple civilizations, underscores its adaptability and enduring value in human agriculture. While

questions about its precise origin and wild progenitor remain unresolved, ongoing research and modern genomic tools are gradually illuminating these aspects of its evolutionary history.

The diversity in faba bean germplasm, manifested through variations in morphological, phenotypic, and end-use traits, represents a valuable genetic resource for crop improvement. This diversity, combined with advancing breeding methodologies, i.e., from traditional approaches to cutting-edge techniques like speed breeding and genomic selection, positions the faba bean as a crucial crop for future food security.

As the world seeks sustainable protein sources and climate-resilient crops, faba bean's nitrogen-fixing ability, nutritional value, and adaptability make it an increasingly important crop for future agricultural systems. The continuing development of improved cultivars through various breeding approaches will be crucial in addressing challenges related to yield, disease resistance, and nutritional quality. Future research should focus on resolving taxonomic uncertainties, expanding our understanding of its genetic diversity, and developing more efficient breeding strategies to harness its full potential as a sustainable food source.

References

Adhikari, K. N., Khazaei, H., Ghaouti, L., Maalouf, F., Vandenberg, A., Link, W. and O'Sullivan, D. M., 2021. Conventional and molecular breeding tools for accelerating genetic gain in faba bean (*Vicia faba* L.). *Frontiers in plant science*, *12*, p. 744259.

Allaire, H. and Brady, T., 2010. Classification and botanical description of legumes. *Retrieved from*.

Augustin, M. A. and Cole, M. B., 2022. Towards a sustainable food system by design using faba bean protein as an example. *Trends in Food Science & Technology*.

Avramidou, E., Ganopoulos, I., Mylona, P., Abraham, E. M., Nianiou-Obeidat, I., Osathanunkul, M. and Madesis, P., 2023. Comparative analysis of the genetic diversity of faba bean (*Vicia faba* L.). *Sustainability*, *15*(2), p. 1016.

Baye, E., 2021. Biosciences and Plant Biology. *Int. J. Curr. Res. Biosci. Plant Biol*, *8*(3), pp. 15-24.

Bhatta, M., Sandro, P., Smith, M. R., Delaney, O., Voss-Fels, K. P., Gutierrez, L., et al. (2021). Need for speed: manipulating plant growth to accelerate breeding cycles. *Curr. Opin. Plant Biol.* 60:101986.

Birch, A. N. E., Tithecott, M. T. and Bisby, F. A., 1985. Vicia johannis and wild relatives of the faba bean: a taxonomic study. *Economic Botany*, *39*(2), pp. 177-190.

Bishop, J., Jones, H. E., O'Sullivan, D. M., and Potts, S. G. (2016). Elevated temperature drives a shift from selfing to outcrossing in the insect-pollinated legume, faba bean (*Vicia faba*). *J. Exp. Bot.* 68, 2055–2063.

Brünjes, L., and Link, W. (2021). Paternal outcrossing success differs among faba bean genotypes and impacts breeding of synthetic cultivars. *Theor. Appl. Genet.* 134, 2411–2427.

Carrillo-Perdomo, E., Vidal, A., Kreplak, J., Duborjal, H., Leveugle, M., Duarte, J., Desmetz, C., Deulvot, C., Raffiot, B., Marget, P. and Tayeh, N., 2020. Development of new genetic resources for faba bean (*Vicia faba* L.) breeding through the discovery of gene-based SNP markers and the construction of a high-density consensus map. *Scientific Reports*, *10*(1), p. 6790.

Cazzola, F., Bermejo, C. J., Gatti, I., and Cointry, E. (2021). Speed breeding in pulses: an opportunity to improve the efficiency of breeding programs. *Crop Pasture Sci.* 72, 165–172.

Cubero, J. I., 1973. Evolutionary trends in *Vicia faba. Theoretical and applied genetics.* 43, 59–65.

Cubero, J. I., 1974. On the evolution of *Vicia faba* L. *Theoretical and applied genetics.* 45, 47–51.

Dangi, P., Chaudhary, N., Paul, A., Prabha, S., Kumar, R. and Poonia, A., 2022. Faba Bean Proteins: Extraction Methods, Properties and Applications. *Faba Bean: Chemistry, Properties and Functionality*, pp. 245-273.

Dhull, S. B., Kidwai, M. K., Noor, R., Chawla, P. and Rose, P. K., 2022. A review of nutritional profile and processing of faba bean (*Vicia faba* L.). *Legume Science*, *4*(3), pp. 129.

Duc, G., Bao, S., Baum, M., Redden, B., Sadiki, M., Suso, M. J., Vishniakova, M. and Zong, X., 2010. Diversity maintenance and use of *Vicia faba* L. genetic resources. *Field Crops Research*, *115*(3), pp. 270-278.

Fouad, M., Mohammed, N., Aladdin, H., Ahmed, A., Xuxiao, Z., Shiying, B. and Tao, Y., 2013. Faba bean. In *Genetic and genomic resources of grain legume improvement* (pp. 113-136). Elsevier.

Gaynor, R. C., Gorjanc, G., Bentley, A. R., Ober, E. S., Howell, P., Jackson, R., et al., (2017). A two-part strategy for using genomic selection to develop Inbred lines. *Crop Sci.* 57, 2372–2386.

Ghaouti, L., and Link, W. (2009). Local vs. formal breeding and inbred line vs. synthetic cultivar for organic farming: case of *Vicia faba* L. *Field Crops Res.* 110, 167–172.

Ghaouti, L., Vogt-Kaute, W., and Link, W. (2008). Development of locally-adapted faba bean cultivars for organic conditions in Germany through a participatory breeding approach. *Euphytica* 162, 257–268.

Gulfishan, M., Bhat, T. A., and Oves, M.. 2015. Mutants as a Genetic Resource for Future Crop Improvement. In: *Advances in Plant Breeding Strategies: Breeding, Biotechnology and Molecular Tools.* Springer International Publishing Switzerland 201.

Hajjar R, Hodgkin T (2007). The use of wild relatives in crop improvement: a survey of developments over the last 20 years. *Euphytica* 156:1-13.

International Atomic Energy Agency (2021).

Karkanis, A., Ntatsi, G., Lepse, L., Fernández, J. A., Vågen, I. M., Rewald, B., Alsiņa, I., Kronberga, A., Balliu, A., Olle, M. and Bodner, G., 2018. Faba bean cultivation–revealing novel managing practices for more sustainable and competitive European cropping systems. *Frontiers in plant science*, p. 1115.

Khazaei, H., Purves, R. W., Hughes, J., Link, W., O'Sullivan, D. M., Schulman, A. H., et al., (2019). Eliminating vicine and convicine are the main anti-nutritional factors restricting faba bean usage. *Trends Food Sci. Technol.* 91, 549–556.

Khursheed, S., Raina, A., Laskar, R. A. and Khan, S., 2018. Effect of gamma radiation and EMS on mutation rate: their effectiveness and efficiency in faba bean (*Vicia faba* L.). *Caryologia*, *71*(4):397–404.

Khursheed, S., Raina, A., Parveen, K. and Khan, S., 2019. Induced phenotypic diversity in the mutagenized populations of faba bean using physical and chemical mutagenesis. *Journal of Saudi Society of Agricultural Sciences*, *18*(2), pp. 113-119.

Kumar, S., Nandana, and Dubey, D. K., 1993. Studies on the effect of gamma rays, diethyl sulphate (DES) on germination, growth, fertility and yield in faba bean. *FABIS Newsletter, 32*:15-18.

Labba, I. C. M., Frokiaer, H. and Sandberg, A. S., 2021. Nutritional and antinutritional composition of fava bean (*Vicia faba* L., var. minor) cultivars. *Food Research International, 140,* p. 110038.

Ladizinsky, G., 1975. On the origin of the broad bean *Vicia faba* L. *Israel J. Bot.* 24, 80–88.

Laskar, R. A. and Khan, S., 2014. Mutagenic effects of MH and MMS on induction of variability in broad bean (*Vicia faba* L.). *Annual Research and Review in Biology, 4* :1129-1140.

Lepse, L., Dane, S., Zeipin¸a, S., Domínguez-Perles, R., and Rosa, E. A. (2017). Evaluation of vegetable-faba bean (*Vicia faba* L.) intercropping under Latvian agro-ecological conditions. *J. Sci. Food Agric.* 97, 4334–4342.

Link, W., Hanafy, M., Malenica, N., Jacobsen, H. J. and Jelenic, S., 2008. Faba bean. *Compendium of Transgenic Crop Plants: Part*, *3*, pp. 71-88.

Maalouf, F., Ahmed, S., Shaaban, K., Bassam, B., Nawar, F., Singh, M., et al. (2016). New faba bean germplasm with multiple resistances to Ascochyta blight, chocolate spot and rust diseases. *Euphytica* 211, 157–167.

Maalouf, F., Hu, J., O'Sullivan, D. M., Zong, X., Hamwieh, A., Kumar, S. and Baum, M., 2018. Breeding and genomics status in faba bean (*Vicia faba*). *Plant Breeding*, *138*(4), pp. 465-473.

Marzinzig, B., Brünjes, L., Biagioni, S., Behling, H., Link, W., and Westphal, C. (2018). Bee pollinators of faba bean (*Vicia faba* L.) differ in their foraging behaviour and pollination efficiency. *Agric. Ecosyst. Environ.* 264, 24–33.

Maxted, N., 1995. An ecogeographical study of Vicia subgenus Vicia. *Systematic and Ecogeographical Studies on Crop Genepools*, vol. 8. IPGRI, Rome.

Mejri, S., Mabrouk, Y., Belhadj, O., and Saidi, M. (2018). Orobanche foetida resistance in two new faba bean genotypes produced by radiation mutagenesis. *Int. J. Radiat. Biol.* 94, 671–677.

Moehring, J., Williams, E. R., and Piepho, H.-P. (2013). Efficiency of augmented prep designs in multi-environmental trials. *Theor. Appl. Genet.* 127, 1049–1060.

Müller, D. (2017). *Genomic selection in synthetic populations (Ph.D. thesis).* University of Hohenheim, Stuttgart, Germany.

Muratova, V. S., 1931. Common beans (*Vicia faba* L.). *Bull. Appl. Bot. Genet. Pl. Breed. Suppl.* 50, 1–298.

Nurmansyah, Alghamdi, S. S., and Migdadi, H. H. (2020). Morphological diversity of faba bean (*Vicia faba* L.) M2 mutant populations induced by gamma radiation and diethyl sulfate. *J. King Saud Univ. Sci.* 32, 1647–1658.

Ofoedu, C. E., Akintayo, O. A. and Zhou, S. J., 2022. Faba Bean Utilization: Past, Present and Future. In *Faba Bean: Chemistry, Properties and Functionality* (pp. 301-329). Cham: Springer International Publishing.

Pietrzak, W., Kawa-Rygielska, J., Król, B., Lennartsson, P. R., and Taherzadeh, M. J. (2016). Ethanol, feed components and fungal biomass production from field bean (*Vicia faba* var. equina) seeds in an integrated process. *Bioresour. Technol.* 216, 69–76.

Rahate, K. A., Madhumita, M. and Prabhakar, P. K., 2021. Nutritional composition, anti-nutritional factors, pretreatments-cum-processing impact and food formulation potential of faba bean (*Vicia faba* L.): A comprehensive review. *Lwt*, *138*, p. 110796.

Shiran, B., Kiani, S., Sehgal, D., Hafizi, A., Chaudhary, M., and Raina, S. N. (2014). Internal transcribed spacer sequences of nuclear ribosomal DNA resolving complex taxonomic history in the genus Vicia L. *Genet. Resour. Crop Evol.* 61, 909–925.

Singh, A. K., Bharati, R. C., Manibhushan, N. C. and Pedpati, A., 2013. An assessment of faba bean (*Vicia faba* L.) current status and future prospect. *African Journal of Agricultural Research*, *8*(50), pp. 6634-6641.

Smykal, P., Coyne, C. J., Ambrose, M. J., Maxted, N., Schaefer, H., Blair, M. W., Berger, J., Greene, S. L., Nelson, M. N., Besharat, N. and Vymyslický, T., 2015. Legume crops phylogeny and genetic diversity for science and breeding. *Critical Reviews in Plant Sciences*, *34*(1-3), pp. 43-104.

Tacke, R., Ecke, W., Höfer, M., Sass, O., and Link, W. (2021). *Zooming into the genomic vicinity of the major locus for vicine and convicine in faba bean (Vicia faba L.).*

Torres, A. M., Avila, C. M., Stoddard, F. L. and Cubero, J. I., 2011. Faba bean. *Pérez de la vega M, Torres AM, Cubero JI, Kole C (eds)* Genetics, genomics and breeding of cool season grain legumes (genetics, genomics and breeding in crop plants). *Science Pubs, New Hampshire*, pp. 50-97.

Tsolakis, N., Wood, T., Privat, I., Srai, J. S., Griffiths, H., Terrier, B. and Roturier, J. M., 2019. *Developing fava beans as a sustainable source of high-quality protein for food supply chains through optimised genetics, farming and processing*. oai:www.repository.cam.ac.uk:1810/298829.

Vandana, Sushil, K. and Dubey, D. K., 1996. Meiotic anomalies induced by EMS and DES in faba bean (*Vicia faba* L.). *Journal of Indian Botanical Society, 75*: 237-240.

Zanotto, S., Vandenberg, A., and Khazaei, H. (2020). Development and validation of a robust KASP marker for zt2 locus in faba bean (*Vicia faba*). *Plant Breed.* 139, 375–380.

Zhao, J., Sykacek, P., Bodner, G., and Rewald, B. (2018). Root traits of European *Vicia faba* cultivars – Using machine learning to explore adaptations to agroclimatic conditions. *Plant Cell Environ.*

Chapter 2

Faba Beans: Origin, Distribution, Production, Uses and Consumption

Saurabh Kumar[1,*]
and Fauzia Naushin[2]

[1]Environmental Botany Laboratory, Department of Botany,
Aligarh Muslim University, Aligarh, Uttar Pradesh, India
[2]Botany Section, Women's College, Aligarh Muslim University, Aligarh,
Uttar Pradesh, India

Abstract

Vicia faba L., commonly known as faba or fava bean, is amongst the oldest legume crops in the world. The faba bean is cultivated in 66 countries worldwide and is ranked 6th among the legume crops. The chief producers of faba beans are Australia, China, Ethiopia, France, and The United Kingdom. Faba bean is primarily utilized as a food source for human consumption and fodder for livestock. Faba bean is a multipurpose crop that is a good source of biomolecules, including proteins, carbohydrates, essential nutrients such as vitamins and minerals, many bioactive compounds, dietary fibre, levodopa, flavonoids, and polyphenols. The faba bean is important in the agricultural system because of its great efficiency in symbiotically fixing atmospheric nitrogen. This way, faba bean improves soil fertility, making it an excellent rotational or mixed crop. In the present era, obtaining food security is a great challenge that demands sustainable production and transformation in food systems. Under present climate change scenarios, legume crops such as faba beans are the top-performing crop because of their distinctive ability to thrive in extreme climatic conditions and varied

* Corresponding Author's Email: s.singh100nl@gmail.com

In: Faba Beans
Editors: Aamir Raina and Samiullah Khan
ISBN: 979-8-89530-447-1

soil regimes. This book chapter describes the origin, distribution, and global trends in faba bean production and consumption.

Keywords: faba bean, legume crop, production, uses, consumption

1. Introduction

The human population is expected to reach 9.6 billion by the end of 2050 (United Nations, 2012), and this will put a lot of pressure on food systems, making the accomplishment of sustainable goals, particularly food security, far from reality. Improving traits in economically important crops such as legumes seems a way to accomplish food and nutrition security. Legumes, crops belonging to the family Leguminosae or Fabaceae, frequently referred to as the bean, pulse, and peas, are ideal crops with huge potential in sustainable agriculture owing to their multiple uses (Lewis et al., 2005; Stagnari et al., 2017). Leguminous crops are considered a sustainable source of nutritional food as they are highly rich in nutrients, easy to preserve, and have a lower cost of production (Rahate et al., 2021). Legume crops are multifunctional and used as a rich source of plant-derived protein for the nourishment of humans and animals and complement protein-deficient cereals (Voisin et al., 2014; Hoehnel et al., 2019). Legumes can fix atmospheric nitrogen symbiotically, which makes it ideal for integration into cost-effective farming techniques and contributes to reducing greenhouse gas emissions (Lemke et al., 2007).

Among legume crops, faba bean is recognized as a versatile crop in terms of nutrition, ecology, and economics and is also considered one of the ancient legumes cultivated globally (Mínguez & Rubiales, 2021). Faba bean is a winter season annual pulse crop (Bilalis et al., 2003) commonly known by different names: faba bean, bell bean, broad bean, field bean, and horse bean (Singh et al., 2010). In India, faba beans are popularly known as Bakala and Kala Matar in their native languages (Singh et al., 2012a; Singh et al., 2013a). Faba bean is chiefly produced in Australia, Afghanistan, China, Egypt, Ethiopia, France, the United Kingdom, Northern Europe, Northern Africa, and India. It is cultivated in 66 different nations globally (Merga et al., 2019). It has been ranked sixth on the global production list among cultivated grain legume crops in 2019, yielding 4.5 metric million tonnes from 2.5 million hectares (Mha) of land.

The faba bean is used as food for humans and animal fodder (monogastric and ruminants) due to its high nutritional quality (Grosjean et al., 2000). Faba bean is considered an outstanding source of proteins (high quantities of lysine and arginine) and also an excellent source of numerous micro as well as macronutrients which includes potassium (K), Calcium (Ca), Magnesium (Mg), Iron (Fe), and Zinc (Zn) (Lizarazo et al., 2015; Longobardi et al., 2015; Neme et al., 2015; Rahate et al., 2021). It is abundantly a storage house of complex carbohydrates, i.e., oligosaccharides such as stachyose, raffinose, and verbascose. Faba beans seeds include a variety of beneficial bioactive chemical molecules such as phenolics, flavonoids, carotenoids, dietary fibre, antioxidants, aminobutyric acid, and non-nutrient secondary metabolites (Khazaei et al., 2019; Liu et al., 2022). Faba bean includes biological elements with health-enhancing characteristics, such as GABA and L-DOPA, and a well-balanced amino acid profile. Peptides of faba bean have been linked to antioxidant, anti-diabetic, anti-hypertensive, lowering of cholesterol, and antiphlogistic properties, showing the high ability of this legume crop to be utilized as a functional food with medicinal properties (Martineau-Côté et al., 2022). Through biological nitrogen fixation and diversification of crops, faba beans offer ecological benefits such as sustainable nitrogen (N_2) addition to crops and soil. It is a great crop for a cropping system that relies heavily on efficient rhizobia populations. The incorporation of legume crops in cropping systems enhances the fertility rate of soil due to their unique ability to symbiotically fix atmospheric nitrogen (N_2) via biological nitrogen fixation (BNF) (Jensen et al., 2010). It may absorb N_2 from the soil and the environment (Rajan and Singh, 2012). Faba bean is also linked to less fertilizer input in agricultural fields and enhanced soil biological activity. Faba beans are a functional crop under climatic conditions because of their ability to thrive in all possible situations of extreme weather and climatic conditions, alongside great resilience to varied soil conditions. Also, they can mitigate climatic variation because they reduce emissions of greenhouse gases and play an important role in the sequestration of atmospheric carbon inside the soil (Jensen et al., 2012; Singh et al., 2013b). Knowing the importance of the faba bean in the agricultural system, it is being utilized as food for consumption by humans and as fodder for feeding livestock due to its high content of protein and nutrients, making it a multipurpose legume crop. It also has medicinal properties and health benefits, ecological services such as N_2 fixation, and is becoming a climate-resilient crop for present and future generations.

2. Origin, History, and Distribution of Faba Bean

An ancient record of the faba bean dates back to the prehistoric period, and different researchers and scientists have different thoughts regarding the origin of the faba bean because of its widespread distribution throughout the world. The centre of origin of the faba bean is thought to be in the Near Middle East (Cubero, 1974), while some researchers claim that the faba bean originated in the far Eastern region of Asia (Hawtin and Hebblethpiait, 1983; Long et al., 1989). Various archaeological surveys in Eurasia and Africa from the earlier Neolithic period indicate that the faba bean was the first cultivated plant (Duc et al., 2015). It is speculated that *V. faba* var. Major might have evolved in North Africa, while *V. faba* var. Minor might have developed in the South Asian Black Sea. However, the researchers found that with the advancement of civilization and trade, faba bean extended to i) Europe, ii) Down the coastline from North Africa to Spain, iii) Down the length of the Nile River to the nation of Ethiopia, and iv) from Mesopotamia into India (Li-Juan et al.,1993). China, Ethiopia, and Afghanistan are believed to be its secondary centre of diversity (Torres et al., 2006; Zong et al., 2009, 2010). It was most likely imported to India as an ornamental crop by Europeans throughout Sultan dynasty (1206 -1555 AD), where the domestication of the faba bean was described (Naqvi, 1984; Akbar, 2000). Faba bean seeds have been discovered in Jericho (a Palestinian city) dated back to 6250 BCE (McVicar et al., 2013). Furthermore, Caracuta et al., 2016 found possible seed of faba bean ancestor on Mount Carmel, which is in Israel, and their remains were Carbon dated to fourteen thousand years ago. Caracuta et al., 2015 revealed that the faba bean was first cultivated in Israel's Lower Galilee approximately 10,200 years ago. Vavilov 1936, identified an ancient variety of faba bean with short fruit and seed near the Himalayan and Hindu Kush mountains and argued that the faba bean came from the region of Central Asia. He discovered a sharp rise in shape and size of the faba bean towards the western region of Central Asian countries across Iranian hills, Turkey, the Mediterranean Sea shore, and Spain. Chinese people used faba beans for food around five thousand years ago. Hebrews domesticated the crops in the Old Testament era (bible time), the Greeks and Romans a little later, and the Egyptians 3,000 years ago. (Mihailovic et al., 2005; Singh and Bhatt, 2012b).

In China, faba bean is originally known as "hu dou", and over 50% of the global faba bean production acreage is in China, where it is utilized for a variety of purposes, such as dried seeds, green leafy vegetables, or processed food items. The dried pods, green haulm (stem of legume plants), and dry

straws are used as feed for livestock. It was most likely first imported to India between 1206 and 1555 AD. Europeans initially cultivated it as a garden crop in India. However, the region under faba bean cultivation is relatively small, and therefore, faba bean continues to be categorized as a minor and underutilized crop.

Faba bean has wide genetic variability, with over 38,000 faba bean accessions preserved in gene storage facilities worldwide, as reported by Duc et al. (2015). According to research undertaken by the EUROLEGUME project, many additional genotypes are possibly accessible regionally, on landowners, and in breeding groups throughout Europe (Lepse et al., 2017). The wild progenitor of faba bean is unknown, but many uncultivated forms, such as *Vicia narbonensis L.* and *V. galilaea Plitmann* and *Zohary,* are systematically similar to domesticated crops; however, these have 14 chromosomes instead of 12 chromosomes in the cultivated crop and have been grouped into three classes on the basis of seed size (Cubero, 1974; Crépon et al., 2010). Faba bean genetic material is categorized as either springtime or wintertime depending on frost tolerance, intended climatic zone, and sowing period, as well as the capacity for adaptation to environments that are either oceanic or continent-wide, i.e., susceptible to dryness (Link et al., 1990).

3. Botanical Description

Faba bean is an annual winter (rabi) season leguminous crop that belongs to Fabaceae or bean and pea family (Bilalis et al., 2003). It has a taproot system out of which many lateral fibrous roots protrude and reach 90 cm deep into the earth (Muehlbauer and Tullu, 1997). Rhizobia-containing nitrogen-fixing nodules can be observed on lateral and tapping roots (Bond et al., 1985). Faba bean grows up to two meters in height and has a rough, upright, hollow stem without branching (Bond et al., 1985; Duc et al., 2015a; Heuzé et al., 2016) (see Figure 1). Tillers grow from the basal nodes of the stem, and stem growth is indefinite, making certain varieties susceptible to lodging. Leaves are long, alternate, and pinnately compound and consist of 2 to 7 rounded or oval leaflets, 6- 8 cm long and 2-4 cm wide. Despite the majority of legumes, the green-colored leaves of this plant lack tendrils that allow them to grow above other vegetation. In faba bean, inflorescences are borne on short axillary racemes consisting of flowers 3-4 cm in length and colorless having purple spots (Figure 1). Flowers are arranged on inflorescences and have a characteristic papilionaceous shape (Bond et al., 1985; Duc et al., 2015a;

Heuzé et al., 2016). Faba bean is an entomophilous plant, and among the insects, *Apis spp.* (honeybee) and *Bombus sp.* (bumblebee) are primary pollinators (Kendall & Smith, 1975; Marzinzig et al., 2018). The fruit is a dehiscent cylindrical pod that can grow up to 10 cm long & 1-2 cm wide. Young pods are green, but when they mature, they turn dark brown or black. The pods contain three to four oblong-oval seeds (beans) with notable hilum. Their colour varies from yellowish to greenish brown to violet, while some seeds are speckled (McVicar et al., 2013; Muehlbauer and Tullu, 1997).

Faba bean thrives well in cool, damp environments with optimum precipitation between 650-1000 mm (Abdel, 2008; Gasim and Link, 2007). Moderately textured sands are suitable for the growth of broad beans. It favors soils with pH levels ranging from neutrality to alkalinity, 6.5 to 8.0 (Rajan and Singh, 2012). An average water supply is required because the crop requires adequate moisture for the best yields. Water requirements are greatest between 9 and 12 weeks after germination (Subash and Priya, 2012).

Figure 1. Specimen of faba bean showing the different parts: a) Faba bean stem with floral appendages, b) Flower, c) Immature Pod, d) Mature Pod, e) Cotyledon, f) Seed's Cross Section.

4. Global Production and Distribution

Faba bean is cultivated in over 60 countries (Merga et al., 2019). It is an important leguminous crop in several nations, including China, Ethiopia, and Egypt, where it is widely grown for consumption by human beings across the entire Mediterranean region and in parts of Latin America (Akbar, 2000). Other notable producers include Northern Europe, Central Asia, East Asia, and Latin America. China, Ethiopia, Australia, France, and the United Kingdom are the top five producers of faba bean, together contributing over 75% of the world's production. China, followed by Ethiopia, is the leading producer of faba beans, which accounts for 60% of total world production (FAO, 2009; FAO, 2016; Torres et al., 2006). In 1961, 5.4 million acres of agricultural land were devoted to faba bean crop production, yielding less than 5 million tons. This estimate reached almost 6 million hectares and 6 million tonnes in 1962. However, faba bean cultivation and yield declined to a minimum of 2 million hectares and 3 million tons in 1990. With the shrinking acreage, production per area will likely rise due to lower vulnerability to environmental stress factors (Link et al., 2010; Sillero et al., 2010; Singh et al., 2012b). From 1991 to 2017, the area allocated to faba beans cultivation stayed roughly constant at about 2 million per hectare, while production climbed steadily from 3 million tons to above 5 million tons. In 2017, the leading five faba beans growers were China with 1,803,019 kg, Ethiopia with 930,633 kg, Australia with 373,605 kg, the United Kingdom with 302,468 kg, and Germany with 188,800 kg. Furthermore, the global average yield of the faba bean is 1,964.3 kg per hectare. In 2019, the global output of faba beans was 5.43 million metric tons, a 25% increase from 4.35 million metric tons in 1990 (Dhull et al., 2022). On a regional basis, Asia accounts for 33% of world faba bean output, which is followed by European (EU) and the African continent, which account for 29.36% and 27.04%, respectively (FAO, 2020). In the European Union, faba beans are ranked second after peas in yield, and they are most commonly utilized for legume production as food and animal livestock feed.

Some of the largest faba bean exporting countries are Australia, which has 201,000 kilograms; the United Kingdom, which has 114,000 kilograms; China, which has 63,000 kilograms; and France, which has 53,000 kilograms. Egypt is the world's highest buyer, importing 309,355 metric tonnes, 40.48% of all imports. It is followed by Norway, Germany, Saudi Arabia, and France (Figure 2a, 2b).

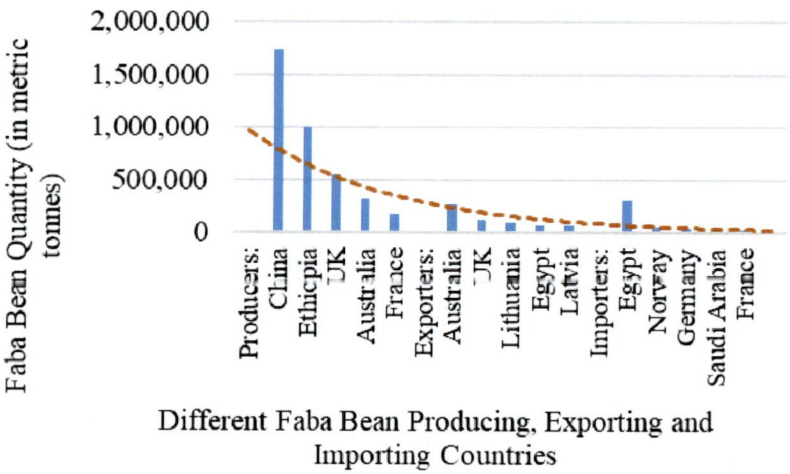

(a)

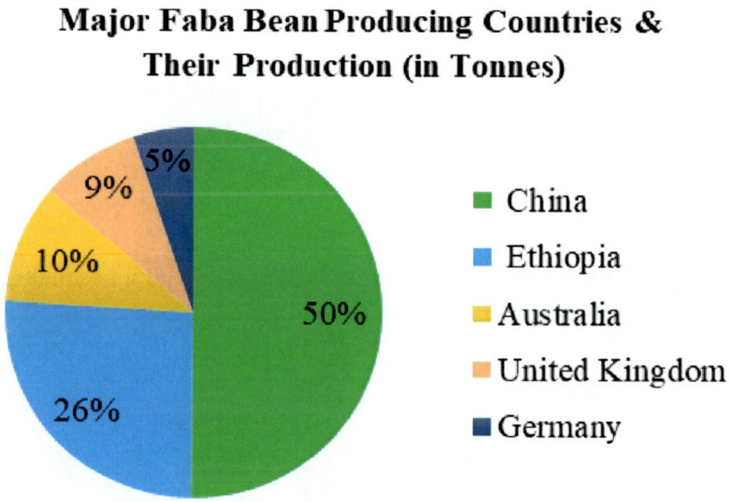

(b)

Figure 2. (a) Major faba bean producers, exporters, and importers, (b) Major faba bean producing countries and their productions (in tonnes).

5. Composition and Nutrition Profiling

Faba bean is a nutrient rich pulse crop with abundant proteins, polysaccharides, roughage, and essential nutrients such as choline lecithin as well as specialized secondary products, or biologically active compounds such

as antioxidants, polyphenols, levodopa, and GABA, i.e., γ-aminobutyric acid (Khazaei et al., 2019; Liu et al., 2022) (see Table 1). It also contains non-nutritive secondary compounds, which include saponins, lectins, tannins, and phytic acid (Dhull et al., 2022).

5.1. Proteins

The protein content of faba bean is 26.1%, nearly twice the amount of protein of cereal grains (Yang et al., 2018). Among leguminous crops, faba bean has higher protein content than other pulses such as chickpeas (19.53%), lentils (22.15%), and beans (22.17%) (Estefania et al., 2018) and lower than soybean and lupin (40.0%) (Raikos et al.,2014). Generally, plant proteins are classified into four classes on the basis of their dissolubility in different solvents (Barłóg et al., 2019). Globulin (solubility in less salt or salt-free liquids), water-soluble albumin, prolamin, which is soluble in 70% alcohol, and glutelin, i.e., soluble in alkali solution (Osborne, 1907). Faba beans have prolamin (8%), glutelin (15%), albumin (20%), and globulin (60%) (Alghamdi, 2009). Most of the faba bean contains globulin-like proteins found in many legumes (Boye et al., 2010). Vicilins (7S) and Legumin (11S) are two principal categories of faba bean globulins based on their sedimentation coefficients (Utsumi, 1992). Most of the globulins found in faba beans are legumins, accounting for more than 50% of the total protein content (Müntz, 1999). Faba bean is rich in amino acids as it has high arginine, histidine, leucine, isoleucine, threonine, and some aromatic amino acids (Millar et al., 2019). However, it has a low amount of amino acids having sulfur (cysteine and methionine) (Małgorzata et al., 2018). The protein portion of faba bean differs mostly due to species and genetic diversity (Ivarsson and Neil, 2018; Nosworthy et al., 2018). Also, the varietal variances and source type, such as the fraction of flour or isolate (Yang et al., 2017), in addition to the pollination technique, growing period, and sowing area (Multari et al., 2015), account for the broad variability. Environmental factors such as the soil composition, nitrogen-fixing rate in the atmosphere, and environmental hardship such as drought influence the protein level of faba bean (Senberga et al., 2017; Ntatsi et al., 2018).

Table 1. Nutritional profile of raw, mature faba bean seeds

Nutrient	Amount	
Water	11	G
Energy	341	Kcal
Energy	1420	kJ
Protein	26.1	G
Total lipid (fat)	1.53	G
Ash	3.08	G
Carbohydrate, by difference	58.3	G
Fiber, total dietary	25	G
Total Sugars	5.7	G
Calcium, Ca	103	Mg
Iron, Fe	6.7	Mg
Magnesium, Mg	192	Mg
Phosphorus, P	421	Mg
Potassium, K	1060	Mg
Sodium, Na	13	Mg
Zinc, Zn	3.14	Mg
Copper, Cu	0.824	Mg
Manganese, Mn	1.63	Mg
Selenium, Se	8.2	µg
Vitamin C, total ascorbic acid	1.4	Mg
Thiamin	0.555	Mg
Riboflavin	0.333	Mg
Niacin	2.83	Mg
Pantothenic acid	0.976	Mg
Vitamin B-6	0.366	Mg
Folate, total	423	µg
Folic acid	0	µg
Folate, food	423	µg
Folate, DFE	423	µg
Choline, total	95.8	Mg
Vitamin B-12	0	µg
Vitamin B-12, added	0	µg
Vitamin A, RAE	3	µg
Retinol	0	µg
Carotene, beta	32	µg
Carotene, alpha	0	µg
Cryptoxanthin, beta	0	µg
Vitamin A, IU	53	IU
Lycopene	0	µg
Lutein + zeaxanthin	0	µg

Nutrient	Amount	
Vitamin E (alpha-tocopherol)	0.05	Mg
Vitamin E, added	0	Mg
Vitamin D (D2 + D3), International Units	0	IU
Vitamin D (D2 + D3)	0	µg
Vitamin K (phylloquinone)	9	µg
Fatty acids, total saturated	0.254	G
SFA 4:0	0	G
SFA 6:0	0	G
SFA 8:0	0	G
SFA 10:0	0	G
SFA 12:0	0.004	G
SFA 14:0	0.002	G
SFA 16:0	0.204	G
SFA 18:0	0.031	G
Fatty acids, total monounsaturated	0.303	G
MUFA 16:1	0.002	G
MUFA 18:1	0.297	G
MUFA 20:1	0	G
MUFA 22:1	0	g
Fatty acids, total polyunsaturated	0.627	g
PUFA 18:2	0.581	g
PUFA 18:3	0.046	g
PUFA 18:4	0	g
PUFA 20:4	0	g
PUFA 20:5 n-3 (EPA)	0	g
PUFA 22:5 n-3 (DPA)	0	g
PUFA 22:6 n-3 (DHA)	0	g
Fatty acids, total trans	0	g
Cholesterol	0	mg
Tryptophan	0.247	g
Threonine	0.928	g
Isoleucine	1.05	g
Leucine	1.96	g
Lysine	1.67	g
Methionine	0.213	g
Cystine	0.334	g
Phenylalanine	1.1	g
Tyrosine	0.827	g
Valine	1.16	g
Arginine	2.41	g
Histidine	0.664	g
Alanine	1.07	g

Table 1. (Continued)

Nutrient	Amount	
Aspartic acid	2.92	g
Glutamic acid	4.44	g
Glycine	1.1	g
Proline	1.1	g
Serine	1.2	G
Alcohol, ethyl	0	G
Caffeine	0	Mg
Theobromine	0	Mg

Source: USDA 2021 (per 100 g).

5.2. Carbohydrates

The primary carbohydrate components in faba bean are starch content, dietary fibres, and sugar types. Carbohydrates range from 51%-68% (510 g per kg to 680 g per kg overall), and starch represents the major proportion of carbohydrates, which is 41-51% (USDA, 2021; Vidal-Valverde et al., 1998). Its seeds contain low molecular weight carbohydrates (LMWC) such as glucose, fructose sucrose (GFS) and raffinose family oligosaccharides (RFO), which include stachyose, raffinose and verbascose. These LMWC and RFO impart flavour and prebiotic nature of faba bean. The primary elements of faba bean starch are amylose and amylopectin (Punia et al., 2019). Scanning electron microscopy has revealed that faba bean granules of starch have uneven shapes (round, oval, elliptical, and irregular forms) and voids on their surfaces (Sofi et al., 2013). A poor soluble rate (9.92 g per 100 g) and high lump capacity (12.67 g/g) characterise faba bean starch, making it suitable for a variety of applications. Faba bean starch is resistant to enzyme hydrolysis due to its elevated resistant starch (RS) level, i.e., 46.7%, along with lower digestible starch (15.7%) and gradually digestible starch (34.5%) levels (Bello-Pérez et al., 2007). Suárez-Diéguez et al. (2021) created and improved a lab-scale method for removing resistance starch from faba bean. A transformation strategy was used to boost the resistant starch amount in broad beans. The earlier investigation demonstrated that faba bean RS may be a potential medicinal ingredient because of its reduced and extended digestion. The faba bean is an excellent source of dietary fibre, containing either soluble or insoluble fibre (Singh et al., 2014). Faba bean-derived flours (FBF) also contain more dietary fibres than flours derived from red kidney, pinto, and

lima beans (Gu et al., 2020). Raw faba beans possess 15 to 30% nutritional fibre, with hemicellulose as its most common kind, followed by celluloid and lignin. The seed coat of the faba bean has an adequate quantity of dietary fibres (82.3%) (Vidal-Valverde et al., 1998; Karatas et al., 2017). Therefore, it is good to have faba bean along with its seed coat due to its high content of dietary fibre, phenolic substances, and minerals (Karatas et al., 2017).

5.3. Essential Nutrients (Vitamins and Minerals)

Faba bean is rich in vitamins, including thiamine (B1), pyridoxine (B6), folate (B9), vitamin K, and vitamin C, as well as different minerals such as K, Cu, Se, Zn, Mg, P, Mn and Ca (Khalil & Mansour, 1995; Luo et al., 2008; Nosworthy et al., 2018; USDA, 2021). These nutrients play an important role in growth and development. In comparison to soybean (250 μg), faba bean contains 423 μg of vitamin B9 (folates). A lack of folate increases the risk of spina bifida (associated with newborn infant spines) as well as anencephaly, a condition in which a newborn child has an underdeveloped brain system (Blancquaert et al., 2010). The elevated level of potassium, i.e., 1062 mg in 100 g, and less salt content, i.e., 13 mg in 100 g of mature seeds, offers a good choice to individuals on a low sodium diet. It is exceptionally rich in iron, an extremely important coenzyme in synthesizing pyrimidines, purines, and amino acids (Hefni et al., 2015). A large portion of the phosphorus in faba bean is inaccessible as it occurs, for example, phytates that can be detrimental to the well-being of humans (Luo et al., 2014).

5.4. Bioactive Compounds

In faba beans, there are several varieties of biologically active phytonutrients such as phenylic substances, flavonoids, lignan, antioxidants, terpenoids, and γ-aminobutyric acid, i.e., GABA (Khazaei et al., 2019). Faba beans contain both free and esterified phenolic chemicals. Abu-Reidah et al., 2014 found 104 phenol compounds in faba bean, primarily flavonoids. The primary compounds that contain flavonoids are flavanol molecules (including catechin, epigallocatechin, and gallocatechol), proanthocyanidins (prodelphynidin and procyanidin), flavonols (glycosylated derivatives of myricetins, quercetins and kaempferols), flavanonols, isoflavones (daidzein and genistein) and flavanones. Additionally, important phenol compounds that

were discovered in faba bean include trans-caffeate, ferulate, 4-hydroxycinnamic acid, protocatechuic acid, vanillic acid, sinapic acid, salvianolic acid (Abu-Reidah et al., 2014; Yao et al., 2011; Kaufman et al., 2007). Polyphenolic compounds in faba beans are variable in their different parts, such as foliage, roots and seeds. Immature faba beans possess more phytochemicals and have stronger antioxidant properties (Chaieb et al., 2011; Boukhanouf et al., 2016). The gross polyphenol composition of faba bean pod extracts ranges between 4.8 to 13 mg of gallic acid equivalents (Valente et al., 2018), while the entire faba bean plant has 2.9 mg GAE/g of overall phenols that gives an antioxidative effect, thus lowering the oxidative stress and decreasing the frequency of chronically degenerative disorders (Turco et al., 2016). Flavonoids, biologically active compounds having anti-inflammation and anti-diabetic action, have been discovered in faba beans (Zhang et al., 2019). Karatas et al. (2017) reported 1.8 mg Trolox equivalent of antioxidative capacity (TEAC) in entire faba bean pods, whereas the husk of seed covering had substantially greater antioxidative function (22.9 mg TE/g). Johnson et al. (2020) examined eleven types of Australian bean varieties and discovered considerable differences in their phenol and anthocyanin content and their potential as antioxidants.

5.5. Antinutritional Factors

Most legumes consist of a particular kind of anti-nutrient. The important antinutrients found in faba bean are convicine, lectins, saponins, vicine, phytates, oligosaccharides such as raffinose, stachyose trypsin as well as protease inhibitors (Labba et al., 2021; Nosworthy et al., 2018). Most of these antinutrients in faba beans are present in their mature seeds and have a detrimental impact on the nutrient quality and affect peptide digestion, the potency of amino acids, and protein quality (Gilani et al., 2012). The inclusion of faba bean in our food has a negative impact on sugar digestion and absorption. The inhibiting activity of trypsin in the faba bean was determined to be 2285 units/g dry weight. Faba bean has lesser trypsin inhibitors (2.24-2.77 TIU/mg) as compared to other legumes, including soybean, chickpea, and lentil (7.65-8.98 TIU/mg) (Sharma & Sehgal, 1992; Khan et al., 2015). Another non-nutritious component in legumes is tannin, which can be localised on seed coverings (Karatas et al., 2017). In faba bean, phosphate is stored as phytic acid, and due to the lack of phytase enzymes, non-ruminants are unable to extract phosphorus from the phytic acid. The anti-nutritional

impact is caused by the six reactive groups of phosphates found in phytic acid. Lectins are carbohydrate protein binders, favin is a type of faba bean lectin protein. Significant progress in lowering vicine tannin in the seed coat of faba bean was acquired through crop improvement programs such as gene editing (Khazaei et al., 2021). It has been demonstrated that the phytic protein present in faba bean modifies the assimilation of protein molecules and decreases mineral absorption. Phytate, considered a detoxifying compound, also reduces the absorption of cations (Luo et al., 2008).

6. Uses of Faba Bean

Faba bean is a versatile crop that has become recognised around the world for its nutritious qualities and wide variety of uses. Because of its high nutritional value, it is a widely used pulse crop for human nutrition and for livestock feed (hay, silage, and straw) (Cazzato et al., 2012; Prolea, 2014; Yasmin et al., 2020). In the central region of the world, it is among the most important rabi crops for human consumption (Yadav et al., 2017). Faba beans are a fantastic supplier of protein, mineral nutrients, organic nutrients, and a variety of biologically active substances. Fresh and dried seeds of faba beans are consumed by humans. They are very nutritious because of their substantial amount of protein (up to 35% in dry seeds) and nutrients, including Potassium, Calcium, Magnesium, Iron, and Zinc (Lizarazo et al., 2015; Longobardi et al., 2015; Neme et al., 2015). Faba bean is utilised not only as a primary food but also in crop rotation and intercropping systems to reduce the prevalence of cyst nematodes (*Heterdera avenae*) and soil-borne diseases (Landry et al., 2016). Soil fertility is improved by integrating faba beans in cropping systems. It has remarkable efficacy in developing symbiotic relations with certain *Rhizobium* bacteria, as well as subsequent biological nitrogen fixation (BNF) integrated with a decrease in fertiliser use in fertile areas and improved biological activity in the soil. Faba bean is also used in crop rotational cycles that include growing legumes with plants that are unable to fix nitrogen (Jensen et al., 2010).

6.1. Ethnobotanical Uses of Faba Bean

In ancient times, faba beans were used for voting in countries like Greece and Rome, in which a "snowy bean" was utilised to give a 'yes' vote while a "dark bean" was used to cast a 'no' vote. Dropping beans on the earth's surface and deciphering a pattern in which they drop was prevalent practice of divination (favomancy) for discovering what will happen in the future in Ubykh culture, and the name for "bean-thrower" has become a generic term for seers and soothsayers in general (Hulse, 1994). Faba beans are commonly planted in Italy on November 2, also known as All Souls Day. Smaller cakes shaped like faba beans (but not made of them) are called "fave dei morti" and "beans of the dead" (Lawes, 1980). As per belief, once Sicily had crop failure except for only one crop left, i.e., beans, which saved the people from hunger, gratitude was offered to Saint Joseph (Acharya and Shrivastava, 2008).

6.2. Medicinal Uses

Faba bean contains a high concentration of levodopa, a type of natural dopamine precursor that is effective in treating Parkinson's disease (Upadhyay et al., 2014; Etemadi et al., 2015). Synthesised levodopa, also known as L-Dopa (L-1,3 dihydroxyphenylalanine), is frequently administered for the treatment of patients who have Parkinson's disease. Levodopa is also a natriuretic (sodium excretion in the urine) drug that might help treat high blood pressure, i.e., hypertension (Jambunathan et al., 1994). Also, it is believed that levodopa is costly and is frequently associated with a number of adverse illnesses, including sickness, hypotension, vomiting, exhaustion, fatigue and uneasiness (Patil et al., 2013). To avoid any adverse effects, consume a natural source of Levodopa (Singh et al., 2013a; Patil et al., 2013). The faba beans can be used to treat hormonal irregularities. Studies have also shown that faba bean levodopa could be used as an organic substitute for Viagra and can enhance libido in humans (Hulse, 1994). The faba bean also possesses anti-inflammatory, anti-oxidative, and anti-diabetic properties, and thus, it could be used to prevent diseases that affect human development (Turco et al., 2016). Young children are often discouraged from eating faba beans uncooked since they might produce bowel movements and signs resembling jaundice. *In vitro* as well as epidemiology investigations reveal that favism-induced haemolysis protects against malarial fever because few types of malaria-causing protozoans (*Plasmodium falciparum*) are vulnerable to oxidative damage

because they lack an enzyme called glucose 6-phosphate dehydrogenase (Hussein and Saleh, 1985). Faba bean and its derivatives help treat diabetes, hypertension and cardiovascular diseases (Madar and Stark, 2002).

6.3. Ecological and Environmental Benefits

Faba bean performs an increased level of biological assimilation of nitrogen due to its capability to fix nitrogen present in the atmosphere through symbiosis using *Rhizobium sp.*, making it an exceptionally good nitrogen-fixing annual legume under a wide range of environmental circumstances (Figure 3). It has been found that faba beans exhibited substantial nitrogen fixation (Neugschwandtner et al., 2015). Nitrogen derived from the atmosphere (NdfA) accounts for up to 96% of the nitrogen taken up by the crop (Lo´pez-Bellido et al., 2006; Paul & Gupta, 2022). NdfA levels of 60% to 80% have been observed multiple times for rich soils in temperate zones or with irrigation. One benefit of faba bean cultivation is that natural strains of *Rhizobium leguminosarum* can be found in most soils that promote crop development. Therefore, faba bean is a great rotational crop, significantly reducing dependability on energy-intensive substances such as nitrogen fertilisers. Faba bean also helps in agroecosystem diversification over time through different crop rotations and spatially via intercropping, therefore increasing the diversity of flora, fauna and animal life and soil-based microorganisms that might influence agricultural system sustainability (Köpke & Nemecek, 2010). Legume-based approaches such as faba bean increase the fertility of the soil through a variety of methods, such as alterations in soil organic carbon and Humic composition and the accessibility of N and P (Jensen et al., 2012).

Faba bean is considered a highly resilient crop in extreme weather and changing climatic conditions due to its distinct ability to thrive in many sorts of climatic conditions and its vast resilience to variable soil regimes (Singh et al., 2013b). Life cycle assessment (LCA) has been used to investigate the environmental implications of legumes at various stages, except for analyzing the early cropping impacts of a single crop. Initially, evaluations of complete crop rotations with and without grain legumes reveal that intensive cereal-rich crop rotation minimizes energy consumption, temperature rise, ozone production, acidity, and ecological toxicity. The substitution of imported soybean meal for locally produced pulses (such as faba bean) might be beneficial. Also, faba bean has the potential to keep human nutrition away

from meat to lessen the environmental loads, but care must be taken to ensure that plant products are not too processed. Despite the several essential advantages of faba beans, their contribution to global warming prevention has been given little attention (Köpke & Nemecek, 2010). On comparing agriculture productivity with chemical-based nitrogen fertilizer, pulses such as faba bean can minimise the emission of greenhouse gases (GHGs), including carbon dioxide (CO_2) and nitrous oxide (N_2O), playing a significant role in storing CO_2 within the soil, as well minimising total system output of fossil energy (Stagnari et al., 2017).

Source: Adapted from Muktadir et al., 2020.

Figure 3. Important ecological and environmental services of faba bean.

7. Faba Bean Consumption

Faba bean is well-known worldwide for its nutritious traits and wide range of uses. The earliest faba bean seeds were discovered in North-Western Syria during the latter part of the 9th century BC (Tanno & Willcox, 2006). Human consuming faba beans dates back to the Viking Age (700-1050 CE) in Northern Europe (Alandia et al., 2020). In various countries throughout the world, studies have shown that faba bean may be utilized as human food and animal feed for pigs, horses, poultry, and pigeons (Singh and Bhatt, 2012a; Crépon et al., 2010). In the Middle East, the faba bean is the most preferred

winter crop for human consumption (Martin et al., 1991). Faba bean can be consumed as a vegetable (fresh or raw, green or dried or cooked) and in various dishes (Gasim and Link, 2007). It is commonly used as a dal (soup) for humans in India and Bangladesh. Faba beans require a considerable boiling time to obtain optimal tenderness and flavour and increased protein digestion, which has limited its use in human diets (Dhull et al., 2022).

7.1. Faba Bean As Food

Faba bean includes huge amounts of proteinaceous peptides, sugars, and essential nutrients with vital health advantages. Faba bean has grown in popularity in the food and feed industries because of its high protein content value. It can be utilized as animal fodder and eaten raw or cooked as a vegetable by humans (Crépon et al., 2010). It has become a popular brunch or morning meal item in countries like the Middle East, the Mediterranean, China, and Ethiopia. Oplinger (1982) explored faba bean as an alternative to meat and also as a substitute for milk. Burnt seeds are eaten like peanuts in India. The most well-known faba bean foods include Medamis (simmered beans), Falafel (deeply oil-fried raw seeds paste mixed veggies and seasonings), Bissara (seed paste), and Nabet soup (made from cooking sprouted faba beans) (Singh et al., 2013a). Many meals are based on the usage of dry faba bean seeds, such as Doubanjiang (a taste of brewed bean paste), Janjetinas bizima (the conventional Dalmatian meat with faba beans), Maccu (the conventional Italian soup). Additionally, a hot meal of fried beans called faba is well-liked in Gilan (a province in North Western Iran), Malaysia, China, Peru, Colombia, Ecuador, Guatemala, Mexico, and Thailand. Research on the beneficial effects of faba bean powder in the creation of gluten-free cakes, extruded appetizers (Giménez et al., 2013), and pasta (Smith & Hardacre, 2011) have all been published.

7.2. Faba Bean As Animal Feed

In developing countries, the Faba bean is mainly used for feeding animals, primarily fowl, birds, cattle, and pigs (Singh and Bhatt, 2012a). On average, faba bean husk raw protein level is 8.7% (w/w), and the required ruminant diet must consist of 7% (w/w) raw protein for optimal rumen microbial activity (Van Soest, 1994). Feeding studies using faba bean straw were conducted on

sheep and rabbits. The metabolized energy (ME) level of faba bean husk has been reported to be between 6.2 and 9.2 MJ kg^{-1} DM for sheep (Abreu & Bruno, 1998; Wegi et al., 2018). Ruminants find faba beans to be highly tasty. According to Fujihara and Nakao (1982), the feedstock could be utilized as fodder for sheep. In Ireland, smaller and rounder faba beans are known as field, horse, or tic beans and are used as animal feed. More precisely, smaller seeds (*Vicia faba* L., var. minor) are tic beans, which are around 500 g per 1000 seeds, while intermediate-sized seeds (*Vicia faba* var. equina) were equine beans (ca. 500-800 g per 1000 seeds) (O'kiely et al., 2022). Because faba bean seeds are high in hemicellulose, cellulose, and fibre, they can be suitable feed for livestock.

7.3. Edible Products Formed from Faba Bean

Faba beans are frequently used to make a variety of nutrient-rich foods. Utilization of faba bean fibre, i.e., FBF, with proteins and glucose isolate has been found in pasta (Rosa-Sibakov et al., 2016; Tazrart et al., 2016), spaghetti (Giménez et al., 2013), bread (Sozer et al., 2019), and tofu or bean curd (Jiang et al., 2020; Zee et al., 1987), yogurt (Jiang et al., 2020), as a substitute of fat in meat or beef patty (Sulaiman et al., 2018), replacing egg yolk in mayonnaise (Ouraji et al., 2020), and product analogue to non-veg food (do Carmo et al., 2021).

7.3.1. Pasta and Spaghetti Products

When rava or semolina was substituted by faba bean in pasta, the fibre and protein quantity increased. However, the texture characteristic was negatively impacted, which depends upon the quantity of faba beans used. To boost protein content, the Food and Agriculture Organisation (FAO) recommends a ratio of 30:70 for legumes and cereal (Abd El-Hady and Habiba, 2003). Fermented bean flour combined with L. plantarum was utilized to strengthen semolina pasta by 30%, which has a huge amount of mineral nutrients. Most nutrient standards got better, and the resistant starch quantity rose by 2.18% without altering the technical or sensory attributes. The faba bean flour, i.e., FBF-fortified pasta, required less cooking time but dried easily. The substantial rise in protein content was 21% higher in faba bean-enriched pasta at 50% against 13.7% in the control sample. Also, the fibre, raw sugar, and mineral (calcium, iron, and zinc) contents of FBF-enriched pasta were enhanced. Rosa-Sibakov et al. (2016) investigated the impact of treated FBF

on the texture, structure and sensory aspects of gluten-free pasta made by FBF, starchy faba bean flour isolate (FBF-S), or fermented faba bean flour (FBF-F). FBF and FBF-F pasta showed considerably increased cooking loss (10.8-11.5%) and poorer water absorption than semolina (control). Giménez et al. (2013) made corn-broad bean spaghetti with a 30:70 corn-broad bean flour ratio.

7.3.2. Bread

The fermentation of lactic acid was explored to enhance nutritive value for making faba bean flour to be used in the creation of sugar-free bread, and the properties of the bread were compared to the bread prepared from soybean (Sozer et al., 2019). Wang et al. (2018) described the inclusion of bean flour (high in protein and biologically active compound) into bread made from wheat, which has been demonstrated to be a protein-enriched food item. Unfermented and fermented faba bean fibre loaves were soft and more porous, with 82% and 72% porosity, respectively, compared to 61% for soybean flour bread. Vioque et al. (2012) reported that protein isolates, optimum nutritional oil, phenolics with a great level of antioxidant activity, and easily solvable polypeptide, which remains rigid with optimum moisture and oil holding capacity obtained from faba bean were analyzed that can be used in the food industry because of their essential and health-forming properties. It was also discovered that faba bean has superior baking qualities when compared to lentils and peas. Dextran (hydrocolloid) integrated through *Weissella confusa* was found to be efficient in improving the standard of wheat bread (dough density characteristics, large volume, and low stiffness) mixed with faba bean flour.

7.3.3. Yogurt and Tofu

The two meals with emulsions of proteins similar to yogurt and bean curd were created by Jiang et al. (2020). Heating pre-treatment of the beans, dehulling, grinding, adding plant oil, mixing, starchy gel formation prevention, and inducement of Gelation of proteins were all part of the processing procedures. These procedures of starch-gelation avoidance, i.e., removing or hydrolyzing starch, resulted in yoghurt and tofu formation having strong emulsifying gel-like effects. Because hydrolysates were demonstrated to enhance density and gelling power, sugar hydrolysis is a better procedure for manufacturing yogurt. Zee et al. (1987) made bean curd using protein isolates of faba beans through several methods.

7.3.4. Fruit Juices

Today, faba bean hydrolysis protein isolates are supplemented in fruit juices and making nutritious beverages is a unique strategy aimed at adding a novel and reliable source of protein and thus increasing economic worth as well as industrial applicability. According to Samaei et al. (2020), apple juice is among the key juices throughout the globe enriched with protein isolates of the faba bean. Apple juice fortified with protein hydrolysate or isolates of faba bean and 3% alcalase may be most suitable and preferred. This acceptability is owing to their nutritional, functional, and bioactive qualities. These faba protein hydrolysates, which are high in amino acids and antioxidants, might be a novel element in manufacturing functional meals. This hydrolysate can improve the beneficial properties of this faba bean protein and can be utilized to generate fortified drinks as functional food components.

Figure 4. Faba bean products.

7.3.5. Isolates of Protein and Other Products

Protein isolates are made in various ways, for example, through dripping or soaking mature seeds of faba beans in water. Functional isolates of faba bean protein characteristics show equivalency to the isolates obtained from soybeans, peas, whey protein, and eggs (Figure 4). Sulaiman et al. (2018) created ground meat patties with 20% faba bean protein isolates, which dramatically boosted product production and dietary fibre content compared to 100% meat patties. Ouraji et al. (2020) extracted faba bean proteins (FBPs) enzymatically for prospective usage in replacement of egg yolk powder (EYP) in dressings such as sauce. Based on quality assessment, production with equal amounts of FBP and EYP (0.37% each), one having 0.50% FBP and the other having 0.25% EYP, was recommended. da Carmo et al. (2021) revealed meat analogue products might be made from FPC produced as a result of dried fractionation. The produced flesh analogue has high persistency, flexibility, fibrous, and stinging, all of which are key neural qualities.

7.3.6. Favism: A Case Study of Excess Consumption of Faba Bean

Intake of faba beans is linked up with the progression of favism, a disorder caused by a genetic defect that occurs after excess consumption of faba bean-based food. It is induced by a high level of aminopyrimidine-derived substances found in faba beans, such as alkaline compounds, aglycones, convicine, and vicine, produced primarily from the broad beans and its metabolic absorption (Oliveira et al., 2000). The two most important anti-nutritional glycosides are vicine and convicine, found in faba bean seeds, which have been responsible for acute haemolytic anaemia (the degeneration of red blood cells), which occurs in favism. Due to this favism, it is also known as haemolytic anaemia. The major cause is a lack of an enzyme called glucose-6-phosphate dehydrogenase (G6PD). If the cells lack G6PD, significant destruction of red blood cells happens, which is entirely hereditary. G6PD regulates NADPH generation in cells via hexose monophosphate shunt (its coenzyme utilized for producing ions (e- and H+) for the process catalyzed through specific enzymes, generally forming big molecules. People with G6PD deficiency are vulnerable to the effects of oxidative stress and cannot regenerate decreased glutathione. As a consequence, natural mortality is greatly increased, and the elimination of biological cells that are chronologically young is intensified (Arese et al., 2012; Rizzello et al., 2016). Faba bean seeds consist of beta-glycosidase activities that change with the development stages, i.e., highest within mature seeds and lowest among immature and old seeds (Arese & De Flora, 1990). Heating and boiling may

inactivate the enzyme, but microbial beta-glycosidase can still create Divicine and Isouramil in the intestine and the cecum (Yannai & Marquardt, 1985). Suitable preparation and manufacturing processes have repeatedly demonstrated that traditional heating methods (steaming or grilling) employed for softening the texture will denature protein, decrease alkaloid content, and reduce favism (Luzzatto & Arese, 2018).

Conclusion

Faba bean is a primitive leguminous crop that ranked sixth in the world. Faba beans are largely consumed worldwide, while India is the leading nation in terms of per-capita use. It has been considered an outstanding multipurpose pulse that could be utilized both for the nourishment of humans as well as a fodder crop for animal feed. Faba bean seeds and pods are consumed (either fresh or dry and cooked). They are considered beneficial for human nutrition because it is a rich source of dietary fibre, proteins, carbohydrates, vitamins, minerals, bioactive compounds and, antinutritional factors etc. Agronomically, faba bean is a potential alternative to food grains since it fixes atmospheric nitrogen, allowing agriculturists to implement green agriculture and achieve environmentally beneficial targets. Furthermore, using faba bean in crop rotation systems improves soil and boosts organic matter. Faba beans are a rich source of lysine-containing amino acids along with multiple essential factors. Levodopa, which is the precursor of dopamine, is found in faba bean and has the potential to be utilized as a biologically active molecule for the cure of Parkinson's disorder. Faba beans are also a good source of polyphenolic substances that have antioxidative and anti-inflammatory properties. Deconstruct, sopping, preparing, disinfecting, growing, exclusion, fermentation, and enzymatic procedures employed to remove antinutritional factors make it more palatable as a nutrient-rich component in various meals. Following the different uses, consumption, and advantages, faba bean has the potential to be employed in making a variety of foods as well as food products, proteinaceous compound as a substitute for an animal-based peptide that matches well with shifting user preferences and current environmental demands. Faba bean is also a climate-resilient crop because of its adaptability to various abiotic and biotic stress factors. Also, the incorporation of faba bean in rotational cropping and intercropping systems is projected towards a vast array of environmental applications.

References

Abd El-Hady, E. A., & Habiba, R. A. (2003). Effect of soaking and extrusion conditions on antinutrients and protein digestibility of legume seeds. *LWT-Food Science and Technology*, 36(3), 285-293.

Abdel, L. Y. I. (2008). Effect of seed size and plant spacing on yield and yield components of Faba bean (*Vicia faba* L.) *Res. J Agric. Biolog.* Sci. 4:146-148.

Abreu, J., Bruno-Soares A. (1998) Characterization and utilization of rice, legume and rape straws. In: Antongiovanni M (ed) *Exploitation of Mediterranean Roughage and by-products.* CIHEAM, Zaragoza, pp 39–51.

Abu-Reidah, I. M. Contreras, M. M. ArráezRomán, D. Fernández-Gutiérrez, A., UHPLCESI-QTOF-MS-based metabolic profiling of *Vicia faba* L. (Fabaceae) seeds as a key strategy for characterization in foodomics. *Electrophoresis*, 35, 2014, pp. 1571–1581. doi: 10.1002/elps.201300646.

Acharya, D., Shrivastava, A. (2008). Indigenous Herbal Medicines: Tribal Formulations and Traditional Herbal Practices, *Aavishkar Publishers Distributor, Jaipur- India.* P. 440.

Akbar, R. (2000). *Nuskha Dar Fanni-Falahat:* (The Art of Agriculture); (Persion Manuscript Compiled In The 17th Century by the Mughal Prince Dara Shikoh) (Agri-History Bulletin No. 3) Munshiram Manoharlal Publishers Pvt Ltd. P. 98.

Alandia, G., Pulvento C., Sellami M. H., Hoidal N., Anemone T., Nigussie E., Agüero J. J., Lavini A., Jacobsen S. E. (2020) Grain legumes may enhance high-quality food production in Europe. In: Hirich A, Choukr-Allah R, Ragab R (eds) *Emerging research in alternative crops,* vol 58. Springer International Publishing, Cham, pp 25–53.

Alghamdi, S. S. Chemical composition of faba bean (*Vicia faba L.*) genotypes under various water regimes. *Pak. J. Nutr.* 2009, 8, 477–482.

Arese, P., & De Flora, A. (1990, January). Pathophysiology of hemolysis in glucose-6-phosphate dehydrogenase deficiency. In *Seminars in hematology* (Vol. 27, No. 1, pp. 1-40).

Arese, P., Gallo, V., Pantaleo, A., & Turrini, F. (2012). Life and death of glucose-6-phosphate dehydrogenase (G6PD) deficient erythrocytes–role of redox stress and band 3 modifications. *Transfusion Medicine and Hemotherapy*, *39*(5), 328-334.

Barłóg, P., Grzebisz, W., Łukowiak, R. The Effect of Potassium and Sulfur Fertilization on Seed Quality of Faba Bean (*Vicia faba L.*). *Agronomy* 2019, 9, 209.

Bello-Pérez, L. A., Islas-Hernández, J. J., Rendon-Villalobos, J. R., Agama- Acevedo, E., Morales-Franco, L., & Tovar, J. (2007). *In vitro* starch digestibility of fresh and sun-dried faba beans (*Vicia faba L.*). *Journal of the Science of Food and Agriculture*, 87(8), 1517–1522. https://doi.org/10.1002/jsfa.2876.

Bilalis, D., Sidiras, N., Economou, G., and Vakali, C. (2003). Effect of different levels of wheat straw soil surface coverage on weed flora in *Vicia faba* crops. *J. Agron. Crop Sci.* 189, 233–241. doi: 10.1046/j.1439-037X.2003.00029.x.

Blancquaert, D., Storozhenko, S., Loizeau, K., De Steur, H., De Brouwer, V., Viaene, J., ... & Van Der Straeten, D. (2010). Folates and folic acid: from fundamental research toward sustainable health. *Critical Reviews in Plant Science*, *29*(1), 14-35.

Bond, D. A., Lawes, D. A., Hawtin, G. C., Saxena, M. C. and Stephens, J. S. (1985). "Faba bean (*Vicia faba* L.)," in *Grain Legume Crops*, eds R.J. Summerfield and E.H. Roberts (London: William Collins Sons), 199–265.

Boukhanouf, S. – Louaileche, H. – Perrin, D., Phytochemical content and *in vitro* antioxidant activity of faba bean (*Vicia faba* L.) as affected by maturity stage and cooking practice. *International Food Research Journal*, 23, 2016, pp. 954–961. ISSN: 1985-4668 (print), 2231-7546 (online).

Boye, J., Zare, F., Pletch, A. Pulse proteins: Processing, characterization, functional properties and applications in food and feed. *Food Res. Int.* 2010, 43, 414–431.

Caracuta, V., Barzilai, O., Khalaily, H., Milevski, I., Paz, Y., Vardi, J., et al., (2015). The onset of faba bean farming in the Southern Levant. *Sci. Rep.* 5:14370. doi: 10.1038/srep14370.

Caracuta, V., Weinstein-Evron, M., Kaufman, D., Yeshurun, R., Silvent, J., and Boaretto, E. (2016). 14,000-year-old seeds indicate the Levantine origin of the lost progenitor of faba bean. *Sci. Rep.* 6:37399. doi: 10.1038/srep37399.

Cazzato, E., Tufarelli, V., Ceci, E., Stellacci, A.M. and Laudadio, V. (2012). Quality yield and nitrogen fixation of faba bean seeds as affected by sulphur fertilization. *Acta Agriculturae Scandinavica Section B-Soil and Plant Science*. 62: 732- 738. https://doi.org/10.1080/09064710.2012.698642.

Chaieb, N. González, J. L. López-Mesas, M. Bouslama, M. Valiente, M., Polyphenols content and antioxidant capacity of thirteen faba bean (*Vicia faba* L.) genotypes cultivated in Tunisia. *Food Research International*, 44, 2011, pp. 970–977. doi: 10.1016/j.foodres.2011.02.026.

Crépon, K., Marget P., Peyronnet C., Carrouée B., Arese P., Duc G. (2010) Nutritional value of faba bean (*Vicia faba* L.) seeds for feed and food. *Field Crop Res* 115:329–339. https://doi.org/10.1016/j.fcr.2009.09.016.

Cubero, J. I. (1974). On the evolution of *Vicia faba* L. *Theor. Appl. Genet.* 45 47–51. 10.1007/BF00283475.

Dhull, S. B., Kidwai, M. K., Noor, R., Chawla, P., & Rose, P. K. (2022). A review of nutritional profile and processing of faba bean (*Vicia faba* L.). *Legume Science*, 4(3), e129.

do Carmo, C. S., Knutsen, S. H., Malizia, G., Dessev, T., Geny, A., Zobel, H., Myhrer, K., Varela, P., & Sahlstrøm, S. (2021). Meat analogues from a faba bean concentrate can be generated by high moisture extrusion. *Future Foods,* 4, 100014.

Duc, G., Aleksic, J. M., Marget, P., Miki ´ c, A., Paull, J., Redden, R. J., et al., (2015a). "Faba bean," in *Grain Legumes,* ed. A.M. De Ron (New York, NY: Springer), 141–178. doi: 10.1007/978-1-4939-2797-5_5.

Estefania, C. M. C., Ivan, V. T. J., Maria, C. T. L., Rosa-Millan, J. D. L., Othona, S. S. S. Physicochemical characteristics, ATR-FTIR molecular interactions and *in vitro* starch and protein digestion of thermally-treated whole pulse flours. *Food Res. Int.* 2018, 105, 371–383.

Etemadi, F., Hashemi, M., Mangan F., Weis, S., (2015). Faba *beans; Growers guide in New England.* http://ag.umass.edu/sites/ag.umass.edu/files/research-reports/faba_bean_guide_2.pdf.

FAOSTAT (2009). *Production stat: crops. FAO statistical databases (FAO stat), food and agriculture organization of the United Nations (FAO),* http://faostat.fao.org.

Food and Agriculture Organization (FAO). (2020). Crop production and trade data.

Food and Agriculture Organization (FAO). (2016). *FAOSTAT Statistical Database of the United Nation Food and Agriculture Organization (FAO) statistical division.* Rome.

Fujihara, T., Nakao, T. (1982) Eating and rumination behaviour in sheep given silage made from the fibrous residue of broad bean (*Vicia faba L.*). *J Agric Sci* 98:237–240. https://doi.org/10.1017/S0021859600041368.

Gasim, S. and Link, W. (2007). Agronomic performance and the effect of soil fertilization on German winter faba bean. *Journal of Central European Agriculture*. 8: 121-127.

Gilani, G. S., Xiao C. W., & Cockell K. A. (2012). Impact of antinutritional factors in food proteins on the digestibility of protein and the bioavailability of amino acids and on protein quality. *British Journal Nutrition*. 108:S315-S332. doi: 10.1017/ S0007114512002371.

Giménez, M. A., González, R. J., Wagner, J., Torres, R., Lobo, M. O., & Samman, N. C. (2013). Effect of extrusion conditions on physicochemical and sensorial properties of corn-broad beans (*Vicia faba*) spaghetti type pasta. *Food Chemistry,* 136, 538–545. https://doi.org/10.1016/j. foodchem.2012.08.068.

Grosjean, F., Bourdillon, A., Rudeaux, F., Bastianelli, D., Peyronnet, C., Duc, G., Lacassagne, L., (2000). Valeur alimentaire pour la volaille de fe´veroles isoge´niques (*Vicia faba L.*) avec ou sans tannins et avec ou sans vicine. *Sci. Tech. Avicoles* 32, 17–24.

Gu, B. J., Masli, M. D. P., & Ganjyal, G. M. (2020). Whole faba bean flour exhibits unique expansion characteristics relative to the whole flours of lima, pinto, and red kidney beans during extrusion. *Journal of Food Science*, 85, 404–413. https://doi.org/10. 1111/1750-3841.14951.

Hawtin, G. C., Hebblethpiait P. D. (1983). Background and history of faba bean production. Pages 3-22 in *The Faba Bean* (*Vicia faba* L.) (Hebblethwaite, P.D., ed.). Buttenvorths, London, U.K.

Hefni, M. E. Shalaby, M. T. Witthöft, C. M., Folate content in faba beans (*Vicia faba L.*) – effects of cultivar, maturity stage, industrial processing, and bioprocessing. *Food Science & Nutrition*, 3, 2015, pp. 65–73. doi: http://dx.doi.org/10.1002/fsn3.192.

Heuzé, V., Tran, G., Delagarde, R., Lessire, M., and Lebas, F. (2016). *Faba bean (Vicia faba). Feedipedia, a Programme by INRA, CIRAD, AFZ and FAO.* Available at: http:// www.feedipedia.org/node/620.

Hoehnel, A., Axel, C., Bez, J., Arendt, E. K., & Zannini, E. (2019). Comparative analysis of plant-based high-protein ingredients and their impact on quality of high-protein bread. *Journal of Cereal Science,* 89, 89.

Hulse, J. H. (1994). Nature, composition and utilization of food legumes. In: F.J. Muehlbauer and W.J. Kaiser (eds.), *Expanding the Production and Use of Cool Season Food Legumes.* Kluwer Academic Publishers, Dordrecht, The Netherlands, pp. 77-97.

Hussein, L. A., Saleh M. (1985). Antinutritional factors in faba bean. In: M. C. Saxena and S. Verma (eds.), Proceedings of the International Workshop on faba bean, Kabuli Chickpeas and Lentils in the 1980s. *ICARDA,* 16-20 May, 1983. Aleppo, Syria, pp. 257-269.

Ivarsson, E., Neil, M. Variations in nutritional and antinutritional contents among faba bean cultivars and effects on growth performance of weaner pigs. *Livest. Sci.* 2018, 212, 14–21.

Jambunathan, R., Blain, H. L., Dhindsa, K. H., Hussein, L. A., Kogure, K., Li-Juan, L. et al., (1994). Diversifying Use of Cool Season Food Legumes through Processing. In: F. J. Muehlbauer and W. J. Kaiser (eds.), *Expanding the Production and Use of Cool Season Food Legumes.* Kluwer Academic Publishers, Dordrecht, The Netherlands. pp. 98-112.

Jensen, E. S., Peoples M. B., Boddey R. M., Gresshof P. M., Hauggaard Nielsen H., Alves J. R. B., Morrison M. J. (2012). Legumes for mitigation of climate change and the provision of feedstock for biofuels and biorefineries A review. *Agron Sustain Dev* 32:329–364. https://doi.org/10.1007/s13593-011-0056-7.

Jensen, E., Peoples, M. B., and Hauggaard-Nielsen, H. (2010). Faba bean in cropping systems. *Field Crops Res.* 115, 203–216. doi: 10.1016/j.fcr.2009.10.008.

Jiang, Z. Q., Wang, J., Stoddard, F., Salovaara, H., & Sontag-Strohm, T. (2020). Preparation and characterization of emulsion gels from whole faba bean flour. *Food,* 9, 755. https://doi.org/10.3390/foods9060755.

Johnson, J. B., Collins, T., Skylas, D., Quail, K., Blanchard, C., & Naiker, M. (2020). Profiling the varietal antioxidative contents and macro-chemical composition in Australian faba beans (*Vicia faba L.*). *Legume Science*, 2, e28.

Karatas, S. Ç., Günay, D., & Sayar, S. (2017). *In vitro* evaluation of whole faba bean and its seed coat as a potential source of functional food components. *Food Chemistry*, 230, 182–188. https://doi.org/10.1016/j.foodchem.2017.03.037.

Kaufman, P. B. Duke, J. A. Brielmann, H. Boik, J. Hoyt, J. E., A comparative survey of leguminous plants as sources of the isoflavones, genistein and daidzein: Implications for human nutrition and health. *Journal of Alternative and Complementary Medicine*, 3, 2007, pp. 7–12. doi: 10.1089/acm.1997.3.7.

Kendall, D. A., Smith, B., D. (1975) The pollinating efficiency of honeybee and bumblebee visits to field bean flowers (*Vicia faba L.*). *Journal of Applied Ecology* 12: 709-717.

Khalil, A. H., & Mansour, E. H. (1995). The effect of cooking, autoclaving and germination on the nutritional quality of faba beans. *Food Chemistry*, 54, 177–182. https://doi.org/10.1016/0308-8146(95)00024-D.

Khan, M. A., Ammar, M. H., Migdadi, H. M., El-Harty, E. H., Osman, M. A., Farooq, M., & Alghamdi, S. S. (2015). Comparative nutritional profiles of various faba bean and chickpea genotypes. *International Journal of Agriculture and Biology*, *17*(3), 449-457.

Khazaei, H., Purves, R. W., Hughes, J., Link, W., O'Sullivan, D. M., Schulman, A. H., Björnsdotter, E., Geu-Flores, F., Nadzieja, M., Andersen, S. U., Stougaard, J., Vandenberg, A., & Stoddard, F. L. (2019). Eliminating vicine and convicine, the main anti-nutritional factors restricting faba bean usage. *Trends in Food Science & Technology,* 91, 549–556. https://doi.org/10.1016/j.tifs.2019.07.051.

Khazaei, H., Hawkins, G., & Vandenberg, A. (2021). Historical review of faba bean improvement in western Canada. *Legume Science*, *3*(4), e92.

Köpke, U., & Nemecek, T. (2010). Ecological services of faba bean. *Field crops research*, *115*(3), 217-233.

Labba, I. C. M., Frøkiær, H., & Sandberg, A. S. (2021). Nutritional and antinutritional composition of faba bean (*Vicia faba L., var. minor*) cultivars. *Food Research International*, 140, 110038.

Landry, E. J., Coyne, C. J., McGee, R. J., & Hu, J. (2016). Adaptation of autumn-sown faba bean germplasm to southeastern Washington. *Agronomy Journal*, *108*(1), 301-308.

Lawes, D. A. (1980). Recent developments in understanding, improvement and use of *Vicia faba*. In: R. J. Summerfield and A. H. Bunting (eds.), *Advances in Legumes Science. Proceedings of the International Legume Conference*, Kew, 31 July-4 August 1978, Royal Botanic Garden, Kew, the Missouri Botanical Garden, and the University of Reading, UK, pp. 625-636.

Lemke, R. L., Zhong, Z., Campbell, C. A., Zentner, R. P. (2007). Can pulse crops play a role in mitigating greenhouse gases from North American agriculture? *Agron J.*;99:1719–25.

Lewis, G., Schrire, B., Mackinder, B., & Lock, M. (2005). *Legumes of the World. Kew: Royal Botanic Gardens.*

Li-Juan, L., Zhao-hai, Y., Zhao-jie, Z., Ming-shi, X., & Han-qing, Y. (1993). Faba bean in China: state-of-the-art review. *International Center for Agricultural Research in the Dry Areas (ICARDA). Aleppo, Syria*, 127-128.

Link, W., Balko C., Stoddard F. L. (2010). Winter hardiness in faba bean: physiology and breeding. *Field Crops Res.* 115 287–296. 10.1016/j.fcr.2008.08.004.

Liu, C., Pei, R., & Heinonen, M. (2022). Faba bean protein: A promising plant-based emulsifier for improving physical and oxidative stabilities of oil-in-water emulsions. *Food Chemistry,* 369, 130879. https://doi.org/10.1016/j.foodchem.2021.130879.

Lizarazo, C. I., Lampi, A. M., Sontag-Strohm, T., Liu, J., Piironen, V., and Stoddard, F. L. (2015). Nutritive quality and protein production from grain legumes in a boreal climate. *J. Sci. Food Agric.* 95, 2053–2064. doi: 10.1002/jsfa.6920.

Lo´pez-Bellido, L., Lo´pez-Bellido, R. J., Redondo, R., Benı´tez, J. (2006). Faba bean nitrogen fixation in a wheat-based rotation under rainfed Mediterranean conditions: effect of tillage system. *Field Crops Res.* 98, 253–260.

Long, J., Lin, L., Hou, X., Duan X., and Duan H. (1989). *Food Legume Crops.* Science Publishing House.

Longobardi, F., Sacco, D., Casiello, G., Ventrella, A., and Sacco, A. (2015). Chemical profile of the carpino broad bean by conventional and innovative physicochemical analyses. *J. Food Qual.* 38, 273–284. doi: 10.1111/jfq.12143.

Luo, Y. Xie, W. Hao, Z. Jin Z. Wang, W. The impact of processing on *in vitro* bioactive compounds bioavailability and antioxidant activities in faba bean (*Vicia faba L.*) and azuki bean (*Vigna angularis L.*). *International Food Research Journal*, 21, 2014, pp. 1031–1037. ISSN: 1985-4668 (print).

Luo, Y., Gu, Z., Han, Y., & Chen, Z. (2008). The impact of processing on phytic acid, *in vitro* soluble iron and Phy/Femolar ratio of faba bean (*Vicia faba L.*). *Journal of Agricultural and Food Chemistry*, 89, 861–866.

Luzzatto, L., & Arese, P. (2018). Favism and glucose-6-phosphate dehydrogenase deficiency. *New England Journal of Medicine*, 378, 60–71. https://doi.org/10.1056/NEJMra1708111.

Madar, Z. and Stark, A.H. (2002). New legume sources as therapeutic agents. *British Journal of Nutrition.* 88: 287–292. https://doi.org/10.1079/BJN2002719.

Małgorzata, S., Jerzy, K., Ewa, H. The Effect of Feeding Native Faba Bean Seeds (*Vicia faba L.*) to Sows and Supplemented With Enzymes to Piglets and Growing Pigs. *Ann. Anim. Sci.* 2018, 18, 1007–1027.

Martin, A., Cabrera, A. and Medina, J.L. (1991). Antinutritional factors in faba bean. Tannin content in *Vicia faba*: possibilities for plant breeding. *Options Méditerranéennes-Série Séminaires.* 10:105-110.

Martineau-Côté, D., Achouri, A., Karboune, S., & L'Hocine, L. (2022). Faba bean: An untapped source of quality plant proteins and bioactives. *Nutrients*, *14*(8), 1541.

Marzinzig, B., Brünjes, L., Biagioni, S., Behling, H., Link, W., Westphal, C. (2018) Bee pollinators of faba bean (*Vicia faba L.*) differ in their foraging behaviour and pollination probability. *Agriculture, Ecosystems and Environment* 264: 24-33.

McVicar, R., Panchuk, D., Brenzil, C., Hartley, S., Pearse, P., Vandenberg, A. (2013). Faba bean. *Gov. Saskatchewan, Agriculture, Crops.*

Merga, B., Egigu, M. C., & Wakgari, M. (2019). Reconsidering the economic and nutritional importance of faba bean in Ethiopian context. *Cogent Food & Agriculture*, *5*(1), 1683938.

Mihailovic, V., Mikic, A., Cupina, B., and Eric, P. (2005). Field pea and vetches in Serbia and Montenegro. *Grain Legumes*. 44: 25-26.

Millar, K. A., Gallagher, E., Burke, R., McCarthy, S., Barry-Ryan, C. Proximate composition and anti-nutritional factors of faba-bean (*Vicia faba*), green-pea and yellow-pea (Pisum sativum) flour. *J. Food Comp. Anal.* 2019, 82, 103233.

Mínguez, M. I., & Rubiales, D. (2021). Faba bean. In V. O. Sadras & D. F. Calderini (Eds.), *Crop Physiology Case Histories for Major Crops* (pp. 452–481). Academic Press. https://doi.org/10.1016/B978-0-12-819194-1.00015-3.

Muehlbauer, F., Tullu, A., (1997). *Vicia faba L.* . Purdue Univ., Cent. New Crops Plants Prod., *New Crop Factsheet* http://www.hort.purdue.edu/newcrop/cropfactsheets/fababean.html.

Muktadir, M. A., Adhikari, K. N., Merchant, A., Belachew, K. Y., Vandenberg, A., Stoddard, F. L. and Khazaei, H., 2020. Physiological and biochemical basis of faba bean breeding for drought adaptation - A review. *Agronomy*, *10*(9), p.1345.

Multari, S., Stewart, D., & Russell, W. R. (2015). Potential of fava bean as future protein supply to partially replace meat intake in the human diet. *Comprehensive Reviews in Food Science and Food Safety*, *14*(5), 511-522.

Müntz, K., Horstmann, C., Schlesier, B. Vicia globulins. In *Seed Proteins; Shewry*, P.R., Casey, R., Eds., Springer: Dordrecht, The Netherlands, 1999; pp. 259–284.

Naqvi, H. K. (1984). Cultivation under the Sultans of Delhi c. 1206–1555. *Indian J. History. Sci.* 19:329–340.

Neme, K., Bultosa, G., and Bussa, N. (2015). Nutrient and functional properties of composite flours processed from pregelatinised barley, sprouted faba bean and carrot flours. *Int. J. Food Sci. Technol.* 50, 2375–2382. doi: 10.1111/ijfs. 12903.

Neugschwandtner, R., Ziegler, K., Kriegner, S., Wagentristl, H., & Kaul, H. P. (2015). Nitrogen yield and nitrogen fixation of winter faba beans. *Acta Agriculturae*

Scandinavica, Section B-Soil & Plant Science, 65(7), 658–666. doi: 10.1080/09064710.2015.1042028.

Nosworthy, M. G., Medina, G., Franczyk, A. J., Neufeld, J., Appah, P., Utioh, A., Frohlich, P., & House, J. D. (2018). Effect of processing on the *in vitro* and *in vivo* protein quality of beans (Phaseolus vulgaris and *Vicia faba*). *Nutrients*, 10, 671. https://doi.org/10.3390/nu10060671.

Ntatsi G., Gutiérrez-Cortines M.E., Karapanos I., Barros A., Weiss J., Balliu A., Rosa E.A.d.S., Savvas D. The quality of leguminous vegetables as influenced by preharvest factors. *Sci. Horticult.* 2018;232:191–205. doi: 10.1016/j.scienta.2017.12.058.

O'kiely, P., McGee, M., & Kavanagh, S. (2022). *Faba beans as a feed for cattle and sheep.*

Oliveira, S., Pinheiro, S., Gomes, P., Horta, A. B., & Castro, A. S. (2000). Favism. *Acta Medica Portuguesa*, 13, 237–240.

Oplinger, E. S. (1982). *Faba beans Field Crops* 32.0 UWEX. Madison, WI 53706.

Osborne, T. B. *The Proteins of The Wheat Kernel.* Carnegie institution of Washington; Washington, DC, USA: 1907. p. 119.

Ouraji, M., Alimi, M., Motamedzadegan, A., & Shokoohi, S. (2020). Faba bean protein in reduced fat/cholesterol mayonnaise: Extraction and physico-chemical modification process. *Journal of Food Science and Technology*, 57, 1774–1785. https://doi.org/10.1007/s13197-019-04211-9.

Patil, S. A., Apine, O. A., Surwase, S. N., Jadhav, J. P., 2013. Biological sources of L-DOPA: an alternative approach. *Adv Parkinson's Dis*, 2: 81–87.

Paul, S. K., & Gupta, D. R. (2021). Faba bean (*Vicia faba* L.), A promising grain legume crop of Bangladesh: A review. *Agricultural Reviews*, *42*(3), 292-299.

Prolea, (2014). Les graines: pois, féveroles, lupin. Prolea. *L'alimentation animale*. http://www.prolea.com/index.php?id=1600.

Punia, S., Dhull, S. B., Sandhu, K. S., & Kaur, M. (2019). Faba bean (*Vicia faba*) starch: Structure, properties, and *in vitro* digestibility - A review. *Legume Science*, 1, e18.

Rahate, K. A., Madhumita, M., & Prabhakar, P. K. (2021). Nutritional composition, anti-nutritional factors, pre-treatments-cum-processing impact and food formulation potential of faba bean (*Vicia faba L.*): A comprehensive review. *Lwt*, *Food Science and Technology 138*, 110796.

Raikos V., Neacsu M., Russell W., Duthie G. Comparative study of the functional properties of lupin, green pea, faba bean, hemp, and buckwheat flours as affected by pH. *Food Sci. Nutr.* 2014;2:802–810. doi: 10.1002/fsn3.143.

Rajan K, Singh AK (2012). Faba bean (*Vicia faba* L.) Carbon sequestration and its role in soil quality. In: Bean (*Vicia faba* L) Eds. Singh A. K. and Bhatt B. P., ICAR, RC for ER, *Patna*, pp. 285-291.

Rizzello, C. G., Losito, I., Facchini, L., Katina, K., Palmisano, F., Gobbetti, M., & Coda, R. (2016). Degradation of vicine, convicine and their aglycones during fermentation of faba bean flour. *Scientific reports*, *6*(1), 32452.

Rosa-Sibakov, N., Heiniö, R. L., Cassan, D., Holopainen-Mantila, U., Micard, V., Lantto, R., & Sozer, N. (2016). Effect of bioprocessing and fractionation on the structural, textural and sensory properties of gluten-free faba bean pasta. *LWT- Food Science and Technology,* 67, 27–36.

Samaei, S. P., Ghorbani, M., Tagliazucchi, D., Martini, S., Gotti, R., Themelis, T., ... Babini, E. (2020). Functional, nutritional, antioxidant, sensory properties and comparative peptidomic profile of faba bean (*Vicia faba L.*) seed protein hydrolysates and fortified apple juice. *Food Chemistry.* Article 127120. https://doi.org/10.1016/j.foodchem.2020.127120.

Senberga, A., Dubova L., Alsina I., Strauta L. *Rhizobium* sp. - A potential tool for improving protein content in peas and faba beans. *Rural Sustain. Res.* 2017; 37:2–9. doi: 10.1515/plua-2017-0001.

Sharma, A., & Sehgal, S. (1992). Effect of processing and cooking on the antinutritional factors of faba bean (*Vicia faba*). *Food Chemistry*, 43, 383–385. https://doi.org/10.1016/0308 8146(92)90311 O.

Sillero, J. C., Villegas-Fernandez A. M., Thomas J., Rojas-Molina M. M., Emeran A. A., Fernandez-Aparicio M., et al., (2010). Faba bean breeding for disease resistance. *Field Crop Res.* 115 297–307. 10.1016/j.fcr.2009.09.012.

Singh, A. K., Bhatt B. P., Upadhyaya A., Kumar S., Sundaram P. K., Singh B. K., et al., (2012b). Improvement of faba bean (*Vicia faba* L.) yield and quality through biotechnological approach: a review. *Afr. J. Biotechnol.* 11 15264–15271.

Singh, A. K., Bhat B. P., Sundaram P. K., Chandra N., Bharati R. C., Patel S. K., (2012a). Faba bean (*Vicia faba L.*) phenology and performance in response to its seed size class and planting depth. Int. J. Agric. Stat. Sci. 8(1):97-109.

Singh, A. K., Bhat B. P., Sundaram P. K., Gupta A. K., Singh D. (2013b). Planting geometry to optimize growth and productivity faba bean (*Vicia faba L.*) and soil fertility. J. Environ. Biol. 34 (1):117-122.

Singh, A. K., & Bhatt, B. P. (2012b). Faba bean: Unique germplasm explored and identified. *HortFlora Research Spectrum*, 1(3), 267–269.

Singh, A. K., Bhardwaj, R., & Singh, I. S. (2014). Assessment of nutritional quality of developed faba bean (*Vicia faba L.*) lines. *Journal of Agri Search*, 1, 96–101.

Singh, A. K., & Bhatt, B. P. (2012a). Faba Bean (*Vicia faba L.*): A potential leguminous crop of India ISBN 978-93-5067-773-5ICAR, RC for ER, *Patna,* P. 518.

Singh, A. K., Bharati, R. C., Manibhushan, N. C., and Pedpati, A., (2013a). An assessment of faba bean (*Vicia faba L.*) current status and future prospect. *African Journal of Agricultural Research.* 8: 6634-6641.

Singh, A. K., Chandra, N., Bharati, R. C., Dimree, S. K. (2010). Effect of seed size and seeding depth on Faba bean (*Vicia faba* L.) productivity. *Environ. Ecol.* 28(3A):1722-1527.

Smith, J., & Hardacre, A. (2011). Development of an extruded snack product from the legume *Vicia faba* minor. *Procedia Food Science, 1*, 1573-1580.

Sofi, B. A., Wani, I. A., Masoodi, F. A., Saba, I., & Muzaffar, S. (2013). Effect of gamma irradiation on physicochemical properties of broad bean (*Vicia faba L.*) starch. LWT-*Food Science and Technology*, 54, 63–72. https://doi.org/10.1016/j.lwt.2013.05.021.

Sozer, N., Melama, L., Silbir, S., Rizzello, C. G., Flander, L., & Poutanen, K. (2019). Lactic acid fermentation as a pre-treatment process for faba bean flour and its effect on textural, structural and nutritional properties of protein-enriched gluten-free faba bean breads. *Food,* 8, 431. https://doi.org/10.3390/foods8100431.

Stagnari, F., Maggio, A., Galieni, A., & Pisante, M. (2017). Multiple benefits of legumes for agriculture sustainability: an overview. *Chemical and Biological Technologies in Agriculture*, *4*(1), 1-13.

Suárez-Diéguez, T., Pérez-Moreno, F., Ariza-Ortega, J. A., Lopez- Rodríguez, G., & Nieto, J. A. (2021). Obtention and characterization of resistant starch from creole faba bean (*Vicia faba L. creole*) as a promising functional ingredient. *LWT*, 145, 111247.

Subash, N., Priya N. (2012). Climatic requirements for optimizing Faba bean (*Vicia faba* L.) production In: Faba Bean (*Vicia faba L.*) *A Potential legume for India*. Eds. Singh and Bhatt. ICAR, RC for ER, Patna, pp. 197-204.

Sulaiman, N., Orfila, C., Ho, P., & Maycock, J. (2018). *Vicia faba*: A cheap and sustainable source of protein and its application in beef products. *Proceedings of the Nutrition Society,* 77(OCE4), E137.

Tanno, K., & Willcox G. (2006) The origins of cultivation of *Cicer arietinum* L. and *Vicia faba* L., early finds from Tell el-Kerkh, north-west Syria, late 10th millennium b.p. *Veget Hist Archaeobot* 15:197–204. https://doi.org/10.1007/s00334-005-0027-5.

Tazrart, K., Lamacchia, C., Zaidi, F., & Haros, M. (2016). Nutrient composition and *in vitro* digestibility of fresh pasta enriched with *Vicia faba*. *Journal of Food Composition and Analysis*, 47, 8-15. https://doi.org/10.1016/j.jfca.2015.12.007.

Torres, A. M., Roman, B., Avila, C. M., Satovic, Z., Rubiales, D., Sillero, J. C., … Moreno, M. T. (2006). Faba bean breeding for resistance against biotic stresses: Towards application of marker technology. *Euphytica*, 147 (1–2), 67–80. doi:10.1007/s10681-006-4057-6.

Turco, I., Ferretti, G. and Bacchet, T. (2016). Review of the health benefits of Faba bean (*Vicia faba L.*) polyphenols. *Journal of Food and Nutrition Research. 55* (4): 283-293.

United Nations: World population prospects: The 2012 revision, key findings and advance tables. *Working paper no.* ESA/P/WP.227; 2013 (United Nations, Department of Economic and Social Affairs, Population Division, New York).

Upadhyay, V., Muntean, B., Hwang, J., Kathem, S., Abou-Alaiwi, W. and Nauli, S. (2014). Roles of dopamine receptor on chemosensory and mechanosensory primary cilia in renal epithelial cells. *Frontiers in Physiology*. 5: 72. https://doi.org/10.3389/fphys. 2014.00072.

USDA (US Dept of Agriculture). (2021). *Food Data Central (Nutrient Database).* https://fdc.nal.usda.gov/ (accessed on January 26, 2021).

Utsumi, S. Plant Food Protein Engineering. In *Advances in Food and Nutrition Research;* Kinsella, J. E., Ed., Academic Press: Kyoto, Japan, 1992; Volume 36, pp. 89–208.

Valente, I. M., Maia, M. R., Malushi, N., Oliveira, H. M., Papa, L., Rodrigues, J. A., Fonseca, A. J., & Cabrita, A. R. (2018). Profiling of phenolic compounds and antioxidant properties of European varieties and cultivars of *Vicia faba L.* pods. *Phytochemistry*, 152, 223–229. https://doi.org/10.1016/j.phytochem.2018.05.011.

Van Soest, P. J. (1994). *Nutritional ecology of the ruminant, 2nd edn.* Cornell University Press, Ithaca.

Vavilov, H. N. (1936). *The Original Centres of the World Crops (Translated by Dong Yusheng).* Agricultural Publishing House, 1980.

Vidal-Valverde, C., Frias, J., Sotomayor, C., Diaz-Pollan, C., Fernandez, M., & Urbano, G. (1998). Nutrients and antinutritional factors in faba beans as affected by processing. *European Food Research and Technology*, 207, 140–145.

Vioque, J., Alaiz, M., & Giron-Calle, J. (2012). Nutritional and functional properties of *Vicia faba* protein isolates and related fractions. *Food Chemistry,* 132(1), 67–72. https://doi.org/10.1016/j.foodchem.2011.10.033.

Voisin, A. S., Guéguen J., Huyghe C., Jeuffroy M. H., Magrini M. B., Meynard J. M., et al., (2014) Legumes for feed, food, biomaterials and bioenergy in Europe: a review. *Agron Sustain Dev.*;34:361–80.

Wang, Y., Sorvali, P., Laitila, A., Maina, N. H., Coda, R., & Katina, K. (2018). Dextran produced *in situ* as a tool to improve the quality of wheat-faba bean composite bread. *Food Hydrocolloids*, *84*, 396-405.

Wegi, T., Tolera A., Wamatu J., Animut G., Rischkowsky B. (2018) Efects of feeding diferent varieties of faba bean (*Vicia faba* L.) straws with concentrate supplement on feed intake, digestibility, body weight gain and carcass characteristics of ArsiBale sheep. *Asian-Australas J Anim Sci* 31:1221–1229. https://doi.org/10.5713/ajas.17.0736.

Yadav, S. K., Verma, N., Singh, A. K., Singh, N., Rana, S. C., Ranga S. S. and Kumar, K. (2017). Diversity and development in faba bean. *Legume Research.* 40(4): 618-623. https://doi.org/10.18805/lr.v0iOF.6200.

Yang, J., Liu, G., Zeng, H., & Chen, L. (2018). Effects of high pressure homogenization on faba bean protein aggregation in relation to solubility and interfacial properties. *Food Hydrocolloids*, 83, 275–286. https://doi.org/10.1016/j.foodhyd.2018.05.020.

Yang, Z., Piironen, V., Lampi, A.-M. Lipid-modifying enzymes in oat and faba bean. *Food Res. Int.* 2017, 100, 335–343.

Yannai, S., & Marquardt, R. R. (1985). Induction of favism-like symptoms in the rat: Effects of vicine and divicine in normal and buthionine sulphoximine-treated rats. *Journal of the Science of Food and Agriculture*, *36*(11), 1161-1168.

Yao, Y. Cheng, X. Wang, L. Wang, S. Ren, G., Biological potential of sixteen legumes in China. *International Journal of Molecular Science*, 12, 2011, pp. 7048–7058. doi: 10.3390/ijms12107048.

Yasmin, W., Paul, S. K. and Anwar, M. P. (2020). Growth, yield and quality of faba bean (*Vicia faba* L.) in response to sowing date and phosphorus fertilization. *Archives of Agriculture and Environmental Science.* 5(1): 11-17. https://dx.doi.org/10.26832/24566632.2020.050102.

Zee, J. A., Bourgeois, M., Boudreau, A., Breton, R., & Jones, J. D. (1987). Use of faba beans in the manufacture of tofu. *Canadian Institute of Food Science & Technology Journal*, 20, 260–266. https://doi.org/10.1016/S0315-5463(87)71197-3.

Zhang, Z., Tian, X., Wang, P., Jiang, H., & Li, W. (2019). Compositional, morphological, and physicochemical properties of starches from red adzuki bean, chickpea, faba bean, and baiyue bean grown in China. *Food Science & Nutrition*, 7, 2485–2494. https://doi.org/10.1002/fsn3.865.

Zong, X., Liu X., Guan J., Wang S., Liu Q., Paull J. G., et al., (2009). Molecular variation among Chinese and global winter faba bean germplasm. *Theor. Appl. Genet.* 118 971–978. 10.1007/s00122-008-0954-5.

Zong, X., Ren J., Guan J., Wang S., Liu Q., Paull J., Redden R. (2010). Molecular variation among Chinese and global germplasm in spring faba bean areas. *Plant Breed.* 129 508–513. 10.1111/j.1439-0523.2009.01718.x.

Chapter 3

Faba Bean: A Protein-Rich Legume

Asif Mohi Ud Din Rather[1,*]
Mohsin A. Hajam[2]
Ummer Atta[3]
Sumati Narayan[4]
Tariq A. Bhat[4]
Sartaj A. Wani[5]
Anil Kumar[6]
and Shahzad A. Bhat[3]

[1]National Horticulture Research and Development Foundation, Nashik, India
[2]Division of Fruit Science, Faculty of Horticulture Shalimar, SKUAST, Kashmir
[3]Faculty of Forestry, SKUAST, Kashmir
[4]Dvision of Vegetable Science, Faculty of Horticulture Shalimar, SKUAST, Kashmir
[5]Division of Soil Science, Faculty of Horticulture Shalimar, SKUAST, Kashmir
[6]Department of Vegetable Science, ANDUAT Kumarganj, Ayodhya, Uttar Pradesh, India

Abstract

Faba bean (*Vicia faba* L.) is a globally significant legume, thriving in Mediterranean and temperate climates with an optimal temperature range of 5°C to 35°C. Though sensitive to extreme temperatures, it is a valuable nitrogen-fixing crop in rotations. The global cultivation area of faba bean has declined from 3.8 million hectares in 1990 to 2.4 million hectares in 2019, but yield per unit area has increased due to improved stress resilience. Grown in over 55 countries, faba beans are high in protein (up

---

* Corresponding Author's Email: itzasif01@gmail.com

In: Faba Beans
Editors: Aamir Raina and Samiullah Khan
ISBN: 979-8-89530-447-1

to 35%) and rich in essential minerals and bioactive compounds. However, they contain antinutritional factors (lectins, saponins, and phytic acid) that can be reduced through cooking or sprouting. While the species is crucial in animal feed, its genetic diversity offers breeding potential to improve yield, winter hardiness, and frost tolerance. However, partial cross-pollination complicates breeding efforts, and faba beans remain vulnerable to biotic stresses. Faba beans, domesticated around 1000 BC in the Near East, are now grown in regions such as East Africa, East Asia, and North Africa, with Ethiopia and China being major producers. In 2021, faba beans accounted for 4% of the global pulse area, with diverse uses in human food, animal feed, and winter cropping. Their genetic variation, particularly in seed size and growth habit, provides opportunities for developing improved cultivars.

Keywords: faba bean, legumes, proteins, food values, food security

Introduction

Faba bean (*Vicia faba* L.), a leguminous family crop having paramount global significance, has witnessed a decline in its worldwide cultivation area from 3.8 to 2.4 million hectares between 1990 and 2019. Notably, country-specific yields exhibit substantial variability, as reported by the Food and Agriculture Organization (FAO, 2020). Despite diminishing acreage, the productivity per unit area exhibited an upward trend. This phenomenon is attributed to a diminished susceptibility to abiotic and biotic stressors, as expounded by Singh et al. (2012).

Across more than 55 nations, faba bean cultivation encompasses both irrigated and rainfed areas. The faba bean seeds, both in their fresh and dry states, find application in human consumption. These seeds are distinguished by their high nutritional value, boasting elevated protein content of up to 35% in dry seeds and serving as a rich source of essential nutrients, including but not limited to magnesium, potassium, calcium, zinc, and iron (Neme et al., 2015). Additionally, the seeds harbor various bioactive compounds, such as carotenoids, polyphenols, and carbohydrates (Landry et al., 2016). Notably, the chemical composition is subject to substantial variation contingent upon factors such as variety, environmental conditions, and management practices (Witten et al., 2015).

From a nutritional standpoint, mature faba bean seeds boast a composition comprising 26.1% proteins, 58.3% carbohydrates, and 25.0% dietary fiber as

per USDA (2021). The bioactive repertoire of faba bean extends to encompass phenolics and flavonoids, which exhibit demonstrable antioxidant activity, as elucidated by Valente et al. (2018). However, the existence of certain antinutritional elements, including saponins, phytic acids, trypsin inhibitors, lectins, condensed tannins, and factors inducing favism, imparts a detrimental impact on its overall biological value (Revilla, 2015). The consumption of faba beans carries the risk of inducing favism, characterized by severe hemolytic anaemia (Minguez & Rubiales, 2021). Furthermore, faba beans contain oligosaccharides such as stachyose, verbascose, and raffinose, which, upon fermentation, can lead to flatulence and abdominal discomfort (Toklu et al., 2021).

It is imperative to acknowledge that lectins are rendered inactive during standard cooking procedures due to the application of high heat. Moreover, Labba et al. (2021) underscore that sprouting constitutes an additional method capable of mitigating numerous antinutritional factors in faba beans, enhancing their nutritional profile for human consumption.

In Mediterranean environments, the faba bean takes center stage in production. However, its cultivation also occurs in cold-temperate and high-altitude subtropical environments. This C_3 species has high productivity, although it exhibits sensitivity to both moisture stress and extreme temperatures, whether high or low. The optimal temperature requirement for photosynthesis is 25°C within a growth range of 5-35°C. Frost threatens flowering, and only a few cultivars can withstand temperatures below -10°C.

Phenologically, faba beans are classified as Mediterranean, spring, or cool weather types based on their varying vernalization requirements for flowering. This differentiation enables adaptation to specific conditions, such as spring sewing in cold and warm regions, areas with mild winters, and regions where the winters are cold but not detrimental to the crop. Faba bean, with its effective N_2 fixation potential, consistently fixes higher rates of approximately 150 kg N ha^{-1}. It is integrated into diverse rotations, primarily among cereals and increasingly in intercrops.

The popularity of the faba bean in Mediterranean climates can be attributed to its development and physiological mechanisms, allowing it to avoid, instead of tolerating, temperature and water stress. Despite its dominance in the animal feed market, the faba bean's domesticated gene pool exhibits significant variability, offering breeding opportunities to enhance yield potential, including the development of determinate cultivars.

Breeding efforts may focus on improving winter hardiness and frost tolerance by utilizing the genetic variation already present for these traits to

extend autumn planting in colder climates. Because of the partial allogamy of the plant, the cultivars that are currently in use are either synthetic or open-pollinated populations. While developing inbred lines and selecting for self-fertility simplifies seed multiplication and breeding, it may not fully exploit the potential benefits of heterosis. The volatility of the cytoplasmic male sterility system prevents hybrids from being produced on a commercial basis, and this has also been a problematic scenario in faba bean. Besides, faba beans are susceptible to a broad array of biotic stresses, including soil-borne pathogens, parasitic weeds, foliar fungal diseases, nematodes, viruses, and insect pests.

Origin

Faba bean, a vital agricultural commodity, was cultivated at the onset of agriculture in the Middle East, signifying the Neolithic era about 9000–10,000 BC. Since then, the cultivation of the crop has proliferated globally, extending beyond its origins (Tanno & Willcox, 2006). The primary center of diversity encompasses Armenia, Syria, Iraq, Azerbaijan, Georgia, Turkey, and Iran (Maxted, 1995). Various routes have been postulated by Cubero (1974), emanating from the Near East to regions like Europe. These routes include Anatolia to Greece, the Nile Delta towards the Maghreb and Iberian lands, alongside the River Nile to Ethiopia, and from Iraq to India. Ethiopia is proposed as a secondary center of diversity, and Afghanistan, although Ladizinsky (1975) proposed Central Asia as the origin.

Trait analyses discern two distinct groups: large-seeded varieties in the west and small-seeded varieties in southwestern Asia. The ancient Eastern group traced back to Neolithic agriculture, exhibits the greatest diversity and numerous specific traits. From Spain to the Himalayas, this group covers a wide region (Muratova, 1931). Evidence from Tell El-Kerkh, an archaeological site in northwest Syria, indicates that the domestication of the crop dates to late in the 10th century BC (Tanno & Willcox, 2006), affirming the southwestern region of Asia as the primary center of origin for *V. faba*. Major types were first discovered in Iraq about 1000 AD (Schultze-Motel, 1972).

The migration of faba bean to South America, and especially the Andes, most likely started in the fourteenth century, aided along by Spanish and Portuguese explorers. This resulted in the emergence of landraces in Bolivia and Peru that showed significant variation in seed form, color, and size (Duc

et al., 2010). In China, historical records suggest that this crop (*V. faba* var. major) was bought from the Middle East through the Silk Road 2100 years ago (Zheng et al., 1997). Fossil evidence indicates over 4000–5000 years of faba bean cultivation in southern China (Ye et al., 2003), making China probably a secondary center of origin, especially for the Chinese winter gene pool, which is distinct from the gene pools of Europe and West Asia (Zong et al., 2009). The nineteenth century witnessed the development of the winter faba beans in Europe, which were developed from small-seeded French and Russian winter-resilient populations. This historical progression underscores the dynamic global dissemination and adaptation of faba beans across centuries and continents.

Distribution

Faba bean cultivation is concentrated in key geographical regions, with East Asia accounting for 34%, East Africa for 20%, North Africa, Central and West Asia for 18.8%, Europe for 12.7%, Australia for 6.3%, and Latin America for 7.3%. Ethiopia plays a prominent role as a faba bean producer in East Africa, sown during the summer, having an area of 0.57 million hectares with a yield of about 0.73 million tons. Morocco, Egypt, Sudan, and Tunisia are important producers in the CWANA region, cultivating faba beans on an area of 0.54 million hectares, generating a total production of 0.87 million tons (FAOSTAT, 2021).

Six of the top ten producing nations are in WANA (West Asia and North Africa), East Asia, and East Africa, where most of the world's faba bean consumption occurs. The genus *Vicia* encompasses 166 species that are either annual or perennial and are primarily found in North America, Europe, and Asia (Allkin et al., 1986). They can also be found in tropical and temperate Southern Africa (Maxted, 1993).

Maxted et al. (1991) categorized it into two subgenera, *Vicia* and *Vicilla*. Legume and canavanine features, style type, keel form, peduncle length, stipule nectary, and other factors distinguish the two subgenera. Studies demonstrating the existence of a partial incompatibility system indicate that *V. faba* exhibits low intraspecific differentiation, which is remarkable. This system is absent in at least one paucijuga group population, weaker in populations in Spain, and stronger in groups found in central Europe. (Cubero, 1974). Cubero (1973) identified four faba bean botanical groups: paucijuga, equina, major, and minor.

In Persoon's seminal work in 1807, the small-seeded category was denoted as *V. faba* minor, encompassing round seeds weighing up to 0.6 or 0.7 grams each. The medium-large-seeded category was designated as *V. faba* equine, with around one gram or less of individual seed weights. In comparison, *V. faba* major included remarkably flat and large seeds that exceeded one gram each. "Field bean" is applied to medium and small-size varieties. In contrast, "broad bean" refers to *V. faba* major. Currently, the term "faba bean" often encompasses the entire species, reflecting its complete variety.

The elusive wild progenitor of the faba bean remains unidentified, and its closest known relative is considered to be the closely related vetch, *V. narbonensis* (Zohary and Hopf, 2000). Recent progress in genetic analysis, particularly utilizing polymerase chain reaction and restriction fragment length polymorphism statistics, has positioned *V. michauxii* and *V. peregrina* in close taxonomic relationship to *V. faba.* These closely related vetches exhibit a notable characteristic of sharing seven chromosomes. *Vica faba*, *V. sativa*, and *V. villosa* are classified within the *Vicia* genus. At the broader taxonomic level, it joins *Pisum sativum* and *Lens culinaris* in the tribe Vicieae. Convergence with the Phaseolus bean, the soybean, and the lupines occurs exclusively at the family level of Fabaceae (Sitte et al., 2002).

Cubero (1974) proposed a center of origin in the Near East that suggests four dispersion routes to Europe: from Mesopotamia to India, from the coast of North Africa to Spain, and along the Nile to Ethiopia (Hawtin and Hebblethwaite, 1983). The secondary centers of diversity are hypothesized to exist in Ethiopia and Afghanistan, although Hajjar and Hodgkin (2007) assert Central Asia as the origin.

The Chinese incorporated faba beans into their diet nearly 5,000 years ago, while the crop found cultivation among the Egyptians 3,000 years ago, during biblical times with the Hebrews, and subsequently by the Greeks and Romans (Singh and Bhatt, 2012). It is likely that during the Sultanic era, Europeans brought it to India as a garden crop (1206–1555); historical records attest to its cultivation (Akbar, 2000). The elusive wild ancestor and precise origin of the faba bean remains unidentified. Although taxonomically akin to the cultivated crop, wild species such as *V. galilaea* and *Vicia narbonensis* L. have 2n = 14 chromosomes, in contrast to the cultivated one, with 2n = 12 chromosomes. Despite numerous attempts, crossbreeding these wild species with cultivated faba beans has proven unsuccessful (Hajjar and Hodgkin, 2007).

V. faba, an annual herb, is distinguished by its upright and coarse unbranched stems, attaining heights of 0.5 to 2.0 meters, with one or more hollow stems originating from the base. The leaves are usually pinnate and alternate, comprised of up to 6 leaflets, measuring up to 8.0 cm in length. In contrast to most other species of this genus, it lacks or has rudimentary tendrils. Faba bean flowers profusely, but only a small percentage of the flowers develop into pods. The blossoms exhibit notable dimensions, characterized by their expansive, white composition adorned with intricate dark purple markings. They are arranged in groups of one to five on each terminal raceme, usually between the fifth and tenth node. Each floral cluster gives rise to 1-4 pods. Approximately 30% of the plant population undergo cross-fertilization, with bumblebees as the primary insect pollinators.

The plant's root system is characterized by a sturdy taproot and abundantly branched secondary roots. Distinguishing between subspecies based on seed size has led to the recognition of two subspecies: faba and paucijuga. *Vicia faba*, specifically, manifests a diploid chromosome count of 12. Among these pairs, five are acrocentric chromosomes, while one pair adopts a metacentric configuration (Alghamdi, 2009).

Economic Importance

The faba bean crop covers about 2.6 million hectares globally, producing 5.7 million tons, accounting for 4% of the total pulse area in 2021 (FAOSTAT, 2022). In Africa, the epicenter of faba bean production is Ethiopia, contributing to 15% of the area worldwide. The aggregate production in 2021 amounted to around 5.7 million tons, with China alone contributing 30%, solidifying its status as the foremost global producer. Faba beans serve diverse culinary purposes, being utilized as a vegetable, whether dried or green, and finding a place in various dishes such as Medamis, Falafel, Bissara, and Nabet soup, each showcasing the versatility of the legume (Singh et al., 2012).

The nutritional value of the faba bean is remarkable, boasting approximately 30% protein content, surpassing field peas and other legumes in certain regions. Particularly significant as a winter crop in the Middle East for human consumption, faba bean has been recognized as a meat extender, a substitute for skim milk, and a popular roasted snack in India akin to peanuts. While the exact proportion of dried faba beans used for human consumption in developing countries remains undefined, available data for green faba beans indicate that they contribute significantly to human nutrition, accounting for

20% of the total production. Notably consumed in regions such as the Middle East, the Mediterranean, Latin America, China, and Ethiopia, the Mediterranean alone contributes 40% to the global green bean production, with the Middle East and Latin America each contributing 19%. Furthermore, the utility of faba bean extends beyond the kitchen, as the haulm derived from harvest is valued as a cash crop in Egypt and Sudan, serving additional functions in brickmaking and as a fuel source in parts of Sudan and Ethiopia.

Conclusion

Faba bean is an important legume with immense potential in terms of food value and ability to withstand the effects of climate change. However, very little attention has been given to its genetic improvement that resulting in few varieties with high yield and better nutritional quality. Owing to its potential of being a climate-smart crop, it is now considered for genetic improvement, and hopefully, much better varieties will be developed in the coming future.

References

Akbar, R. (Tr.) (2000). Nuskha Dar Fanni-Falahat: (The Art of Agriculture); (Persion Manuscript Compiled In The 17Th Century by the Mughal Prince Dara Shikoh) *(Agri-History Bulletin No. 3)* Munshiram Manoharlal Publishers Pvt Ltd. P. 98.

Alghamdi, S. S. (2009). Chemical composition of faba bean (*Vicia faba* L.) genotypes under various water regimes. Pakistan Journal of Nutrition, 8, 477-482.

Allkin, R., Goyder, D. J., Bisby, F. A., & White, R. J. (1986). *Names and synonyms of species and subspecies in the Vicieae [Issue 3]*. Southampton, UK: Biology Department, University of Southampton.

Cubero, J. I. (1973). Evolutionary trends in *Vicia faba*. *Theoretical and Applied Genetics*, *43*(2), 59–65.

Cubero, J. I. (1974). On the evolution of *Vicia faba* L. *Theoretical and Applied Genetics*, *45*, 47–51.

Duc, G., Bao, S. Y., Baum, M., Redden, B., Sadiki, M., Suso, M. J., et al., (2010). Diversity maintenance and use of *Vicia faba* L. genetic resources. *Field Crops Research*, *115*, 270–278.

FAO (Food and Agriculture Organization). (2020). *Crop Production and Trade Data.* http://www.fao.org/faostat/en/#data.

FAOSTAT (2022). *https://www.fao.org/faostat/en.*

FAOSTAT, (2021). *World statistics on faba bean.* Rome, Italy: Food and Agriculture Organization of the United Nations. Available at: <http://faostat.fao.org/>.

Hajjar, R., Hodgkin T. (2007). The use of wild relatives in crop improvement: a survey of developments over the last 20 years. *Euphytica* 156:1-13.

Hawtin, G. C., Hebblethpiait P. D. (1983). Background and history of faba bean production. *Pages 3-22 in The Faba Bean* (*Vicia faba* L.) (Hebblethwaite, P. D., ed.). Buttenvorths, London, U.K.

Labba, I. C. M., Frøkiær, H., & Sandberg, A. S. (2021). Nutritional and antinutritional composition of fava bean (*Viciafaba* L., var. minor) cultivars. *Food Research International*, 140, 110038.

Ladizinsky, G. (1975). Seed protein electrophoresis of the wild and cultivated species of section *faba* of *Vicia*. *Euphytica*, *24*, 785–788.

Landry, E. J., Fuchs, S. J., and Hu, J. (2016). Carbohydrate composition of mature and immature faba bean seeds. *J. Food Compos. Anal.* 50, 55–60.

Maxted, N. (1993). A phenetic investigation of *Vicia* L. subgenus *Vicia* (Leguminosae-Vicieae). *Botanical Journal of the Linnean Society*, *111*, 155–182.

Maxted, N. (1995). An ecogeographical study of *Vicia* subgenus *Vicia*: *Systematic and ecogeographic studies on crop genepools 8*. Rome, Italy: International Plant Genetic Resources Institute.

Maxted, N., Callimassia, M. A., & Bennett, M. D. (1991). Cytotaxonomic studies of eastern Mediterranean *Vicia* species (Leguminosae). *Plant Systematics and Evolution*, *177*, 221–234.

Mínguez, M. I., & Rubiales, D. (2021). Faba bean. In V. O. Sadras & D. F. Calderini (Eds.), *Crop Physiology Case Histories for Major Crops* (pp. 452–481). Academic Press.

Muratova, V. S. (1931). Common beans (*Vicia faba* L.). *Bulletin of Applied Botany, Genetics and Plant Breeding Supplement*, *50*, 1–298.

Neme, K., Bultosa, G., and Bussa, N. (2015). Nutrient and functional properties of composite flours processed from pre-gelatinised barley sprouted faba bean and carrot flours. *Int. J. Food Sci. Technol.* 50, 2375–2382.

Revilla, I. (2015). Impact of thermal processing on faba bean (*Vicia faba*) composition. *Processing and impact on active components in food.* Academic Press.

Schultze-Motel, J. (1972). Die archaologischen reste der ackerbohne, *Vicia faba* L. und die genese de art. *Kulturpflanze*, *19*, 321–358.

Singh, A. K., Bhat B. P., Sundaram P. K., Chndra N., Bharati R. C. and Patel S. K. (2012). Faba bean (*Vicia faba* L.) phenology and performance in response to its seed size class and planting depth. *International Journal Agriculture Stat Sci.* 8(1):97-109.

Singh, A. K., Bhatt B. P. (2012). Faba Bean (*Vicia faba* L.): A potential leguminous crop of India ISBN 978-93-5067-773-5 ICAR, RC for ER, Patna, P. 518.

Sitte, P., Ziegler, H., Ehrendorfer, F. and Bresinsky, A. (2002) *Strasburger, Lehrbuch der Botanik*. 34. Auflage. Spektrum, Gustav Fischer Verlag.

Tanno, K., & Willcox, G. (2006). The origins of cultivation of *Cicer arietinum* L. and *Vicia faba* L.: Early finds from Tell el-Kerkh, north-west Syria, late 10th millennium B.P.. *Vegetation History and Archaeobotany*, *15*, 197–204.

Toklu, F., Gupta, D. S., Karaköy, T., & Özkan, H. (2021). Bioactives and Nutraceuticals in Food Legumes: Nutritional Perspective. In D. S. Gupta, S. Gupta, & J. Kumar (Eds.), *Breeding for Enhanced Nutrition and Bio-Active Compounds in Food Legumes* (pp. 229–245). Springer.

USDA (US Dept of Agriculture). (2021). *Food Data Central* (Nutrient Database). https://fdc.nal.usda.gov/.

Valente, I. M., Maia, M. R., Malushi, N., Oliveira, H. M., Papa, L., Rodrigues, J. A., Fonseca, A. J., & Cabrita, A. R. (2018). Profiling of phenolic compounds and antioxidant properties of European varieties and cultivars of *Viciafaba* L. pods. *Phytochemistry*, 152, 223–229.

Witten, S., Böhm, H., and Aulrich, K. (2015). Effect of variety and environment on the contents of crude nutrients, lysine, methionine and cysteine in organically produced field peas (Pisum sativum L) and field beans (*Vicia faba* L). *Landbauforsch Volkenrode* 65, 205–216.

Ye, Y., Lang, L., Xia, M., & Tu, J. (2003). *Faba beans in China*. Beijing, China: China Agriculture Press, in Chinese.

Zheng, Z., Wang, S., & Zong, X. (1997). *Food legume crops in China*. Beijing, China: China Agriculture Press, in Chinese.

Zohary, D. and Hopf, M. (2000). *Domestication of Plants in the Old World*, 3rd eds. Oxford: Clarendon Press.

Zong, X, Liu, X., Guan, J., Wang, S., Liu, Q., Paull, J. G., et al., (2009). Molecular variation among Chinese and global winter faba bean germplasm. *Theoretical and Applied Genetics*, *118*, 971–978.

Chapter 4

Germplasm Diversity of Faba Beans (*Vicia faba*)

Mahima Gotame[1,*]
Santosh BK[2]
and Abhisek Shrestha[3]
[1]Mahendra Ratna Multiple Campus, Department of Horticulture, Ilam, Nepal
[2]Agriculture Knowledge Center, Sankhuwasabha, Nepal
[3]College of Natural Resource Management, Bardibas, Nepal

Abstract

Faba bean (*Vicia faba L.)* is an important legume crop worldwide. It has been important since ancient times as a food source in Asia and Africa. Faba bean represents a valuable climate-resilient legume under changing climatic scenarios. *Vicia faba* accessions comprising over 38,000 have been preserved globally, though the conservation of species has declined in recent years. The purpose of domesticated *Vicia* species cultivation is to provide food, while some wild species are also cultivated for fodder, and certain *Vicia* species are also noxious weeds. Most faba bean varieties are grown for human consumption, while equine varieties are mostly used as animal feed. South Mediterranean regions are rich in large, seeded types, while small, seeded beans are native to North Europe, and equine types are grown throughout the Middle East, North Africa, and Australia. Genetic diversity among beans reflects from the Arctic Circle to the equator. Near East, Mediterranean, Afghanistan, India, and China are speculated as centre of diversity for faba bean. Although it has high genetic potential and diversity, limited information

* Corresponding Author's Email: mahimagotame40@gmail.com

In: Faba Beans
Editors: Aamir Raina and Samiullah Khan
ISBN: 979-8-89530-447-1

creates difficulties in conservation and their use in future breeding strategies. Germplasm is a basic requirement for any crop improvement program. So, resources should be conserved and utilized more effectively with information on domesticated species and their wild relatives. Thus, this chapter provides information on the diversity of *Vicia* species.

Keywords: *Vicia faba*, diversity, legumes, germplasm

Introduction

Faba bean *(Vicia faba L.)* is an important legume crop worldwide. According to FAOSTAT, faba bean ranks seventh among the legumes in production (Oliveira et al., 2016). Faba bean, commonly known as the broad bean, horse bean, and field bean, belongs to the family Fabaceae/Leguminosae. It is one of the earliest cultivars of legumes documented (Zohary et al., 2012). Faba beans are mainly grown as human food in developing countries, primarily those in Asia and Africa (Duc, 1997). It is an excellent source of protein, fiber, and essential amino acids (Rahate et al., 2021). It is possible to cultivate *Vicia faba* in a kitchen garden, rooftop garden, or on marginal terrain. Growers benefit from integrating faba bean with other crops or sole cropping systems, and intercropping with faba bean is considered excellent for maintaining soil health. The pod of faba bean is consumed as vegetables or pulses, and vegetative parts provide fodder to livestock. It is also used as green manure. Faba bean is mostly grown as a winter season crop as sole cropping, mixed cropping, or in rotation with rice. They are valued culturally due to their ancient symbolic significance in connection with rituals. It is used on auspicious days and to make different food items for functions. Being a major food in developing countries, its cultivation is continuously declining (Gresta et al., 2009). Genetic erosion of crops is visible, and it is believed that 50% of traditional varieties have disappeared from farmers' fields (Bhatta, 2012; Paudel et al., 2016). Species diversity must be characterized, and knowledge of their wild relatives must be gathered for crop development as the first approach. Hence, it is necessary to study the diversity of *Vicia faba* for the conservation of resources, which have great significance in crop improvement programs.

Vicia faba Germplasm Diversity

Several plant genetic resources were diversified into ecological zones cultivated by farmers over the years. There are 650 genera and 18,000 species in the family Fabaceae, among which 100 genera and 379 species are found in Terai to the alpine region of Nepal (Neupane et al., 2017). Out of the 379 legume species, including grain, vegetables, and forage legumes, 262 are native, and 20 species belonging to the subgroup Papilionaceae are used as food legumes (Neupane et al., 2017). *Vicia* comprises 140-190 species primarily found in Europe, Asia, and North America to temperate South America and tropical East Africa (Kupicha et al., 1981). Since Linnaeus, more than 20 significant classifications of *Vicia* have been proposed (Maxted, 1993). ICARDA preserves around 6,000 accessions of *Vicia* species, particularly over 3,000 accessions of wild *Vicia* species (genebanks.org, 2009). Germplasms were collected and conserved in national gene banks and research centers in different countries. AARI Turkey, NGB Egypt, INRA Morocco, CGSAR Syria, NARC Pakistan, and MOA Tunisia are the largest national collection centers for faba beans. The region includes the world's largest faba bean collection held at ICARDA, Syria (genebanks.org, 2009). Germplasm is requisite for crop improvement programs. Numerous crop species are underutilized, and their closely related species were not given enough attention. Local landraces and wild species are more resistant to biotic and abiotic stresses. Attention has switched from landraces/cultivars with traits of adaptability to particular ecological niches with agronomic methods and socio-cultural value to modern cultivars, with a major focus in recent days. *Vicia faba* has a large genetic diversity around the globe. Faba bean germplasm is conserved in more than 38,000 accessions. The largest collection of 10,045 accessions worldwide from 71 countries with specific characteristics was in The International Center for Agricultural Research in Dry Areas (ICARDA) (Fouad et al., 2013). However, winter grain legumes are more significant than summer legumes in area and production (Pokhrel et al., 2018). Asian regions have negligible area and production (Malik et al., 1988). Researchers and policymakers are keeping one of the crop species behind the eye: the faba bean (Paudel and Joshi, 2018). The total number of accessions conserved in different countries based on different sources is presented in Table 1.

Table1. Conserved accessions of *Vicia faba* indifferent genebanks of the world

S.N.	Country	Collection center	Accession number	Source
1.	China	Institute of Crop Science (CAAS), Beijing	5900	Duc et al., 2015
2.	Nepal	NAGRC, Khumaltar	60	Poudel & Joshi, 2018
3.	Australia	Australian Temperate Field Crops Collection, Victoria	2665	CGIAR, 2009
4.	Syria	ICARDA/Aleppo	10045	Duc et al., 2015
5.	Germany	Gene Bank IPK/Gatersleben	1920	Duc et al., 2015
6.	France	INRA/Dijon	1900	Duc et al., 2015
8.	Ecuador	Instituto de Ciencias Naturales Universidad Central del Ecuador (ICN)	1650	CGIAR, 2009
9.	Ethiopia	Biodiversity Conservation and Research Institute (BCRI)	1208	CGIAR, 2009
10.	Ethiopia	PGRC/Addis Ababa	1118	Duc et al., 2015
11.	Spain	IFAPA/Cordoba	1091	Duc et al., 2015
12.	Netherlands	Centre for Genetic Resources, the Netherlands (CGN	728	CGIAR, 2009
13.	Netherlands	DLO/Wageningen	726	Duc et al., 2015
14.	India	National Bureau of Plant Genetic Resources (NBPGR)	554	CGIAR, 2009 Duc et al., 2015
15.	USA	USDA/Pullman	750	Duc et al., 2015
16.	Uk	John Innes Centre (JIC) Norwich	156	CGIAR, 2009
17.	Egypt	Field Crops Research Institute, Agricultural Research Center (NGB)	125	CGIAR, 2009
18.	Canada	PlantGene Resources of Canada (PGRC	274	CGIAR, 2009
19.	Morocco	Rabat INRA	1715	Fouad et al., 2013

S.N.	Country	Collection center	Accession number	Source
20.	Turkey	Aegean Agricultural Research Institute (AARI)	351	CGIAR, 2009
21.	Pakistan	Plant Genetic Resource Institute, National Agriculture Research Center (PGRI)	181	CGIAR, 2009
22.	Russia	VIR	1881	Fouad et al., 2013
23.	Poland	IOPG-PAS	1258	Fouad et al., 2013
24.	Portugal	INRB-IP	788	Fouad et al., 2013

Table 2. Gene pool of faba bean based on classification proposed by Maxted & Kell (2009)

Primary gene pool	*V. faba subsp. paucijuga, V. faba subsp. faba, V. faba subsp. faba var. minor, V. faba subsp. faba var. equina, V. faba subsp. faba var. faba*
Secondary gene pool	No secondary wild relatives
Tertiary gene pool	*V. kalakhensis, V. johannis, V. johannis var. ecirrhosa, V. johannis var. procumbens, V. johannis var. johannis, V. galilaea, V. galilaea var. galilaea, V. galilaea var. faboidea, V. serratifolia, V. narbonensis, V. narbonensis, V. narbonensis var. salmonea, V. narbonensis var. jordanica, V. narbonensis var. affinis, V. narbonensis var. aegyptiaca, V. narbonensis var. narbonensis, V. hyaeniscyamus, V. bithynica, V. eristalioides.*

Source: Yadav et al., 2017.

Ecological Distribution of *Vicia faba*

Vicia species is native to Southeastern Europe and southwestern Asia (Fouad et al., 2013) and is the only grain legume widely cultivated in Europe for a long time (Gresta et al., 2009). The distribution of cultivated faba bean is in more than 55 countries, either in rainfed or irrigated conditions (Maalouf et al., 2021). Bond et al. (1985) reported that genetic diversity among beans reflects from the Arctic Circle to the equator (Torres et al., 2006). Based on trait studies, the faba bean was divided into two groups. Beans with small seed forms have been reported in southwest Asia, and the western group consists

of beans with larger seeds. The Eastern group was characterized by diverse traits and distributed more widely (Fouad et al., 2013). The mid-east region is the center of faba bean domestication (Caracuta et al., 2015). According to Perrino et al. (1991), the near East, Mediterranean, Afghanistan, India, and China are centers of diversity for the faba bean (genebanks.org, 2009). Although the parent of the faba bean is still unknown, many species closely related to the seeded faba bean predominate in the Middle East region (Smither-Kopperl, 2019).

It is grown as a winter-season crop from Terai to the high hills of Nepal. Cultivated varieties may vary from region to region, whereas wild species are grown as weeds in crops and pastureland from Terai to high hills. *Vicia rigidula royle*, *Vicia anugustifolia* L., *Vicia hirsuta* (L.), *Vicia tetrasperma* (L.) and *Vicia tenuifolia.* Roth are wild species grown as weeds in crop and pasture lands from Terai to the high hills of Nepal (Neupane et al., 2017). Large-seeded type bean *Vicia faba var* major is commonly grown in kitchen gardens in Kathmandu Valley and adjoining districts, whereas small-seeded bean *Vicia faba var.* minor with green or black color testa are grown as a field crop or in a kitchen garden in the Terai region of Nepal. Large pods are primarily consumed for green vegetable purposes, dry seeds as roasted beans, and small seeds are consumed as soup (Neupane et al., 2017). Major and minor type beans are cultivated for edible use. In contrast, horse beans are mostly used as animal feed, and minor types are used mainly for crop rotation and livestock nutrition (Pasqualone et al., 2020). Numerous primary and tertiary wild relatives of *Vicia faba* were identified, though there was no secondary wild relative; the crop is closely related to *Vicia narbonensis* species (Yadav et al., 2107). Table 2 highlights the wild relatives/gene pool of the faba bean. Southern and northern Europe, Ethiopia, and southern China have all been identified as secondary domestication regions for this crop. South Mediterranean regions are rich in major-type beans, minor types are common in North Europe, and equine types are widespread throughout the Middle East to North Africa and Australia (Alghamdi et al., 2012). The purpose of cultivation varies with varieties among the regions. Vicia species are mainly used for green manure and forage purposes. Some species are used as pulses, vegetables, roasted snacks, and soups in the human diet.

Morphological Diversity in Faba Bean Genotypes

In developing countries, more than 80% of cultivation has been performed traditionally (Torres et al., 2010). Central Europe and Northwest European gene pools consist of V. faba var. minor and V. faba var. major types and a Mediterranean gene pool, including the former types and *V. faba var. equina* (Terzopoulos and Bebeli, 2008). Large-seeded cultivars were developed in Europe and southern China (Torres et al., 2006). *Vicia faba* var. equine has an intermediate seed size, whereas *Vicia faba* minor possesses small-sized seeds (Pasqualone et al., 2020). Morphological characteristics of major faba bean types (*Vicia faba var.* major) include a thick non-dehiscent pod having 2-3 seeds per pod, whereas *Vicia faba var.* equine have plate, thick, nondehiscent pods with 3-5 seeds and a small cylindrical pod with round to cylindrical seed with seed weight less than number of seeds are major characteristics of *var.* minor. Small seeded bean subsp. faba var. minor or subsp. Paucijuga was the most ancient variety of *V. faba* (Maxted and Bennett, 2001). Five thousand years ago (BP), the medium-sized variety became known in the Iberian Peninsula, including Portugal, Spain, and Central Europe. Flattened larger types were unknown before 1,500 BP (Ladizinsky, 1989). The greatest global producer of *V. faba* is China, followed by Ethiopia, UK, Australia, Germany, Morocco, France, Egypt, Italy, and Sudan, and despite having a lower yield, *V. faba* is the most extensively cultivated plant in Algeria and Tunisia (fao.org, 2017). It is one of the traditional staple foods throughout the Mediterranean Basin and the Middle East, India, Pakistan and China (Torres et al., 2010). Faba bean is a major crop for dryland farming systems in East Africa, West Asia, and North Africa (Maalouf et al., 2021). Narbon vetch (*V. narbonensis L.*) is a *Vicia* species that can cope with drought better than *Vicia faba* and thrives well in the Mediterranean basin (Sanz et al., 2007). Faba beans are susceptible to a variety of diseases, including chocolate spot, rust, Stem and root rots, damping-off, downy mildew, leaf and pod spots or blight, foot rots, bean yellow mosaic virus, and bean leaf roll virus (genebanks.org, 2009). Some *Vicia* species possess abilities to cope with changing climatic conditions. Several *V. narbonensis* species showed attributes of winter hardiness and resistance to black aphid and chocolate spot disease (Birch et al., 1985).

Conclusion

Faba bean is one of the potential legume crops grown throughout the world. Broad variation among faba beans has been studied with peculiar characteristics for their adaptability. The distribution of *Vicia faba* and their wild relatives in distinct regions plays a significant role in the conservation and utilization of resources for breeding programs. Declining genotypes of beans and shifting to modern cultivars should be greatly alarming for the loss of germplasm. Faba bean is grown traditionally in different regions of Nepal. Since it is a minor crop, not much attention is given to the documentation of the faba bean. Attention should be given to the collection, documentation, characterization, and conservation of faba beans.

References

Alghamdi, S. S., Migdadi, H. M., Ammar, M. H., Paull, J. G., & Siddique, K. H. M. (2012). Faba bean genomics: current status and future prospects. *Euphytica, 186*(3), 609-624.

Bhatta, M. R. (2012). Status of plant genetic resources (PGR) management system in Nepal. *AFACI Pan-Asia Project Meeting, 19*-22 March 2012. NAAS, RDA, Korea. Pp. 112-124.

Birch, A. N. E., Tithecott, M. T., & Bisby, F. A. (1985). Vicia johannis and wild relatives of the faba bean: a taxonomic study. *Economic Botany, 39*(2), 177-190.

Bond, D. A., Lawes, D. A., Hawtin, G. C., Saxena, M. C. and Stephens, J. S. (1985). "Faba bean (*Vicia faba* L.)," in *Grain Legume Crops*, eds R.J. Summerfield and E.H. Roberts (London: William Collins Sons), 199–265.

Caracuta, V., O. Barzilai, H. Khalaily, I. Milevski., Y. Paz, J. Vardi, L. Regev, and E. Boaretto. (2015). The onset of faba bean farming in the Southern Levant. *Sci. Rep.* 5, 14370; doi: 10.1038/srep14370. https://www.nature.com/articles/srep14370#ref-link-section-2.

CGIAR GENEBANKS (2009). Global strategy for the Ex Situ conservation of Faba Bean (*Vicia faba*).

Duc, G. (1997). Faba bean (*Vicia faba* L.). *Field crops research*, 53(1-3), 99-109.

Duc, G., Aleksić, J. M., Marget, P., Mikic, A., Paull, J., Redden, R. J., & Torres, A. M. (2015). *Faba bean. Grain legumes,* 141-178.

FAO (2017). *Data of Crops Production*. http://www.fao.org/faostat/en/#data/QC. Accessed 7 October 2019.

Fouad, M., Mohammed, N., Aladdin, H., Ahmed, A., Xuxiao, Z., Shiying, B., & Tao, Y. (2013). Faba bean. In *Genetic and Genomic Resources of Grain Legume Improvement* (pp. 113-136). Elsevier.

Gresta, F., Albertini, E., Raggi, L., &Abbate V (2009). A study of variability in the Sicilian faba bean landrace 'Larga di Leonforte.' *Genet. Res. Crop. Evol.* 57: 523-531.

https://www.genebanks.org/resources/publications/faba-strategy-2009.

Kupicha, F. K., Polhill, R. M., & Raven, P. H. (1981). Advances in legume systematics. *Vicieae*, 377-381.

Ladizinsky, G., 1989. Origin and domestication of the Southwest Asian grain legumes. In: *Foraging and Farming* (eds. Harris, D. H. and Hillman, G. C.). pp. 376–389. Unwin Hyman, London

Maalouf, F., Ahmed, S., & Bishaw, Z. (2021). Faba bean. In *The Beans and the Peas*. pp. 105-131.

Malik, B. A., Verma, M. M., Rahman, M. M., & Bhattarai, A. N. (1988). Production of chickpea, lentil, pea and faba bean in South-East Asia. In *World crops: Cool season food legumes* (pp. 1095-1111). Springer, Dordrecht.

Maxted, N. (1993). A phenetic investigation of Vicia L. subgenus Vicia (Leguminosae, Vicieae). *Botanical Journal of the Linnean Society, 111*(2), 155-182.

Maxted, N., & Bennett, S. J. (2001). Conservation, diversity and use of Mediterranean legumes. In *Plant genetic resources of legumes in the Mediterranean* (pp. 1-32). Springer, Dordrecht.

Maxted, N., & Kell, S. P. (2009). Establishment of a global network for the in situ conservation of crop wild relatives: status and needs. *FAO Commission on Genetic Resources for Food and Agriculture, Rome, Italy, 266.*

Neupane, R. K., Shrestha, R., & Darai, R. (2017). Food Legumes: Diversity, utilization and conservation status. In Conservation and Utilization of Agricultural Plant Genetic Resources in Nepal (B. K. Joshi, H. B. K. C. and A. K. Acharya, eds). *Proceedings of 2nd National Workshop* (pp. 22-23).

Oliveira, H. R., Tomás, D., Silva, M., Lopes, S., Viegas, W., & Veloso, M. M. (2016). Genetic diversity and population structure in *Vicia faba* L. landraces and wild related species assessed by nuclear SSRs. *PloS one,* 11(5), e0154801.

Pasqualone, A., Abdallah, A., & Summo, C. (2020). Symbolic meaning and use of broad beans in traditional foods of the Mediterranean Basin and the Middle East. *Journal of Ethnic Foods*, 7(1), 1-13.

Paudel, M. N., & Joshi, B. K. (2018) Agriculture plant genetic resources behind the eyes of researchers and policy makers. *Workshop of Agricultural Plant Genetic Resources (APGRs)* in Nepal during (June 21-22, 2018) 7-8 Ashad 2075 in the Seminar Hall of the Agronomy Division, NARC, Khumaltar, Lalitpur, Nepal.

Paudel, M. N., Joshi, B. K., & Ghimire, K. H. (2016). Management status of agricultural plant genetic resources in Nepal. *Agronomy Journal of Nepal,* 4, 75-91.

Perrino, P., Maruca, G., Lester, R. N., Linsalata, V., Lattanzio, V., & Hanelt, P. (1991). Chemotaxonomic relationship among species of Vicia section Faba. *Feddes Repertorium, 102*(5-6), 319-334.

Pokhrel, A., Aryal, L., & Poudel, P. P. (2018). A review on research work of grain legumes research program, NARC. *Government of Nepal, Nepal Agricultural Research Council, Grain Legume Research Program, Banke, Nepal.*

Rahate, K. A., Madhumita, M., & Prabhakar, P. K. (2021). Nutritional composition, anti-nutritional factors, pretreatments-cum-processing impact and food formulation potential of faba bean (*Vicia faba* L.): A comprehensive review. *Lwt, 138*, 110796.

Sanz, A. M., Gonzalez, S. G., Syed, N. H., Suso, M. J., Saldaña, C. C., & Flavell, A. J. (2007). Genetic diversity analysis in Vicia species using retrotransposon-based SSAP markers. *Molecular Genetics and Genomics*, *278*(4), 433-441.

Smither-Kopperl, M. 2019. Plant Guide for fava bean (*Vicia faba*). USDA-Natural Resources Conservation Service, Lockeford Plant Materials Center. *Lockeford,* CA 95237.

Terzopoulos, P. J., & Bebeli, P. J. (2008). Genetic diversity analysis of Mediterranean faba bean (*Vicia faba* L.) with ISSR markers. *Field Crops Research*, *108*(1), 39-44.

Torres, A. M., Avila, C. M., Gutierrez, N., Palomino, C., Moreno, M. T., & Cubero, J. I. (2010). Marker-assisted selection in faba bean (*Vicia faba* L.). *Field crops research*, *115*(3), 243-252.

Torres, A. M., Roman, B., Avila, C. M., Satovic, Z., Rubiales, D., Sillero, J. C., & Moreno, M. T. (2006). Faba bean breeding for resistance against biotic stresses: towards application of marker technology. *Euphytica*, *147*(1), 67-80.

Yadav, S. K., Verma, N., Singh, A. K., Singh, N., Rana, S. C., Ranga, S. S., & Kumar, K. (2017). Diversity and development in faba bean. *Legume Research: An International Journal,* 40(4).

Zohary, D., Hopf, M., & Weiss, E. (2012). *Domestication of Plants in the Old World: the origin and spread of domesticated plants in Southwest Asia, Europe, and the Mediterranean Basin*. Oxford University Press.

Chapter 5

Breeding Methods: Limitations and Achievements in Faba Beans

Satvinder Singh[1]
Shubham Kumawat[1]
S. G. P. Karthikeya Reddy[1]
Amit Singh[1]
and Vivek Thapliyal[2]
[1]PhD Research Scholar, Genetics and Plant Breeding, College of Agriculture, Govind Ballabh Pant University of Agriculture and Technology, Pantnagar, Udham Singh Nagar, Uttarakhand, India
[2]PhD Research Scholar, Department of Vegetable Sciences, College of Agriculture, Govind Ballabh Pant University of Agriculture and Technology, Pantnagar, Udham Singh Nagar, Uttarakhand, India

Abstract

Faba bean breeding programs strive to develop improved cultivars with traits like high yield, desirable seed characteristics, nutritional content, and stress resilience. Various breeding methods, including mass selection, recurrent selection, and traditional pedigree approaches, are employed to develop modern cultivars. Challenges such as genetic complexity, environmental influences, and limited genetic diversity hinder breeding progress. Despite these challenges, notable achievements include the development of Orobanche-tolerant and disease-resistant varieties. Exotic varieties have also contributed to genetic diversity. Molecular markers and genomic selection offer promising avenues for accelerating breeding cycles, although challenges in understanding complex resistance mechanisms persist. The progress

In: Faba Beans
Editors: Aamir Raina and Samiullah Khan
ISBN: 979-8-89530-447-1

made in the model legumes aids in addressing gaps in knowledge, and the application of biotechnological tools holds potential for targeted genetic improvements in faba beans.

Keywords: faba bean, breeding, genetics, diversity

Introduction

Breeding Methods Limitations and Achievements

Breeding Goals

- A high and consistent yield
- Desired seed size and color following consumer preferences
- Methionine and cysteine, two types of high-seed protein
- Earliness and stiff straw, low tannins, low pyrimidine glucosides (vicine and convicine)
- Tolerant to abiotic stresses, such as dryness and salt
- Resistance to bean leaf roll virus, bean YMV, rust, and chocolate spot.

Faba Bean Breeding Methods

Different methods, such as mass selection, recurrent selection, and traditional methods, like pedigree selection and single-seed descent line breeding, are used to develop the modern cultivars of faba bean. Parental lines are selected for these approaches based on their phenotypic characteristics and pedigree. As a largely allogamous plant (Figure 1), the faba bean is, in theory, open to hybrid, population, and line breeding. One of the main difficulties in breeding the crop is that the level of cross-pollination in faba bean extends greatly among genotypes (10–70%) and is strongly influenced by environmental factors, notably the level of heat stress (Link, 1990; Rowland, 1987; Gasim et al., 2004). Outcrossing has a low heritability and is significantly influenced by the inbreeding coefficient and the type of insect pollinators. However, there are significant differences in faba bean genotypes in terms of their capacity for spontaneous self-fertilization, i.e., without the help of their pollinators, notably honeybees and bumblebees (Link, 1990; Brunjes and Link, 2021). As a result,

breeding pure lines in insect-proof cages results in disaster. These genotypes completely rely on insect pollinators and cannot set seeds without mechanical flower tripping in their absence. The plant's ability to self-fertilize and set seed afterwards without being tripped is called auto fertility (Drayner, 1956; Suso et al., 1999).

In the faba bean, there is genetic variation for auto fertility, and F_1 hybrids have been found to have higher levels of auto fertility than inbred lines (Link, 1990; Bishop et al., 2020). Due to their genetic ancestry in largely auto-fertile ICARDA germplasm, Australian-bred cultivars are extremely self-fertile. Also, they are originally raised and chosen inside insect-proof containers. Breeding for *Vicia* can be done under-monitored selfing environment with the help of cages that are insect-proof or with less control and in cases of low outcrossing, based on emerging lines under open field settings (Gharzeddin et al., 2019). In the faba bean, heterosis, or hybrid vigor, is evident. The yield of heterozygous F_1 hybrids is 40 to 70% higher than that of their homozygous parents (Dieckmann and Link, 2010; Zeid et al., 2004). David Bond in Cambridge, UK, and Pierre Berthelem in Rennes and France in 1967 were the first to use the cytoplasmic male sterility (CMS) technique in *Vicia faba* hybrid breeding (Pfeiffer, 1993; Link et al., 1997). Due to the fragility of the current CMS systems, hybrid breeding has not yet been established in this crop. Synthetic cultivar development is a substitute for hybrid breeding (i.e., a population breeding method). For the enhancement of faba beans, breeding lines, synthetic and composite cultivars are often utilized products (Abo Mostafa et al., 2014).

Figure 1. Flowers of faba bean.

Figure 2. Field layout of line breeding.

Heterosis Breeding

Vicia faba hybrid development has three notable challenges.

1. A functioning Cytoplasmic Male Sterility mechanism must be accessible. The CMS that is accessible in faba bean has revealed insufficient male sterility. The sterility of pollen is unreliable, and recurrent sudden return to fertility is common (Link et al., 1997; Maalouf et al., 2019).
2. The accessibility of suitable pollinators poses the second barrier. Nectar is frequently "stolen" by pollinators, who remove it on the outside, depositing pollen on the flower's stigma. By gathering pollen, pollinator insects learn to avoid mother plants devoid of pollen, which reduces seed set during the development of hybrid seed (Bond 1989a; 1989b; Brunet et al., 2019; Duc, 1997; Marzinzig et al., 2018).
3. Seed increase through three or more generations from propagation at manually managed single row level is necessary due to the low faba bean propagation coefficient to produce the necessary number of seeds for certified seed production. Therefore, for hybrid seed production to be effective, all operational processes and tools must be

carried out successfully, including the dependability of the CMS mechanism. The breeders have used monogenic characteristics to encourage the production of hybrid seeds due to the non-availability of CMS and gametocides. These qualities include markers for testa color, hilum color, cotyledon color, blossom color, and other plant morphological traits, such as dominant and recessive nuclear genetic pollen sterility. Although seed size is a multigenic variable, it was used as well. Over the years, breeders have demonstrated remarkable innovation in the hybrid breeding of faba beans. However, a breakthrough has not yet occurred (Bond et al., 1966; Berthelem and Guen, 1967; Bond, 1989a; 1989b; Duc and Stoddard, (2018).

Line Breeding

Line breeding cultivars are subject to rigorous selection (Figure 2). As lines are required as components in synthetic cultivars, line variety can be generated promptly. Line cultivars are more commonly approved than synthetic cultivars based on standards of uniqueness, homogeneity, and stability. The SSD approach can be replaced by testing, selecting, and releasing new cultivars employing F_2-derived lines (i.e., partial bulks) at F_3 or advanced generations; nevertheless, such cultivars can be very diverse. This strategy relies on rapid seed multiplication rates and short breeding cycles. F_5- or F_6-derived lines are frequently used as the foundation for new line cultivars and their maintenance breeding. Regardless of the method used to create the lines, they must be multiplied under isolation or controlled self-fertilization, whether the replication is started as a line variety, entered into a polycross, or created as a Syn-0. The experimental trial, which focuses on the breeding of synthetic cultivar, may be set up as a top cross consisting of few replicates and a joint, constant pollinator genotype constituting half or less of the area or as an easy way of a polycross with several rather than one plant per label and thus fewer replicates (Ederer and Link, 1992a; 1992b). In comparison to other crops, the faba bean has a low propagation coefficient, which restricts breeding and, in particular, yield testing. For instance, a single faba bean plant only yields enough quantity of seed to create a field plot measuring 1 m^2 (about 20–40 seeds). Hence, one SSD pipeline plant produces only enough quantity of seed for a short observation trial following regulated self-fertilization; harvesting that plot would produce the required quantity of seed for a test in an area of 20 to 40 m^2. This study might be conducted at a single site with two replicates,

or it could be conducted at two sites with one replicate and some checks as part of an enhanced design (Moehring et al., 2013). In traditional faba bean breeding, selecting the best possible parents is important. Concentrating on line cultivars, it may be said that around 50% of the genetic diversity pertaining to the breeding material is found among the lines resulting from crossings that could theoretically be made. These crosses still contain the other half of the genetic diversity, emphasizing the significance of selecting parents (Falconer and Mackay, 1996). Hence, before spending too much money testing their progeny, rejecting complete crosses that promise to produce lines that perform worse than other crosses is crucial.

Synthetic Breeding

A cultivar typically consists of four or more founder lines in synthetic breeding. Hence, the cultivar can always be "resynthesized" from these lines. Since the synthesized population won't be as similar as its inbred components, selecting these inbred lines based only on how well they perform per se is probably not the best course of action. Using as much heterosis as possible for agronomic performance or reducing inbreeding is the goal of opting for synthetic breeding over line breeding (Ghaouti and Link, 2009; Ghaouti et al., 2008). Inbred lines should be chosen primarily for their purported varietal potential when used as synthetic components. This factor affects breeding value and overall combining ability. The performance of a candidate line's polycross or top-cross progeny roughly represents the varietal aptitude of that line. We might be able to more accurately anticipate the execution of the superior synthetics if we analyze per se the performance of the lines in addition to their level of cross-fertilization and paternal success. Since the seed of the lines must be multiplied under regulated selfing circumstances for yield testing, it is more costly than the progenies (from a top cross or the poly cross). Depending on each candidate line's particular extent of outcrossing, the tester (pollen parent) of a top cross may be laid out in strips to cover 50% or less of the field area. Cross-pollination of the candidate lines would thus mostly be done with the common pollinator. In addition to adjustments in the field and experimental layout (Figure 3), behavior of the pollinator, and differences in the success of the paternal outcrossing, several other factors contribute to the discrepancy between the performance of the line as a whole and also that of their offspring produced by open pollination (polycross progenies; top cross-progenies). General combining ability, a line-specific performance factor of

the progenies through polycross/top cross progenies, is the main cause of this disparity. The same holds for the candidate lines' level of self-fertilization and the success of paternal outcrossing. Inbreeding levels and the actual makeup of the progeny are, hence, line-specific. The seed production of open-pollinated offspring occurs in an open condition, but the seed production of inbred lines occurs in cages, creating two different environments. Hence, it is expected that the size and quality of the seed, as well as the subsequent standing crop, will vary for non-genetic reasons when the performance of the lines is compared with that of their progenies derived in open-field conditions (Bond and Pope, 1984).

The decline caused by synthetic generations may vary frequently and may be the result of the increasing heterozygosity and the heterogeneity among the (quartets of) homozygous lines and synthetic generations that succeed them. These arguments suggest that field-phenotyping the candidate lines' derived offspring from open fields should take priority over further evaluation of line performance. It is not obvious whether different synthetic populations must be created and tested to choose amongst them, or certainly, those synthetics should be built to be further put through experimental field trials.

Mutagenesis in Faba Bean

New mutations on specific faba bean genes will improve the variety available for breeding. Chemical mutagenesis or radiation exposure may be used to create these mutations. Both chemical and physical mutagens, such as gamma rays and ethyl methane sulphonate-EMS, have been effective in inducing mutations in faba beans (Khursheed et al., 2018, 2019). It has also been noted that ion beam irradiation is a powerful and original method for causing mutations in faba beans. The use of mutation in faba bean breeding developed novel varieties, particularly in terms of protein content, biotic resistance, plant growth habit, and early days to maturity. X-rays, neutrons, ethyl methane sulphonate (EMS), methyl methane sulphonate (MMS), ethylene imine (EI), diethyl sulphonate (DES), and colchicine, used separately or in combination, are the most frequently employed mutagenic agents (Abdalla, 1982). The first faba bean mutations observed with the description of determinate mutants drastically decreased the number of flowering branches, resulting in the plant morphology being "topless" (Tacke et al., 2021). Later, the dwarf mutant dw1 had an almost 50% reduction in internode length (van Norel and Hoogendoorn, 1989). Five nodulation mutants were obtained using an EMS

mutagen; one was a super nodulating mutant (f 32), with 3-5 times nodulation as the healthy type (Duc, 1995). Several imazapyr resistance variants were developed using the EMS, leading to Australia's first imidazolinone herbicide-resistant variety (Ramsay et al., 1991). Several mutation breeding studies conducted worldwide have demonstrated the effectiveness of mutagenesis in seed legumes to enhance genetic variability. The faba bean mutants' resistance to herbicides would be particularly helpful. Similar to this, mutants with thinner pod walls will produce more grain. Other chemicals or fungicides might cause mutations. Some generated mutations may contribute to developing disease resistance or Orobanche tolerance. The effects of mutagens depend on the genetic constitution being treated, the quantity used, and whether the agent is being used alone or in combination with other agents. Some induced variants, or determinate types, resemble those that emerge naturally. There are now 20 mutants of *Vicia faba* that have been created from different nations and are recorded in the IAEA mutant cultivar database (https://bit.ly/3yVEReM) (International Atomic Energy Agency, 2021). These mutant varieties may be employed in breeding efforts.

Speed Breeding

Any breeding effort to increase genetic gains annually must use speed breeding to reduce breeding cycles. As a method for expediting legume crop breeding, progress achieved has been significant towards cutting the reproductive cycle short and consequently the whole process of development of cultivar. For the very first time, a faba bean *in vivo* speed breeding protocol has recently been published (Mills et al., 2020). This can be a useful technique for producing superior cultivars and diversified germplasm in a short amount of time. The breeding cycle of the faba bean from seed to seed has been reduced by 22 days, according to their findings, by using cytokinin or cold therapy to promote pollen viability and seed setting. Faba bean generation time can be further reduced by controlling light and temperature and adopting expedited development of faba bean to the flowering stage as part of a trustworthy speed breeding strategy. In addition to being coupled with other front-line breeding technologies like marker-assisted breeding and genomic selection based on anticipated breeding values, speed breeding can hasten the creation of faba bean lines (Bhatta et al., 2021). Since the seed produced in a typical speed breeding program is very low and the cost incurred in maintaining a controlled environment is high, speed breeding protocols are

suitable only for primary cross/backcross and inbreeding generations until more research related to speed breeding expansion is carried out. A significant seed growth, most often under open pollination conditions, is a prerequisite for all steps in the phenotyping of agronomic performance. Eventually, phenotypic assessments must be performed in "natural" circumstances.

Marker-Assisted Selection (MAS)

Marker-assisted selection is the most promising application of DNA markers in breeding for cultivar development. The foundation of this strategy is the identification of markers in populations that segregate to identify advantageous alleles at target loci in a given genome. To aid in phenotypic screening, markers connected to desired features are found and employed in MAS (Torres et al., 2010). Phenotypic selection of phenotypes involves specialized, costly procedures and produces findings that continue to depend on the genotypic interaction. Vicine and convicine (v-c), the two main antinutritional components restricting the use of faba bean seed (Duc et al., 1989), provide a good illustration of an application of DNA markers in breeding of faba bean. Because the v-c decreases feed conversion efficiency among monogastric animals and causes favism in susceptible people, its genetic elimination is unquestionably desired (Khazaei et al., 2015; 2017). A diagnostic SNP assay for selection was created after mapping the gene on chromosome 1. It is now understood that faba bean cultivars with low v-c levels and the vc gene are reliable for people with favism. Everything is substituting for a simple foreground selection of genotypes with low v-c, removing the need for costly, time-consuming, destructive, and biochemical study of seeds (Khazaei et al., 2019; van Norel and Hoogendoorn, 1989). In Australia, faba bean breeding efforts frequently employ the vc marker and different herbicide tolerance markers (imidazolinone). The biosynthetic process for v-c was just recently discovered. Similar to this, strong DNA markers for rust resistance genes (Uvf1 and Uvf2), low seed coat tannins (Zt1 and Zt2), growth habit (Vf-TFL1), and other traits have been developed. Moreover, it has been noted that closely related genetic markers for the resistance of host plants to ascochyta blight exist (Ijaz et al., 2021). Creating useful genetic markers for lowering susceptibility to chocolate spot would be very helpful for faba bean breeding efforts.

Figure 3. Field layout for successful outcrossing.

Genomic Selection

High-throughput phenotyping tools and low-cost genome-wide genotyping create further opportunities for accelerating selection for complicated characters. Genomic *selection (GS) is a specialized form of MAS in which information from genotype data on marker alleles covering the entire genome forms the basis of selection.* Rounds of intermating and selection based on genomic predicted breeding values make up the recombination cycle. (GEBVs). Updates to the model and line evaluation happen simultaneously. Selected lines are inbred and then re-selected based on GEBVs or perhaps phenotypes after at least one recombination cycle for line evaluation and updating of the training population. The model is updated using genotypic and phenotypic information from the chosen lines. Bringing fresh germplasm into the recombination cycle can preserve diversity in the breeding program. To update the prediction model, genotypic and phenotypic information from the new germplasm should be added to the training population (Rutkoski et al., 2011). In base populations of cross-pollinated species created as synthetic populations, genomic selection has been demonstrated to be an effective technique for recurrent selection (Muller, 2017). Grain yield accuracy values are often less accurate than heritability values (or their square roots) from numerous yield experiments. Yet, genomic predictions are made for specific cases. Breeding values for grain yield or agronomic performance that are genomically determined for individual plants are additionally more accurate than the performance of such individual plants' heredity. This makes it possible to use single plants to implement genomic selection. It also makes breeding cycles shorter as the laborious process of creating and phenotyping

superior inbred lines can be moved from the selection process followed by recombination breeding to the cultivar-making phase. Breeding faba beans can benefit greatly from genomic selection, provided that a reliable technology for genome-wide genotyping becomes accessible. By achieving a crossing cycle from 6 months to a year and enhancing genetic gain per unit of time, genomic selection may speed up the breeding process for faba beans. Using GS primarily in the first (population improvement) parts of the method is one potential implementation of genomic selection that would execute well in *Vicia faba* (Gaynor et al., 2017). Breeding germplasm, or the breeding population, must be genetically superior and diversified. The individuals are more or less heterozygous as a result of partial allogamy. This group has a high proportion of genotyped (base) individuals. Instead of plants, seeds are used for genotyping, and genotypic information is used to forecast line performance or varietal aptitude (Mills, 2020). The appropriate selection is made from the chipped seeds. Only the chosen chipped seed is then planted as the breeding population's new generation, and it is immediately permitted to continue reproducing following its natural, partially cross-pollination mode, under natural pollination, without following any fragmentary stages of inbreeding, multiplication, or phenotyping. The progeny is again significantly genotyped, and selection is done based on genomic estimated values. Two cycles each year can be achieved with this method, even without the help of speed breeding. "Recurrent genomic selection" may be used to describe this quick turnover. It is organized as a distinct pipeline called "cultivar development" or "product development" that produces inbred lines from the most promising, chosen plants (Gaynor et al., 2017). During cultivar development, phenotypic information is obtained to direct and modify the algorithm's training for the genomic prediction. The faba bean's largely allogamous nature allows for the recombination required for rapid-cycling recurrent selection and the selfing needed to get inbred material for varietal development combined and at the cheapest cost. The only work required to analyze the inbreeding levels of the mating population's individuals is to identify the other superior inbred ones using genotyping to develop an inbred variety. The progenies in this rapid-cycling population need to be genotyped due to genomic selection, and as a by-product, the progenies are found to have high inbred levels. The inbred individual acts as a bridge to transfer the level of genetic improvement that is obtained in the recurrent population improvement in the varietal development pipeline. Both phenotypic and genetic selection will be used while developing cultivars. An index value selection will be made based on the genomic estimated value and phenotypic

value. The scheme would be more or less relatable to synthetic and traditional line breeding. Maintenance breeding should start when the inbred progenies enter into a varietal development programme.

Phenotypic Recurrent Selection of Populations

The above-mentioned selection technique modified for faba bean has not yet been used in actual practice. The high expense of removing a genetically subpar progeny plant by genome-wide genotyping using currently accessible genotyping technology is still out of reach for most private or governmental breeding projects. A recurrent selection method for use in faba bean was evaluated over a 10-yr period. Growing superior lines in open-pollinated random mating nurseries (RMN) and bulk samples of the seed of each of the retained lines in an RMN are used for seeding the RMNs the following year. Every year, both RMNs produced single plant selections of early maturing legumes, and single rows of each of these selections were cultivated the following year under an allogamous environment to assess the selection's overall agronomic merit and seed material replicate for yield trials in year 2. The choices were assessed in yield experiments using a lattice design with three replications in year 2. Two rows (2R), each 5 m long with a 30 cm space, are grown in the plot. Each plot was observed for seed production, seed protein content, maturity, height, and 1000-seed weight. Based on how well a selection performed in these yield tests, it was either eliminated, bulked up, and planted in the appropriate RMN or yield tested the next year in a layout of three replications with four rows (4R) per plot yield test to further assess a line's potential as a cultivar (Rowland, 1987). What would happen if the phenotypic selection was carried out based on each single plant's seed yield? This strategy has historically been ignored since it is thought that a single plant's agronomic performance, such as yield, has a low heritability and is complicated by selection on the likely linked trait of competitive ability in heterogeneous crop stands (Fasoula and Tokatlidis, 2012). Only 5–10% of the best-performing progenies will be selected from the population (Figure 4). The selected progenies would be threshed and weighed; the remaining population could be removed from the unthreshed plants, which have lesser weight, followed by evaluation and bulking. As a result, genotyping can be done on an individual plant basis at a low expense. Fast recurrent selection based on phenotypes has undergone practical testing, and the results are encouraging.

Figure 4. Plant population under selection.

Major Breeding Achievements

The agronomy department of Cairo University developed 4 varieties (Cairo 1, C. 375, C. 241 & C. 2). Three of them except (C. 375) was developed for Orobanche – tolerance (Darwish and Abdalla, 1997). Eleven improved varieties are under distribution (Giza 2, G. 3, G. 461, G. 714, G. 716, G. 717, G. 643, G. 402, G. 429, G. 674, and Giza blanca). Three of them are reported to be Orobanche tolerants (G. 402, G. 429; G. 674). The rest, except G. 2, are recorded as foliar disease resistant.

Some Exotic Varieties Introduced from Abroad

Long Pod Type

- Aquadule Claudin, Imperial White Long Pod, Masterpiece Green Long Pod, Imperial Green Long Pod, Red Epicure.

Windsor Type

- Imperial White Windsor, Giant four-seeded Green Windsor, Imperial Green Windsor.

Vicia faba breeding has advanced thanks to numerous individual efforts. They include Taylor's and numerous other cytogeneticists' research on the DNA of faba bean root tips, Darwin's studies on the effects of outcrossing, and the verifiable mass mating activities of nearby farmers before 1950. Research on faba beans was also greatly aided by the foundation of the ICARDA in 1977. ICARDA maintains more than 9000 germplasm lines. The third outline of genetic variability within the faba bean contains descriptions of more than 200 traits, making it the largest collection of its kind in the world. Many breeding programs use this dataset. Additional collections are kept at ZIGuK Gatersleben in Germany, CNR-Bari in Italy, and ETSIA-Cordoba in Spain, and they span a wide geographic range of accessions.

Limitations

It has been observed that the breeding work to develop the resistance in plants to biotic and abiotic stress is cumbersome due to the complex nature of the resistance mechanism. Despite significant accomplishments of QTL's studies related to *Ascochyta* blight, rust and chocolate spot resistance, orobanche resistance, and the formation of consensus maps in *Vicia*, emerging molecular tools, such as MAS, must be used to speed up breeding cycles. These gaps will be filled, and the process will be sped up with the information developed in the model legumes *Medicago truncatula* and *Lotus japonicus* (Dita et al., 2006). Understanding plant/pathogen interactions will be aided by phenotyping at a large scale accelerated by functional assessments carried out in the models. Although relevant development was made through biotechnological tools, the change in the genetic makeup of the faba bean can be applied to appropriate genes governing stress tolerance or resistance.

Conclusion

The study provides insights into faba bean breeding based on objectives, methods, challenges, and accomplishments. Improving yield and seed quality, increasing the protein content, and growing tolerance to adverse environmental factors and disease are the main objectives for breeding. Various methods have been described: heterosis breeding, line breeding, synthetic breeding, mutagenesis, speed breeding, marker-assisted selection,

and genomic selection. Each technique presents unique benefits and has limitations, especially challenges such as obtaining stable hybrid seeds, managing genetic diversity, and accelerating breeding cycles. The other crucial aspects include major breakthroughs in breeding and newly developed notable faba bean varieties with specific traits highlighted in the study. However, limitations abound in developing resistance to such stresses, although molecular advances provide avenues for future developments.

References

Abo Mostafa, R. A. I., Sarhan, E. A. D., Atwa, M. A., and Ghareeb, Z. E. (2014). Evaluation of some breeding methods on faba bean improvement. *Journal of Plant Production*, *5*(7), 1225-1237.

Abdalla, M. M. F., (1964). Variation of some agronomic characters in different collections of *Vicia faba L. M. Sc. Thesis. Fac. Agric.* Cairo University.

Berthelem, P., and Le Guen, J. (1967). *Rapport d'activite* 1967. Rennes: INRA.

Bhatta, M., Sandro, P., Smith, M. R., Delaney, O., Voss-Fels, K. P., Gutierrez, L., and Hickey, L. T. (2021). Need for speed: manipulating plant growth to accelerate breeding cycles. *Current Opinion in Plant Biology*, *60*, 101986. doi: 10.1016/j.pbi.2020.101986.

Bishop, J., Garratt, M. P. D., and Breeze, T. D. (2020). Yield benefits of additional pollination to faba bean vary with cultivar, scale, yield parameter and experimental method. *Sci. Rep.* 10:2102. doi: 10.1038/s41598-020-58518-1.

Bond, D. A. (1989a). Prospects for commercialisation of F_1 hybrid field beans *Vicia faba* L., *Euphytica*. 41, 81–86. doi: 10.1007/BF00022415.

Bond, D. A. (1989b). A short review of research on male sterility and prospects for F_1 hybrid varieties in field beans (*Vicia faba* L.). *Euphytica* 41, 87–90. doi: 10.1007/BF00022416.

Bond, D. A., and Pope, M. (1984). "Effect of seed source on performance of faba bean varieties," in *Vicia faba: Agronomy, Physiology and Breeding. World Crops: Production, Utilization, Description*, Vol. 10, eds P. D. Hebblethwaite, T. C. K. Dawkins, M. C. Heath, and G. Lockwood (Dordrecht: Springer), 127–133. doi: 10.1007/978-94-017-3647-3-13.

Bond, D. A., Fyfe, J. L., and Toynbee-Clarke, G. (1966). Male sterility in field beans (*Vicia faba* L.). III. Male sterility with a cytoplasmic type of inheritance. *J. Agric. Sci.* 66, 369–377. doi: 10.1017/S002185960006367X.

Brunet, J., Zhao, Y., and Clayton, M. K. (2019). Linking the foraging behavior of three bee species to pollen dispersal and gene flow. *PLoS ONE.* 14: e0212561. doi: 10.1371/journal.pone.0212561.

Brunjes, L., and Link, W. (2021). Paternal outcrossing success differs among faba bean genotypes and impacts breeding of synthetic cultivars. *Theor. Appl. Genet.* 134, 2411–2427. doi: 10.1007/s00122-021-03832-z.

Darwish, D. S., & Abdalla, M. M. F. (1997). Faba bean breeding in Egypt. *Egypt. J. Plant Breed, 1*, 115-139.

Dieckmann, S., and Link, W. (2010). Quantitative genetic analysis of embryo heterosis in faba bean (*Vicia faba* L.). *Theor. Appl. Genet.* 120, 261–270. doi: 10.1007/s00122-009-1057-7.

Dita M. A., Rispail N, Prats E, Rubiales D and Singh K. B. (2006). Biotechnology approaches to overcome biotic and abiotic stress constraints in legumes. *Euphytica* 147:1–24. doi: 10.1007/s10681-006-6156-9.

Drayner, J. M. (1956). Regulation of outbreeding in field beans (*Vicia faba*). *Nature* 177, 489–490. doi: 10.1038/177489b0.

Duc, G. (1995). Mutagenesis of faba bean (*Vicia faba* L.) and the identification of five different genes controlling no nodulation, ineffective nodulation or supernodulation *Euphytica* 83, 147–152. doi: 10.1007/BF016 78042.

Duc, G. (1997). Faba bean (*Vicia faba* L.). *Field Crops Res.* 53, 99–109.

Duc, G., and Stoddard, F. (2018). David Bond and Jean Picard: two pivotal breeders of faba bean in the 20th century. *Plant Genet. Resour.* 16, 483–487. doi: 10.1017/S 147926211800031X.

Duc, G., Sixdenier, G., Lila, M., and Furstoss, V. (1989). "Search of genetic variability for vicine and convicine content in *Vicia faba* L.," in Proceedings of the Conference of a First Report of a Gene which Codes for Nearly Zero-Vicine and Zero-Convicine Contents (Wageningen), 305–313.

Ederer, W., and Link, W. (1992a). "How shall we select the components for open pollinated faba bean (*Vicia faba* L.) varieties?," in Proc. *1st European Conference on Grain Legumes, Angers, France* (Paris: Association Européenne des Protéagineux), 71–72.

Ederer, W., and Link, W. (1992b). "The Polycross-test as step in breeding partially allogamous crops. Theoretical considerations," in *XIIIth EUCARPIA Congress*, Book of Abstracts, France (Angers: EUCARPIA), 543–544.

Falconer, D. S., and Mackay, T. F. C. (1996). *Introduction to Quantitative Genetics, 4th* Edn, Harlow: Addison Wesley Longman.

Fasoula, V., and Tokatlidis, I. (2012). Development of crop cultivars by honeycomb breeding. *Agron. Sustain. Dev.* 32, 161–180. doi: 10.1007/s13593-011-0034-0.

Gasim, S., Abel, S., and Link, W. (2004). Extent, variation and breeding impact of natural cross-fertilization in German winter faba beans using hilum colour as marker. *Euphytica* 136, 193–200. doi: 10.1023/B:EUPH.0000030669.75809.dc.

Gaynor, R. C., Gorjanc, G., Bentley, A. R., Ober, E. S., Howell, P., Jackson, R., Mackay, I. J., and Hickey, J. M. (2017). A two-part strategy for using genomic selection to develop inbred lines. *Crop Science, 57*(5), 2372-2386. doi: 10.2135/cropsci 2016.09.0742.

Ghaouti, L., and Link, W. (2009). Local vs. formal breeding and inbred line vs. synthetic cultivar for organic farming: case of *Vicia faba* L. *Field Crops Res.* 110, 167–172. doi: 10.1016/j.fcr.2008.07.013.

Ghaouti, L., Vogt-Kaute, W., and Link, W. (2008). Development of locally-adapted faba bean cultivars for organic conditions in Germany through a participatory breeding approach. *Euphytica* 162, 257–268. doi: 10.1007/s10681-007-9603-3.

Gharzeddin, K., Maalouf, F., Khoury, B., Khater, L. A., Christmann, S., and El Dine, N. A. J. (2019). Efficiency of different breeding strategies in improving the faba bean productivity for sustainable agriculture. *Euphytica* 215:203. doi: 10.1007/s10681-019-2521-3.

Ijaz, U., Sudheesh, S., Kaur, S., Sadeque, A., Bariana, H., Bansal, U., & Adhikari, K. (2021). Mapping of two new rust resistance genes Uvf-2 and Uvf-3 in faba bean. *Agronomy*, *11*(7), 1370. 10.3390/agronomy11071370.

International Atomic Energy Agency (2021). Available online at: https://bit.ly/3yVEReM (accessed on September 9, 2021).

Khazaei, H., O'Sullivan, D. M., Jones, H., Pitts, N., Sillanpaa, M. J., Parssinen, P., Manninen, O., and Stoddard, F. L. (2015). Flanking SNP markers for vicine–convicine concentration in faba bean (*Vicia faba* L.). *Molecular Breeding*, *35*, 1-6.doi: 10.1007/s11032-015-0214-8.

Khazaei, H., Purves, R. W., Hughes, J., Link, W., O'Sullivan, D. M., Schulman, A. H., Bjornsdotter, E., Geu-Flores, F., Nadzieja, M., Andersen, S. U., and Stoddard, F. L. (2019). Eliminating vicine and convicine are the main anti-nutritional factors restricting faba bean usage. *Trends in Food Science & Technology*, *91*, 549-556. doi: 10.1016/j.tifs.2019.07.051.

Khazaei, H., Purves, R. W., Song, M., Stonehouse, R., Bett, K. E., Stoddard, F. L., and Vandenberg, A. (2017). Development and validation of a robust, breeder-friendly molecular marker for the vc-locus in faba bean. *Molecular Breeding*, *37*, 1-6. doi: 10.1007/s11032-017-0742-5.

Khursheed, S., Raina, A., Laskar, R. A., and Khan, S. (2018). Effect of gamma radiation and EMS on mutation rate: their effectiveness and efficiency in faba bean (*Vicia faba* L.). *Caryologia* 71, 397–404. doi: 10.1080/00087114.2018.1485430.

Khursheed, S., Raina, A., Parveen, K., and Khan, S. (2019). Induced phenotypic diversity in the mutagenized populations of faba bean using physical and chemical mutagenesis. *J. Saudi Soc. Agric. Sci.* 18, 113–119. doi: 10.1016/j.jssas.2017.03.001.

Link, W. (1990). Autofertility and rate of cross-fertilization: crucial characters for breeding synthetic varieties in faba beans (*Vicia faba* L.). *Theor. Appl. Genet.* 79, 713–717. doi: 10.1007/BF00226888.

Link, W., Ederer, W., Gumber, R. K., and Melchinger, A. E. (1997). Detection and characterization of two new CMS systems in faba bean (*Vicia faba*). *Plant Breed.* 116, 158–162. doi: 10.1111/j.1439-0523.1997.tb02171.x.

Maalouf, F., Hu, J., O'Sullivan, D. M., Zong, X., Hamwieh, A., Kumar, S., and Baum, M. (2019). Breeding and genomics status in faba bean (*Vicia faba*). *Plant Breeding*, *138*(4), 465-473.doi: 10.1111/pbr.12644.

Marzinzig, B., Brunjes, L., Biagioni, S., Behling, H., Link, W., and Westphal, C. (2018). Bee pollinators of faba bean (*Vicia faba* L.) differ in their foraging behaviour and pollination efficiency. *Agric. Ecosyst. Environ.* 264, 24–33. doi: 10.1016/j.agee.2018.05.003.

Mills, A. M., Allsman, L. A., Leon, S., and Rasmussen, C. G. (2020). Using seed chipping to genotype maize kernels. *Bio* 101:e3553. doi: 10.21769/BioProtoc.3553.

Moehring, J., Williams, E. R., and Piepho, H.-P. (2013). Efficiency of augmented p1rep designs in multi-environmental trials. *Theor. Appl. Genet.* 127, 1049–1060. doi: 10.1007/s00122-014-2278-y.

Muller, D. (2017). *Genomic selection in synthetic populations (Ph.D. thesis)*. University of Hohenheim, Stuttgart, Germany.

Pfeiffer, P., Jung, J. L., Heitzler, J., and Keith, G. (1993). Unusual structure of the double-stranded RNA associated with the "447" cytoplasmic male sterility in *Vicia faba*. *J. Gen. Virol.* 74, 1167–1173. doi: 10.1099/0022-1317-74-6-1167.

Rutkoski, J. E., Heffner, E. L., and Sorrells, M. E. (2011). Genomic selection for durable stem rust resistance in wheat. *Euphytica*, *179*, 161-173.

Ramsay, G., Griffiths, D. W., and Dow, N. D. (1991). Spontaneous and induced variation in levels of vicine and convicine in faba beans. *Asp. Appl. Biol.* 27, 43–47.

Rowland, G. G. (1987). A recurrent selection scheme for faba bean. *Canadian journal of plant science*, *67*(1), 79-85.

Suso, M. J., Moreno, M. T., and Melchinger, A. E. (1999). Variation in outcrossing rate and genetic structure on six cultivars of *Vicia faba* L. as affected by geographic location and year. *Plant Breed.* 118, 347–350. doi: 10.1046/j.1439-0523.1999. 00389.x.

Tacke, R., Ecke, W., Hofer, M., Sass, O., and Link, W. (2021). Zooming into the genomic vicinity of the major locus for vicine and convicine in faba bean (*Vicia faba* L.). *bioRxiv*. doi: 10.1101/2021.02.19.431996.

Torres, A. M., Avila, C. M., Gutierrez, N., Palomino, C., Moreno, M. T., and Cubero, J. I. (2010). Marker-assisted selection in faba bean (*Vicia faba* L.). *Field crops research*, *115*(3): 243-252.

van Norel, A., and Hoogendoorn, J. (1989). Faba bean dwarf selections outyield Dutch top varieties, *FABIS Newsl.* 25, 10–13.

Zeid, M., Schon, C. C., and Link, W. (2004). Hybrid performance and AFLP-based genetic similarity in faba bean. *Euphytica* 139, 207–216. doi: 10.1007/s10681-004-3156-5.

Chapter 6

Conventional Breeding Approaches in Faba Beans

Rumana Khan[1,*]
Shivam Yadav[1]
Pawan Kumar Prajapati[2]
Vijay Sharma[2]
and Sushil Kumar Chaturvedi[3]
[1]Department of Genetics and Plant Breeding,
Rani Lakshmi Bai Central Agricultural University, Jhansi, India
[2] Department of Genetics and Plant Breeding,
Banda University of Agriculture and Technology, Banda, India
[3] Director Research, Rani Lakshmi Bai Central Agricultural University, Jhansi, India

Abstract

Faba bean (*Vicia faba* L.) is an important food and feed legume crop that has been cultivated for centuries. The main objectives of faba bean breeding programs include increasing yield, improving resistance to biotic and abiotic stresses, and enhancing nutritional quality. Conventional breeding methods such as mass selection, pedigree selection, and single-seed descent have been employed to develop improved faba bean cultivars. These methods involve selecting superior plants based on desirable traits and self-pollination and progeny testing to stabilize the desired characteristics. Hybridization between genetically diverse parents is also a common approach to create new genetic variations and combine favorable traits. Screening for disease resistance,

* Corresponding Author's Email: shivam00789@gmail.com

In: Faba Beans
Editors: Aamir Raina and Samiullah Khan
ISBN: 979-8-89530-447-1

insect tolerance, and abiotic stress tolerance, such as drought and salinity, is integral to faba bean breeding programs. Breeders also focus on improving the nutritional profile of faba beans, including protein content, amino acid composition, and the reduction of anti-nutritional factors. This chapter discusses the conventional breeding approaches used to improve faba bean varieties.

Keywords: faba bean, legumes, breeding strategies, *Vicia faba,* mutation breeding

Introduction

The faba bean (*Vicia faba L.*) exhibits partial allogamy, i.e., it can undergo both self-pollination and cross-pollination (Suso et al., 1999). Within a typical population of faba beans, approximately one-third of the plants exhibit hybrid characteristics, while the remaining two-thirds are classified as self-pollinated or inbred plants. Multiple reports from different countries have presented evidence indicating that the degree of cross-pollination ranges from 10 to 70% in field conditions. The prevailing range of estimates often is within the interval of 20 to 40% (Filippetti and Pace, 1982; Picard, 1960; Poulsen, 1975). The breeding system of *Vicia faba* is intermediate between autogamy and allogamy, and seed production is largely controlled by the presence of pollinators (Filippetti, 1979; Free, 1966; Fyfe & Bailey, 1951). The primary pollen carriers in this context are the long-tongued bumblebees, specifically *Bombus hortorum* and *Bombus agronim*. On the other hand, Honeybees are typically regarded as less efficient in this role (Bishop et al., 2020; Link, 1990; Torres et al., 1993). Bumblebees with short tongues, such as *B. Terrestris* and *B. Lucorum* bees, extract nectar from the bottom of the flower and aid in triggering cross-pollination by collecting pollen (Hawtin et al., 1985). The heritability of the degree of outcrossing is low and is significantly affected by the inbreeding coefficient of a genotype and the specific kind of insect pollinators (Brünjes & Link, 2021; Link, 1990).

Domestication

Vicia faba L. (2n = 12) is closely related to Narbonensis species (2N = 14) within the subgenus Vicia despite having distinct chromosome counts and

nuclear DNA content (Kew Science's data and resources, 2017). The Narbonensis species display advantageous agronomical characteristics such as resilience to winter conditions and disease tolerance (Birch et al., 1985). The faba bean has a long cultivation history and is known for its capacity to adapt to many environments. It also exhibits diverse responses to human selection, leading to a wide range of variations in plant features and seed traits (Maalouf et al., 2013). The species exhibits a notable variety in plant structure, leaf form, seed color, shape, and size, with seed sizes ranging from 20 to 250 g/100 seeds (Maalouf et al., 2013).

The variation in faba bean has resulted in its classification into three primary groups according to seed size: small-seeded forms in south-western Asia, large-seeded forms in the west, and medium-seeded ancient kinds found across a wide geographical range spanning from Spain to the Himalayas (Muratova, 1931). Archaeological findings indicate that the domestication of faba beans can be traced back to the late 10th millennium BC. This is supported by the discovery of specimens that are 14,000 years old in the Mount Carmel region. These specimens have been recognized as the ancient ancestor of faba beans (Tanno & Willcox, 2006; Caracuta et al., 2016). According to Tanno and Willcox (2006), the bigger varieties of faba bean seeds are thought to have been deliberately chosen by humans. The medium-sized kind can be traced back to the Iberian Peninsula and Central Europe, around 5,000 years ago, as noted by Ladizinsky (1998). Although there are differences in seed size and proposed subdivisions of subspecies based on seed size, these groups have no genetic or sterility barriers. As a result, the faba bean is classified as a single species with diverse botanical groups (Cubero, 1973).

Breeding Methods

Faba bean breeders employ techniques adapted from those utilized for self-pollinated crops, such as the introduction and germplasm collection, as well as the processes of selection and hybridization. These approaches are adjusted to accommodate the specific breeding system and the objective of preserving partial heterozygosity within the plant population. Aside from conventional breeding methods, a methodology has been developed to reintroduce cytoplasmic male sterility in faba beans, which enables the creation of hybrid variants. A population that has undergone recurrent selection procedures to enhance a specific quantitative inherited characteristic can be created and

utilized to create open-pollinated cultivars. It can also serve as a source of variability for further enhancements through other methods, such as developing inbred lines for use in synthetic varieties. A high degree of cross-fertilization in faba bean, above 0.5, is crucial for developing synthetic varieties and enhancing open-pollinated populations (Metz et al., 1994). The optimal approach is to create artificial variations by utilizing self-fertilizing strains to guarantee a minimal yield when insect pollinators are absent and to capitalize on their presence by investigating heterosis to enhance yield and yield consistency (Hebblethwaite et al., 1984).

Mass Selection

Mass selection is a process in which a significant number of individual plants are chosen based on their superior phenotype and compositing the harvested seeds without progeny-testing (Allard, 1999). Mass selection, whether conducted within local populations or in subsequent generations after hybridization, has been a common breeding technique in faba bean breeding programs. However, it has not resulted in significant yield enhancement when selection has been solely based on yield alone (Sjodin, 1977). Enhancing traits with high heritability is highly beneficial, particularly when these traits are closely linked to yield. Nevertheless, the primary benefit of mass selection may reside in enhancing uniformity (Bond, 1971). The study conducted by Sahabuddin in 1984 examined the effectiveness of mass selection in faba beans. The results revealed substantial differences in variance for all traits between the progenies, but no significant differences were observed between the control groups. However, the average yield of the selection progenies was not higher than the controls, and the regression of selection progenies on their parents was not statistically significant; it may be argued that mass selection for yield was not successful. The analysis revealed that the mean grain yields were substantially linked with pod and seed numbers per plant while only showing a small correlation with weight per seed and seeds per pod. All the characters, except for seeds per pod, exhibited a positively skewed frequency distribution.

On the other side, a research project was conducted to enhance the winter hardiness of faba beans by utilizing bulk-harvested micro plots drawn from 466 accessions of the National Plant Germplasm System (NPGS) (Landry et al., 2017). The authors generated four large populations by specifically choosing plants that had successfully survived in different winter conditions

in Southeastern Washington. After undergoing three cycles of mass selection for winter survival, these breeding materials exhibited a level of winter hardiness comparable to that of northern European populations. Significantly, it was demonstrated that northern European populations displayed a lower yearly percentage survival growth rate (about 5%) in comparison to the NPGS bulks. Furthermore, there was a decline in populations with large seeds over time, indicating a correlation between seed size and survival throughout the winter season. The breeding materials produced have the potential to improve the ability of faba beans to tolerate cold conditions in future cultivars.

Hybridization

Hybridization is an essential breeding technique employed in crops that undergo self-pollination. This method entails mating two distinct parental genotypes, which is subsequently followed by the identification and isolation of pure lines in subsequent generations. Various selection processes, including pedigree selection, bulk population, single seed descent, and backcrossing, are frequently utilized to discern favorable genotypes within segregating progenies after hybridization. When it comes to faba bean breeding, the pedigree method, typically supplemented with early testing, is a favored option after the original hybridization process.

The bulk population method is commonly used due to its simplicity and cost-effectiveness. This approach promotes bulk development during the initial generations, followed by selecting individual plants typically occurring in the F6 generation or subsequent generations. Nevertheless, it is possible to incorporate alterations to improve its efficacy. Despite the numerous benefits it offers, a significant obstacle in the field of faba bean breeding is the substantial duration necessary for the development of cultivars, which normally extends over a period of 10 to 12 years following the process of hybridization.

As an alternative selection method, the single seed descent approach (Hawtin et al., 1985) has been utilized in faba bean breeding by certain individuals. However, it has not achieved universal acceptance among field bean breeders. The approach entails progressing each F2 offspring within the population through the process of single seed selection till reaching subsequent generations (F5-F6), enabling the opportunity to pick from predominantly homozygous lineages. The primary advantages of this approach encompass the preservation of genetic diversity for traits with

limited heritability, the possibility to achieve many generations within a single year, and the facilitation of streamlined data management during successive generations of segregation (Figure 1). In addition, the backcrossing practice has proven valuable in numerous faba bean breeding initiatives, specifically in the context of incorporating desirable characteristics from a donor parent into an established cultivar. These characteristics may include resistance to diseases or a determinate growth pattern. It is worth mentioning that the development of haploid culture in *Vicia faba* has not been achieved, and the utilization of line breeding through recombination from biparental crossings has not yielded significant advancements in the enhancement of faba bean.

Historically, the breeding protocols employed for self-pollinating crops have been characterized by a constrained gene pool, mostly resulting from using two-parent crosses and the limited scope for disrupting linkage blocks over successive generations. To address these constraints, breeders have investigated various alterations, including augmenting the number of crosses, doing early yield testing, and implementing measures for population improvement.

On the other hand, compared to other crops, the faba bean's low propagation coefficient presents special difficulties in breeding and yield assessment. The seed production of a single faba bean plant is sufficient to cultivate a field plot measuring 1 square meter, hence imposing constraints on the extent of observation trials and experimentation. Another aspect to consider in traditional faba bean breeding is optimizing parental selection. When considering line cultivars, it is important to note that approximately 50% of the genetic variation within the breeding material is attributed to the differences between the lines produced through prospective crossings. The remaining portion of the genetic variance persists within these crosses, underscoring the significance of parental selection (Falconer, 1996). Consequently, it is crucial to discard crosses that exhibit inferior performance compared to other crosses before allocating excessive resources toward evaluating their progeny. In addition to conventional techniques such as single seed descent, pedigree-based selection enables the evaluation of complete crosses (families) during initial generations, facilitating the exclusion of less favorable crosses. The implementation of marker-assisted selection can play a crucial role in facilitating well-informed decision-making both before and following the process of crossbreeding.

The breeding of faba beans encounters distinct problems attributable to its poor propagation coefficient. Consequently, meticulous parent selection and investigation of diverse breeding techniques are imperative to enhance genetic

advancements and expedite the cultivar development procedure. The aforementioned factors are crucial in effectively handling the distinctive attributes and constraints associated with cultivating faba beans.

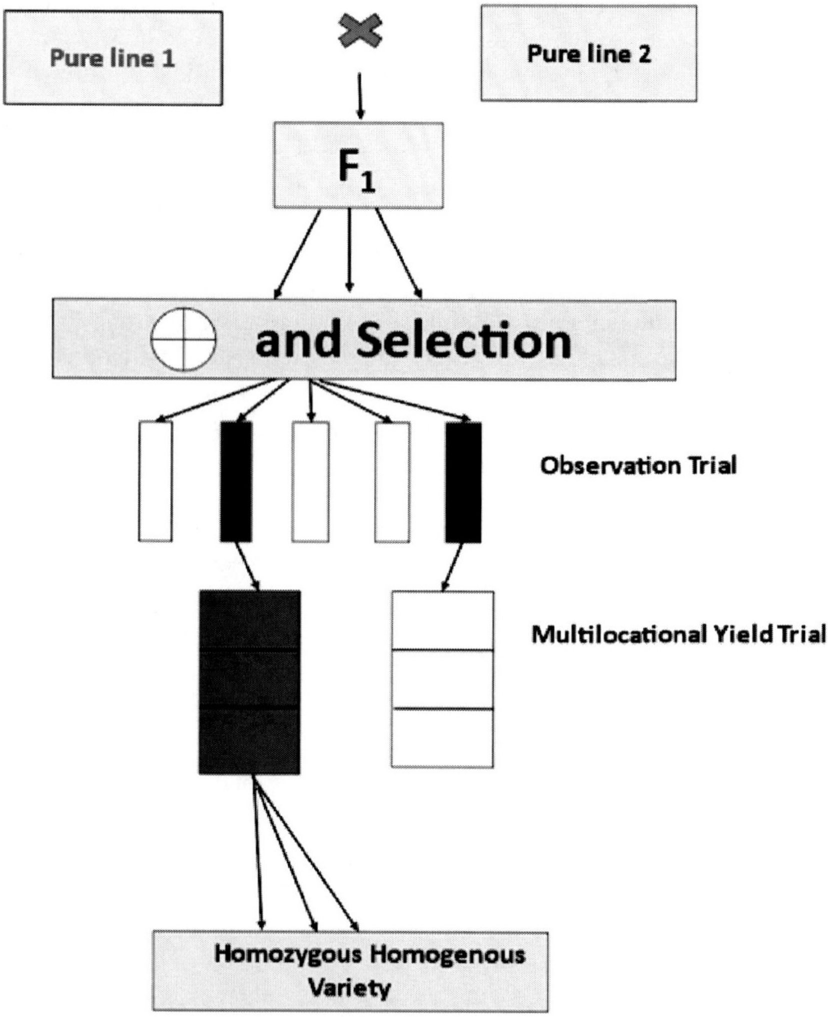

Figure 1. Methods of handling segregating generations adapted from Poehlman, 2013.

Synthetic Breeding

A synthetic variety is a population that has been constituted from hybridization among a limited number of distinct and well-evaluated lines. The term was originally defined by (Allard, 1999) as "being synthesized from genotypes which have been tested for combining ability." However, it is now commonly used to describe a variety maintained through open pollination after being created through hybridization using selected genotypes based upon general combining ability in various combinations. A revised definition is required for faba beans since not all hybridizations occur through bee pollination, and there may be a lack of prior testing to determine the combining ability. In this case, the performance of the synthetic variety itself indicates the combining ability of its components, or the components' performance is assumed to be linked to their overall combining ability (Bond, 1982). In crops with partial allogamy, such as faba bean, enhancing the proportion of heterosis in a synthetic cultivar can enhance yield and yield stability. Greater levels of cross-fertilization lead to an increased proportion of heterosis in synthetic cultivars (Brünjes & Link, 2021).

Synthetic varieties can be classified into two main categories: (i) those that are derived from inbred or partially inbred lines and (ii) those that are derived from populations or other synthetic varieties. Synthetic varieties derived from inbred lines usually undergo national testing and are commercially utilized from the syn-4 to syn-6 generations. In contrast, synthetic materials derived from population sources can be added at earlier stages, such as syn-2 or syn-3, to reduce the need for scaling up parent populations.

Typically, synthetic varieties containing four to ten or more lines are favored. However, it is most common to choose six to eight components (Kinman & Sprague, 1945), as this strikes a balance between the risk of excessive inbreeding and the inclusion of lower-ranked lines if the number exceeds seven to nine. In the case of the faba bean, the highest-yielding synthetic varieties were those with six and seven parents. The superiority of the synthetics with six and seven parents has been remarkable for all the features, except for seeds per pod, hundred seed weight, and podding nodes in the main stem (Maalouf et al., 2002).

The cultivation of synthetic faba bean varieties began with "Throws MS of RHM (Agriculture) Ltd." marking the first commercially grown synthetic faba beans in the United Kingdom. The recommendation was made in 1960, and the program had four component populations with recurrent selection.

Maris Beagle, a significant synthetic variety, was introduced in 1974 and consists of five inbred lines. Currently, there is ongoing research on several synthetic variations, such as Herra from Franck (GFR) with over six lines and two PBI synthetics with three and six components. These varieties are currently undergoing national testing in the U.K.

In synthetic cultivar development, it may not be advisable to depend only on assessing pure lines to develop synthetic cultivars. This discrepancy arises since the overall level of homozygosity found in pure lines does not provide an adequate representation of the degree of inbreeding within a synthetic population. In synthetic breeding, it is customary to employ a minimum of four founder lines as essential constituents of a cultivar. Hence, it is feasible to reconstitute the crop using these genetic lines at any given juncture. Choosing inbred lines exclusively based on their individual performance may not produce optimal outcomes. This phenomenon arises from the observation that the synthetic population generated will not exhibit an equivalent degree of consanguinity as its individual constituents. The primary reason for choosing synthetic breeding instead of line breeding is to optimize the exploitation of heterosis to improve agronomic performance, reducing the adverse consequences of inbreeding (Ghaouti et al., 2008; Ghaouti and Link, 2009). An efficient strategy in faba bean entails the production of offspring through the utilization of top cross or poly cross methods employing pure lines. The assessment of progeny and subsequent identification of superior lineages is carried out as an alternative to directly evaluating the parental lines. The offspring produced through polycross breeding demonstrate a similar degree of heterozygosity and heterogeneity as reported in a synthetic cultivar. Progeny exhibits differences between their paternal lineages in three dimensions: (1) individual performance, (2) capacity to interact with other individuals, and (3) degree of cross-fertilization (Link & Ederer, 1993). The intentional preservation of genetically homogeneous parental lines facilitates the regulated reconstitution of cultivars according to specific preferences.

A commonly employed approach for developing a synthetic cultivar entails allowing a series of around four to eight generations to undergo reproduction within an allogamous setting. The original combination is commonly known as Syn-0, whereas subsequent generations are labeled Syn-1, Syn-2, and so on. The achievement of minimal heterosis in Syn-1 depends on cross-fertilization rates being less than 100% and can only be observed in the following generations.

Busbice, (1970) presented a comprehensive account of the projected production yield of a synthetic material in subsequent generations. In a generation t the expected performance Y is

$$Yt=A + (1-Ft) B$$

where A is the performance of the homozygous parent, Ft is the coefficient of inbreeding in generation, and r and B are the amount of heterosis (average differences between the mean of parents and their F! crosses). This formula assumes a linear relationship between Yt and Ft.

The level of consanguinity in the first generation (Syn-1) resulting from a population of unrelated parents is influenced by various parameters, including the degree of polyploidy (k), the coefficient of inbreeding displayed by the parents (F0), and the degree of self-fertilization (s) (Busbice, 1970). The numerical value of 22 is the focal point of the present discourse. The theory about the behavior of synthetic variations during the multiplication process is particularly pertinent to allogamous species or those that consistently demonstrate a self-pollination ratio. Therefore, the practicality of its use to faba bean is constrained, as the degree of interbreeding in this particular species fluctuates on a yearly basis and relies on the presence and behavior of bees. Bond, 1981 suggested implementing selective breeding techniques to develop artificial variations in the faba bean species. Bond, 1987 presented a revised definition of synthetic faba bean based on two primary factors. To begin with, it should be noted that not all hybridizations in faba bean parents are exclusively facilitated by bee pollination. Additionally, it is worth noting that there is no prerequisite for conducting preliminary tests to assess combining abilities. Alternatively, the performance of the synthetic variety might be utilized as a measure of the combining ability exhibited by its constituent components. Conversely, it is postulated that the performance of the individual components is linked to their collective combining capacity (Bond, 1987).

The concerns are contingent upon the presumption of a sustained degree of cross-pollination, as posited by Busbice in 1970. In the event of environmental heterogeneity concerning outcrossing, it becomes imperative to emphasize the attainment of the highest degree of cross-fertilization during the concluding phase of seed development. This methodology guarantees reduced inbreeding and elevated expression of heterosis, resulting in advantageous outcomes for farmers in terms of increased vigor, yield, and stability in their agricultural plots.

A study by Gasim and Mohamed (2018) aimed to assess the importance of synthetic varieties compared to line cultivars in cultivating faba beans in semi-arid zones. Five inbred lines, which had undergone at least six generations of inbreeding, were created and utilized to produce F1 hybrid offspring. All feasible combinations of these lines were used, except for reciprocal crosses. Additionally, blends of these lines (Syn-0) and subsequent synthetic generations of Syn-1 were constructed. The evaluation of the entries revealed that the lines have a significant level of both general and specific combining ability, resulting in a high yield and a substantial average degree of cross-fertilization (0.36). The heterosis, relative to the mid-parent, for yield was 67%. The combination of four inbred lines (Hudeiba/93, Bassabier, Ed-Damar, and Shabah) resulted in the maximum yield of 3.40 t/ha for Syn-0 and 3.96 t/ha for Syn-1. The average yield of the pure stand of the four lines was 3.11 t/ha. The yield increase was 9% in Syn-0 and 27% in Syn-1. The yield per hectare of the mixed crop, Syn-0, was 14% higher than that of the individual pure stands of the Hudiaba/93 and Bassabier lines. For the lines Ed-Damar and Shabah, the yield of Syn-0 was 4% and 7% higher, respectively. Similarly, the yield of Syn-1 was 32%, 25%, and 21% higher than the yield of the pure stands for Hudiaba/93, Bassabier, and Ed-Damar, respectively. The results validate the existing understanding that yield in faba beans increases with subsequent generations due to the influence of heterogeneity and heterozygosity (Figure 2).

Male Sterility and Hybrid Faba Bean

In *Vicia faba*, male sterility is necessary to produce hybrid variants. There have been reports of two forms of male sterility (Bond et al., 1964a).

1. Genetic male sterility (Male sterility under the control of the nucleus only).

An experimental study utilized genetic male sterility to demonstrate the extent of hybrid vigor in faba beans and the potential of hybrid varieties. Commercial production of hybrid seed with this recessive gene is challenging due to the maintenance scheme, which results in only half of the seed being predicted to be male sterile. To enhance its efficiency, it is necessary to develop intricate systems (Bond et al., 1964b).

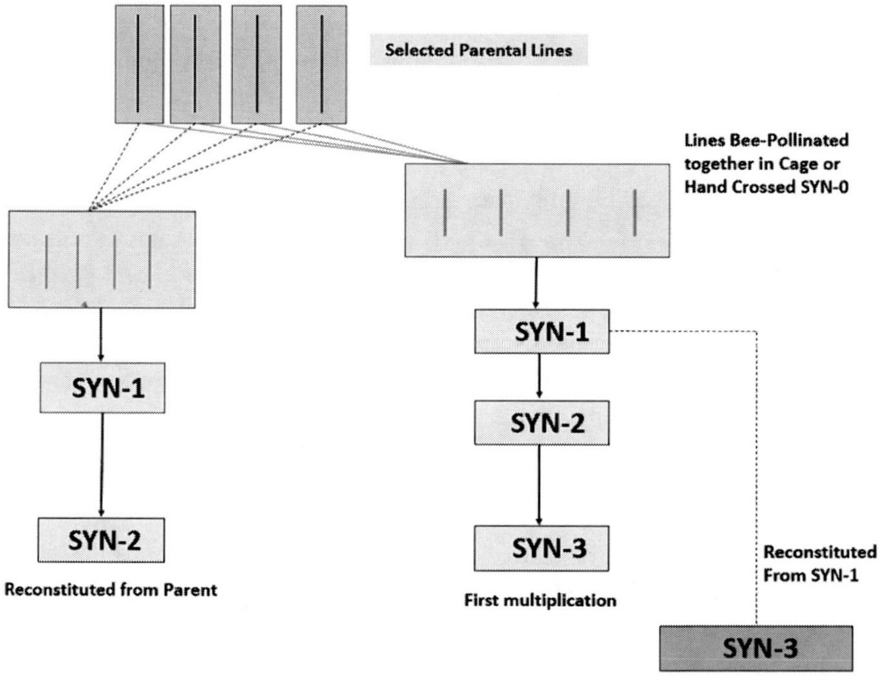

Figure 2. Synthetic breeding scheme adapted from D. A. Bond, 1981.

2. Male sterility controlled through an interaction of the cytoplasm and nucleus (cytoplasmic genetic male sterility, commonly named CGMS Bond, 1989a).

Cytoplasmic Male Sterility in Faba Bean

Two sources of CMS are known; the first one, designated 447, was discovered by Bond et al. in 1957. The second one, termed 350, was discovered by Berthelem & Le Guen in 1967. These individuals were found separately within populations and do not appear to come from crosses between other species, which is often the cause of CMS.

This finding emerged from the observation of a significant rise in the rate of crossbreeding in the progeny of a plant that was allowed to be pollinated naturally, particularly referred to as plant No. 447, compared to other plants. Nevertheless, it was observed that both the 447 and 350 cytoplasms exhibited instability and vulnerability to external influences, resulting in sporadic and

unpredictable instances of reversion to fertility. Consequently, these cytoplasms were considered unsuitable for creating F1 hybrids.

Classical Scheme for the Utilisation of the CMS System in Hybrid Breeding

The illustration in Figure 3 depicts techniques for transferring a specific genetic makeup into a particular cytoplasm, a process crucial for maintaining cytoplasmic male sterile (CMS) lines and ultimately producing fertile single-cross hybrids. From these methods, it's straightforward to derive strategies for generating three-way crosses and potentially other hybrid types, following conventional breeding principles. However, what sets apart the comparison involving 447 faba bean cytoplasm and other species is a distinct feature: once male sterility in faba beans has been restored, whether by a dominant restorer gene or any reversal mechanism, subsequent generations resulting from self-pollination will never exhibit male sterility again. (Bond et al., 1966).

Different Parameters Controlling CMS Instability in Faba Bean

External and internal factors affect phenotypic instability of expression of cytoplasmic male sterility (Thiellement et al., 1979). The initial occurrence of genetic male sterility was spontaneous; however, its practical application for large-scale F1 hybrid production was hindered by its genetic linkage with semi-lethal characteristics (Bond et al., 1964b). However, the utilization of genetic male sterility has proven to be advantageous for plant breeders. This is primarily due to its ability to eliminate the need for emasculation in hybrid production, enhance natural cross-pollination and recombination in populations undergoing recurrent selection, and facilitate the production of commercial hybrid seeds. The application of hybrid vigor in faba bean via F1 hybrids has been advocated, with several ways being recommended, such as manual pollination, the use of chemical gametocides, natural cross-pollination, and the implementation of genetic or cytoplasmic male sterility.

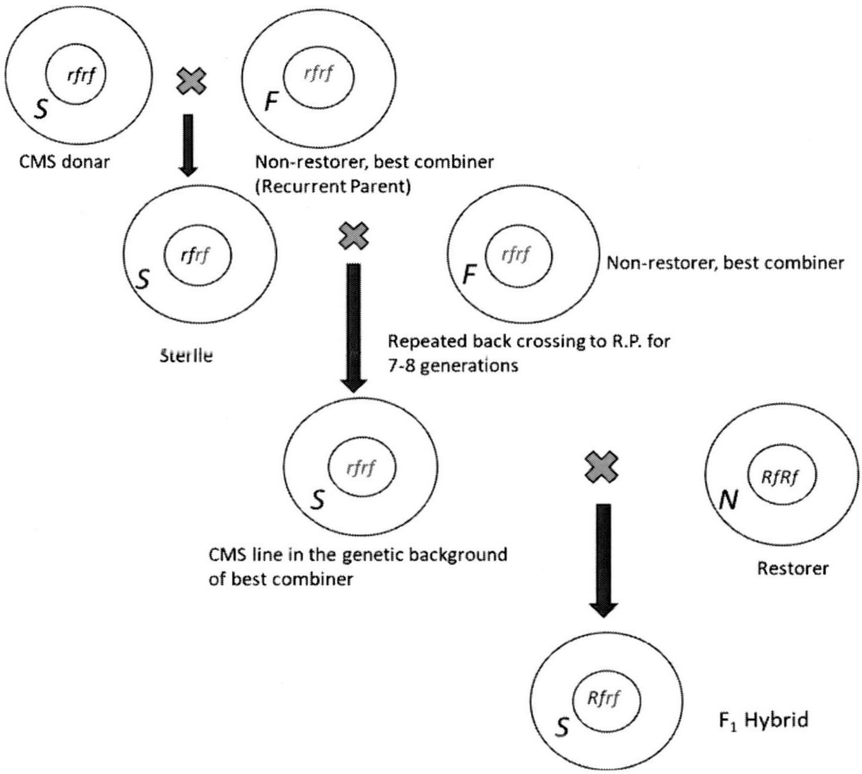

Figure 3. Classical Scheme for Utilisation of CMS System in Hybrid production.

Nevertheless, the economic viability of hand pollination in nations with high labor costs was considered impractical, and the use of chemical gametocides had not resulted in adequate levels of male sterility. The task of effectively managing field plantings and insect pollinators continues to pose a significant problem, particularly in the context of faba bean cultivation, where cross-pollination heavily relies on the presence and activity of insects. To tackle this issue, a suggestion was put out to include honeybee colonies in seed production areas to enhance seed set, hence mitigating the insufficiency of indigenous pollinators.

In the absence of cytoplasmic male sterility and gametocides, breeders resorted to utilizing monogenic characteristics to facilitate hybrid seed production. The observed qualities encompassed both recessive and dominant nuclear genetic pollen sterility (Duc et al., 1985), alongside a range of morphological characteristics such as colors for testa, hilum, cotyledon, and blossom. (Link, 2009) These morphological characteristics were utilized as

markers in the study. Breeders have also investigated the potential utilization of seed size despite its complex nature as a feature influenced by multiple genes. Over the course of several decades, breeders have demonstrated ingenuity in their efforts to develop hybrid varieties of faba beans. Despite numerous attempts, however, a significant breakthrough in this area has not yet been achieved. (Bond, 1989b; Bond et al., 1966; Link et al., 1994).

Mutation Breeding

Vicia faba has been extensively utilized as a test system for evaluating and comparing the mutagenic effects of chemical compounds and radiations. While there have been numerous studies in this regard, few have focused on comparing mutagenic agents for their capacity to induce beneficial genetic variations in *Vicia faba* (Abdalla and HAS, 1977; Sjödin, 1971). Some studies have highlighted the potential of mutagenesis to expand genetic diversity in available seed stocks (Abdalla and HAS, 1977; Chapman, 1981). However, these studies typically used cultivars from the "minor" botanical group, limiting their immediate applicability in regions like the Mediterranean basin, where cultivars from the "major" botanical variety are predominant.

To address this limitation, a breeding program aimed at enhancing the faba bean through mutagenesis was initiated in 1978 at the Plant Breeding Institute in Bari (Filippetti, 1986; Filippetti & Pace, 1983; Filippetti & Ricciardi, 1988). Various mutagenic agents have been employed in this program, including X-rays, neutrons, ethyl methane sulfonate (EMS), methyl methane sulfonate (MMS), ethylene imine (El), and diethyl sulfonate (DES), either individually or in combination. Some induced mutations have shown promise in enhancing yield (Filippetti & De Pace, 1986) mutants with determinate habits and terminal inflorescences, offering opportunities to revolutionize plant architecture and develop new faba bean varieties (Abdalla & HAS, 1977; Sjödin, 1971). Other easily identifiable variants have been created, which can serve as markers in cytogenetic analyses and potentially aid in genetic transformation, benefiting breeders in the long term.

Furthermore, there is a pressing need to identify mutants that show resistance against root parasitic weeds like broomrape and stem parasitic dodders (Fernández-Aparicio et al., 2016; Mejri et al., 2018; Rubiales et al., 2016; Rubiales and Fernández-Aparicio, 2012), which are increasingly problematic in the Mediterranean basin. Developing faba bean varieties with resistance to herbicides targeting these weeds is particularly valuable.

Similarly, mutants that reduce pod wall thickness have the potential to boost grain yield (Bhatia et al., 2001). Inbred lines with new diversity, whether from recombination or mutagenesis, can be used as a starting point in breeding programs.

Presently, there are 20 faba bean mutant varieties from various countries listed in the International Atomic Energy Agency (IAEA) mutant variety database. These mutants can potentially be valuable in breeding initiatives. Combining mutagenesis with biotechnology tools has the potential to expedite the release of novel faba bean germplasm and improved cultivars, further contributing to faba bean production in the Mediterranean basin and beyond.

Breeding for Quality in Faba Bean

Faba bean, being a leguminous crop, has both nutritional and environmental benefits. It is a good source of plant-based protein, nutrients, and dietary fiber (Warsame et al., 2018; Multari et al., 2015). Its nutritional profiling (Figure 4) revealed that it contains protein, amino acid, Iron, Zinc, and dietary fibre as nutritional factors, whereas as antinutritional factors, it has vicine, convicine, phytate, saponin, oligosaccharides, phenols, lectins, and trypsin inhibitors (Labba et al., 2021). Moreover, it fixes nitrogen symbiotically from the atmosphere, which improves the fertility of the soil, and it also helps in breaking the cycle of soil-borne disease by not being the host for cereal pathogens (Köpke and Nemecek, 2010). Being a substitute for animal protein, it could also help in reducing greenhouse gas emissions. Breeding for quality in faba beans is mainly dependent on two key factors: producer and consumer preferences. Producer preferences involve resistance to biotic stress (fungal, viral, bacterial diseases, restriction caused by parasitic weeds) and abiotic stress (frost, waterlogging, heat, drought) harmful to crops, whereas consumer preferences involve attractive consumable parts, e.g., pods and seeds with minimum anti-nutritional and higher nutritional content. It has also been found to have curing properties in chronic diseases like Parkinson's disease (Apaydin, 2000; Rabey et al., 1992; Liu et al., 2000), liver cirrhosis, kidney failure, and hypertension (Randhir and Shetty, 2004; Kwok et al., 1997; Randhir et al., 2002), inflammation, diabetes, and lower the level of cholesterol (Labba et al., 2021) due to the wide range of its levo-dihydroxy phenylalanine (L-dopa) content. Different nutritional and anti-nutritional factors for their content range in faba beans are given in Table 1.

Considering all these facts, we can say that breeding for quality in faba beans should include methods that can enhance the protective capacity of plants against different biotic and abiotic factors, which ultimately leads to healthy and attractive consumable parts and helps in bridging the demand and supply gap. Likewise, such methods should be chosen or developed that can have the potential to minimize anti-nutritional properties and enhance their nutritional properties by exploiting their present genetic variation in cultivars or by transferring them from other rich sources by using them as donors for the traits. The breeding approaches that can be exploited to achieve the goals are germplasm screening and characterization, hybridization, and mutation.

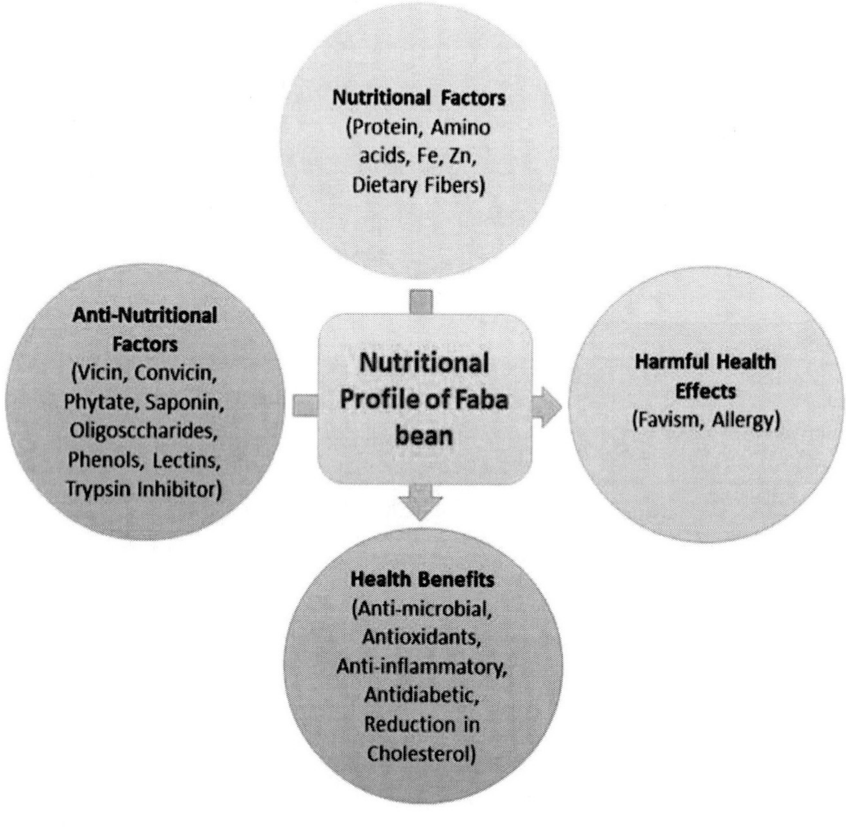

Figure 4. Nutritional and anti-nutritional quality of faba bean with their effects on human health.

Table 1. Content range of different nutritional and anti-nutritional factors in faba bean

S.N.	Factors		Content range	References
Nutritional Factors				
1.	Protein		26-33%	Labba et al., 2021; Makkar et al., 1997; Griffiths and Lawes, 1978; De Angelis., 2021
2.	Amino acids	Leucine	50.8-72.1 mg/g protein	Labba et al., 2021; Malgorzata., et al., 2018; Boye et al., 2010
		Lysine	44.8-74.8 mg/g protein	Labba et al., 2021; Malgorzata., et al., 2018; Boye et al., 2010
3.	Dietary fiber	Soluble fraction	0.55-1.06%	Labba et al., 2021; Singh et al., 2014
		Insoluble fraction	10.7-16.0%	Labba et al., 2021; Singh et al., 2014; Vidal-Valverde et al., 1998
4.	Iron		1.8-21.3 mg/100 g	Labba et al., 2021; Khan et al., 2015
5.	Zinc		0.9-5.2 mg/100 g	Labba et al., 2021; Cabrera et al., 2003
Antinutritional factors				
1.	Lectin		0.8-3.2 HU/mg	Labba et al., 2021; Jiang et al., 2010
2.	Trypsin inhibitor		1.2-23.1 TIU/mg	Labba et al., 2021; Millar et al., 2019; Shi et al., 2017; Hern´ Andez-Infante et al., 1998
3.	Saponin		18-109 µg/g	Labba et al., 2021
4.	Phytate		112-1281 mg/100 g	Labba et al., 2021; Lazarte et al., 2015; Luo et al., 2012
5.	Total Phenolic content		1.4-5 mg GAE/g	Labba et al., 2021; Saleh et al., 2019
6.	Vicine		403-7014 µg/g	Labba et al., 2021; Khazaei et al., 2019
7.	Convicine		35.5-3121 µg/g	Labba et al., 2021; Khazaei et al., 2019
8.	Raffinose		1.1 3.9 g/kg	Labba et al., 2021; Landry et al., 2016

Table 2. Some quality traits and the description of work done

S. No.	Quality traits	Description	Reference
1.	Tannin	zt-1 and zt-2 controlling the absence of tannin in faba beans were studied	Gutierrez et al., 2008
2.	Drought, Protein and sulphur	Studies have shown that drought helps in increasing the level of protein and sulfur content of faba bean	Alghamdi, 2009; Singh, 2007; Rharrabti et al., 2001

S. No.	Quality traits	Description	Reference
3.	Protein and tannin	It was observed that protein and tannin make soluble complexes at initial binding, but as the process progresses, it converts into insoluble precipitates.	Kosińska et al., 2011
4.	Protein, Phenol and Tannin	Screening of germplasm revealed that they have protein content varies from 19.5 to 33.3%, phenol 0.95 to 2.56%, and tannin 0.426 to 1.59%	Kumar et al., 2015
5.	Tannin, vicine, and convicine	Linkage of single gene controlling seed quality traits (zero tannin and low content of vicine or convicine or both) with markers has been analysed using bulk segregant analysis.	Gutierrez et al., 2006; Gutierrez et al., 2017
6.	Carbohydrates, protein, minerals (nitrogen, phosphorus, and potassium), tryptophan, lysine, arginine, and phenylalanine	The application of fulvic acid @ 9 g/L has significantly improved the content of Carbohydrates, protein, minerals (nitrogen, phosphorus, and potassium), tryptophan, lysine, arginine, and phenylalanine.	Abdel-Baky et al., 2019
7.	Vicine and convicine	QTL mapping for gene vc and vcr has been done	Khazaei et al., 2019
8.	Tannin, Ca, Mg, Mn, Cd and Protein	Low tannin varieties of faba bean have higher content of Ca, Mg, Mn, Cd and Protein in comparison to higher tannin varieties.	Khazaei and Vandenberg, 2020
9.	Polyphenolic compounds	Quantification done for polyphenolic compounds revealed that p-coumaric acid (42.1 mg/100 g), chlorogenic acid (32.6 mg/100 g), and epicatechin (31.2 mg/100 g) were in plenty.	Castaldo et al., 2022
10.	Tannin and amino acids	Breeding line FLIP08-016FB was found to have a higher yield with low tannin and higher amino acids (lysine, methionine, and cysteine)	Akgun et al., 2023

In faba beans, there is much variation available for quality traits (Pradhan et al., 2014). However, this variation is unraveled, and there is a need for intense screening for the quality traits to utilize the potential variation. Kumar et al. (2015) screened 11 germplasm for their nutritional and anti-nutritional attributes and observed plenty of variation for protein, amino acid, sugar, phenol, and tannin. After screening, morphological, biochemical and molecular characterization of the germplasm is necessary to validate the presence and extent of the quality traits. It has been observed that F_1 hybrids have 40-70% higher hybrid vigour than homozygous parents for different quality traits, including yield (Dieckmann and Link, 2010; Zeid et al., 2004). An operational cytoplasmic male sterility (CMS), appropriate pollinators, and a higher propagation coefficient are three basic things that are needed for a successful crossing program not only in the faba bean but in all other crops. Unfortunately, these systems are not well developed in the faba bean and have emerged as a potential obstacle. The reversion of male sterility (Maalouf et al., 2019; Link et al., 1997), nectar stealing by pollinators (Marzinzig et al., 2018), and low propagation coefficient (Link and Stutzel, 1995) are the common problems for a dynamic and successful breeding program in the crop. Breeders have tried to conquer these obstacles by using different creative ideas, but an ultimate solution has yet to occur (Duc and Stoddard, 2018; Link, 1998; Berthelem, 1970). Studies done so far for some quality traits have been given in Table 2.

Biotic and Abiotic Stresses

Faba bean is an important legume crop grown worldwide for its nutritious seeds and forage (Duc, 1997). However, the productivity of the faba bean is often constrained by various biotic and abiotic stresses. Biotic stresses, including diseases caused by fungi, bacteria, viruses, nematodes, and pests, can significantly reduce yield and quality (Stoddard et al., 2010). Abiotic stresses, including drought, salinity, extreme temperatures, and nutrient deficiencies, also pose major challenges to faba bean production and productivity. Understanding the impact of these stresses and developing effective management strategies are crucial for improving the sustainability and resilience of faba bean cultivation. Some of the major stresses, their genetics, and the degree of damage are discussed in Tables 3 and 4.

Table 3. Some major biotic stresses, resistance mechanism and nature of damage in faba bean

Biotic Stress	Causal Agent	Yield loss severity	Resistance Mechanism	Nature of Damage	References
Chocolate Spot	*Botrytis fabae*	Heavy yield losses	Quantitative resistance, additive with some dominance	Lesions on leaves, defoliation, spread in humid regions	Robertson & Saxena, 1993
Ascochyta Blight	*Ascochyta fabae*	Average yield losses	Resistance has been observed in tannin-free as well as tannin-containing cultivars, but seed transmission may be more frequent in tannin-free cultivars.	Leaf spots, stem damage, pod damage, varying susceptibility	Jellis et al., 1991
Foot Rot & Wilt	*Fusarium solani, Fusarium oxysporum*	Yield loss dependent on rotation frequency	Resistance can be selected in zero-tannin lines	Root rot, wilting, yield loss, influenced by soil conditions and concurrent stresses	Van Loon et al., 1989; Villalobos & Jellis, 1990
Stem Rot	*Sclerotinia sclerotiorum*	Major stress with susceptible crop rotation	Resistance has been reported in Lines KU 189, 190, and 191	Rotting of stems, impact on crop rotation, need for further field observations.	Jellis et al., 1990

Table 3. (Continued)

Biotic Stress	Causal Agent	Yield loss severity	Resistance Mechanism	Nature of Damage	References
Viruses	Bean yellow mosaic virus (BYMV), Pea enation mosaic virus (PEMV), Bean leaf roll virus (BLRV), True broad bean mosaic virus (TBBMV), Broad bean stain virus (BBSV)	Bean yellow mosaic virus (BYMV) 13%, Pea enation mosaic virus (PEMV) 5%, Bean leaf roll virus (BLRV) 3%	Challenging to breed for resistance to all viruses	Symptoms vary based on virus type, including mosaic patterns, leaf roll, and transmission	Sillero et al. (2010)
Rust	*Uromyces fabae*	Up to 80% yield loss in China, 27-32% in the UK	Quantitative "slow rusting" resistance is more durable than race-specific immunity	Lesions on leaves, defoliation, earlier appearance with earlier maturing cultivars	Conner & Bernier, 1982
Stem Nematode	*Ditylenchus dipsaci*		Single gene resistance controlled maternally through 29H cytoplasm	Stunting, wilting, yield loss	Caubel & Le Guen, 1992
Broomrape	*Orobanche crenata*	Major stress in the Mediterranean	Resistance from Giza 402 and other sources, additive inheritance	Parasitic attachment, yield loss, need for integrated control measures	Cubero et al., 1988
Aphids	*Aphis fabae*, *Aphis craccivora*	Considerable stress, yield loss	Resistance in some lines like Rastatt and Line 14	Sap-sucking, population buildup, yield loss	Muller, 1968; Bond & Lowe, 1978

Table 4. Some major abiotic stresses, resistance mechanism and nature of damage in faba bean

Abiotic Stress	Resistance Mechanism	Yield Loss Severity	Nature of Damage	References
Drought	-Osmotic adjustment -Stomatal regulation -Increased root growth -Reduced leaf area	Not quantified	-Wilting -Reduced growth -Yield loss	Robertson & Saxena, 1993
High Temperatures	-Transportation cooling -Antioxidant production -Heat shock proteins	Not quantified	-Leaf scorch/Flower/pod abortion -Reduced growth -Yield loss	Graff, 1988, Hanelt, 1972
Low Temperatures/Frost	-Supercooling - Antifreeze proteins - Membrane lipid changes	Not quantified	-Leaf/stem necrosis -Flower/pod damage -Reduced growth - Yield loss	Lin et al., 1987
Waterlogging	-Aerenchyma formation -Anaerobic respiration	Up to 48% yield loss	-Leaf chlorosis -Root rot -Reduced growth -Yield loss	Dawkins & Brereton, 1984
Salinity	-Ion exclusion -Osmotic adjustment -Antioxidant production	Not quantified	-Leaf burn/Reduced growth/Yield loss	El-Karouri, 1979

Limitations and Future Breeding Strategies

Faba beans are a partially cross-pollinated crop; it is very difficult to multiply pure lines under insect-proof environments due to their dependence on mechanical tripping of flowers by their insect pollinators. However, auto fertility has also been reported in faba beans with a high level of genetic variability. Adaptation to the target environment should be studied by characterizing morpho-physiological and molecular aspects of the traits and suitable QTLs to select quality traits. Emphasis should be given to functional genomics and reverse genetics approaches. That is how we can better understand the genetic control of quality traits. There is a need to develop low-cost genotype-by-sequencing approaches, which have commonly been developed in other important crops (Adhikari et al., 2021). Once easy and cost-effective genotype-by-sequencing is developed, genomic selection will be revolutionized, and this knowledge can be exploited in other arenas. The gene editing tools can strengthen the "breeding by design" approach to obtain an ideal plant with high nutritional value and low anti-nutritional factors (Jacquier et al., 2020). Low-cost resequencing may make in silico mutagenesis a reality. The study of genetic control of the mutual relationship between the crop and their pollinators is more demanding in understanding the attraction and function of pollinators as well as the traits of the crop (Bailes et al., 2018). Moreover, the relationship between rhizobia and crop needs a thorough study of its genetic control to understand the efficiency of both the microbiome and crop in fixing nutrients in the soil.

References

Abdalla, M. M. F., & HAS, H. (1977). Effects of single and combined treatments of gamma-rays and EMS on M2-quantitative variation *in Vicia faba L. Zeitschrift fur Pflanzenzuchtung,* 78(1), 57-64

Abdel-Baky, Y. R., Abouziena, H. F., Amin, A. A., Rashad El-Sh, M., & Abd El-Sttar, A. M. (2019). Improve quality and productivity of some faba bean cultivars with foliar application of fulvic acid. *Bulletin of the National Research Centre,* 43(2). https://doi.org/10.1186/s42269-018-0040-3.

Adhikari, K. N., Khazaei H, Ghaouti L, Maalouf F, Vandenberg A, Link W and O'Sullivan DM (2021) Conventional and Molecular Breeding Tools for Accelerating Genetic Gain in Faba Bean (*Vicia faba* L.). *Front. Plant Sci.* 12:744259. doi: https://doi.org/10.3389/fpls.2021.744259.

Akgun, D., Canci, H. (2023). Selection of Faba Bean (*Vicia faba* L.) Genotypes for High Yield, Essential Amino Acids and Low Anti-Nutritional Factors. *Agriculture*, 13, 932. https://doi.org/10.3390/agriculture13050932.

Alghamdi, S. S. (2009). Chemical composition of faba bean (*Vicia faba* L.) genotypes under various water regimes. *Pakistan Journal of Nutrition*, 8, 477–482.

Allard, R. W. (1999). *Principles of Plant Breeding*. John Wiley & Sons.

Apaydin, H. (2000) Broad bean (*Vicia faba*) – A natural source of L-DOPA prolongs 'on' periods in patients with Parkinson's disease who have 'on-off' fluctuations. *Mov Disord* 15:164–166.

Bailes, E. J., Pattrick, J. G., and Glover, B. J. (2018). An analysis of the energetic reward offered by field bean (*Vicia faba*) flowers: nectar, pollen, and operative force. *Ecol. Evol.* 8, 3161–3171. doi: https://doi.org/10.1002/ece3.3851.

Berthelem, P. (1970). *Rapport d'activité* 1967-1970. Rennes: INRA.

Berthelem, P., & Le Guen, J. (1971). Rapport d'activite de Station d'Amelioration des Plantes. *Le Rheu, Rennes, France, 74*.

Bhatia, C. R., Maluszynski, M., Nichterlein, K., & Zanten, L. van. (2001). *Grain legume cultivars derived from induced mutations, and mutations affecting nodulation*. https://www.osti.gov/etdeweb/biblio/20169549.

Birch, A. N., Tithecott, A. A., & Bisby, F. A. (1985). The agronomic importance of vicia species and their breeding for pest and disease resistance. *Aspects of Applied Biology,* 10, 111-118.

Bishop, J., Garratt, M. P. D., & Breeze, T. D. (2020). Yield benefits of additional pollination to faba bean vary with cultivar, scale, yield parameter and experimental method. *Scientific Reports*, *10*(1), Article 1. https://doi.org/10.1038/s41598-020-58518-1.

Bond, D. A. (1981). *Development and performance of synthetic varieties of Vicia faba L.* Paper presented at International Faba Bean Conference.

Bond, D. A. (1982). The Development and Performance of Synthetic Varieties of *Vicia faba* L. In G. Hawtin & C. Webb (Eds.), *Faba Bean Improvement* (pp. 41–51). Springer Netherlands. https://doi.org/10.1007/978-94-009-7499-9_4.

Bond, D. A. (1987). Recent Developments in Breeding Field Beans (*Vicia faba* L.). *Plant Breeding*, *99*(1), 1–26. https://doi.org/10.1111/j.1439-0523.1987.tb01144.x.

Bond, D. A. (1989a). A short review of research on male sterility and prospects for F1 hybrid varieties in field beans (*Vicia faba* L.). *Euphytica*, *41*(1), 87–90. https://doi.org/10.1007/BF00022416.

Bond, D. A. (1989b). Prospects for commercialisation of F1 hybrid field beans *Vicia faba* L. *Euphytica*, *41*(1), 81–86. https://doi.org/10.1007/BF00022415.

Bond, D. A., & Lowe, H. J. B. (1975). Tests for resistance to Aphis fabae in field beans (*Vicia faba*). Annals of Applied Biology, 81(1), 21-32.

Bond, D. A., Drayner, J. M., Fyfe, J. L., & Toynbee-Clarke, G. (1964a). A male sterile bean (*V. faba* L.) inherited as a Mendellian recessive character. *J. Agri. Sc*, *63*, 229–234.

Bond, D. A., Fyfe, J. L., & Toynbee-Clarke, G. (1966). Male sterility in field beans (*Vicia faba* L.). IV. Use of cytoplasmic male sterility in the production of F1 hybrids, and their performance in trials. *The Journal of Agricultural Science*, *66*(3), 369–377. https://doi.org/10.1017/S002185960006367X.

Bond, D. A., Jellis, G. J., Rowland, G. G., Le Guen, J., Robertson, L. D., Khalil, S. A., & Li-Juan, L. (1994). Present status and future strategy in breeding faba beans (*Vicia faba* L.) for resistance to biotic and abiotic stresses. In: Muehlbauer, F.J., Kaiser, W.J. (eds) Expanding the Production and Use of Cool Season Food Legumes. Current Plant Science and Biotechnology in Agriculture, 19, 592-616. Springer, Dordrecht. https://doi.org/10.1007/978-94-011-0798-3_36

Bond, D., Drayner, J. M., Fyfe, J., & Toynbee-Clarke, G. (1964b). Male sterility in field beans (*Vicia faba* L.). I. A male-sterile bean inherited as a Mendelian recessive. *The Journal of Agricultural Science*, *63*(2), 229–234.

Bond, D., Fyfe, J., & Toynbee-Clarke, G. (1964a). Male sterility in field beans (*Vicia faba* L.). II. Yield trials of F1 hybrid winter beans produced with the aid of male sterility. *The Journal of Agricultural Science*, *63*(2), 235–243.

Boye, J., Zare, F., Pletch, A. (2010). Pulse proteins: Processing, characterization, functional properties and applications in food and feed. *Food Res. Int.*, 43, 414–431.

Brünjes, L., & Link, W. (2021). Paternal outcrossing success differs among faba bean genotypes and impacts breeding of synthetic cultivars. *Theoretical and Applied Genetics*, *134*(8), 2411–2427. https://doi.org/10.1007/s00122-021-03832-z.

Busbice, T. H. (1970). Predicting Yield of Synthetic Varieties1. *Crop Science*, *10*(3), cropsci https://doi.org/10.2135/cropsci1970.0011183X001000030017x.

Cabrera, C., Lloris, F., Gim´enez, R., Olalla, M., & Lopez, ´ M. C. (2003). Mineral content in legumes and nuts: Contribution to the Spanish dietary intake. *Science of the Total Environment,* 308, 1–14.

Caracuta, V., Weinstein-Evron, M., Kaufman, D., Yeshurun, R., Silvent, J., & Boaretto, E. (2016). 14,000-year-old seeds indicate the Levantine origin of the lost progenitor of faba bean. *Scientific Reports,* 6, 37399.

Castaldo, L., Izzo, L., Lombardi, S., Gaspari, A., De Pascale, S., Grosso, M., Ritieni, A. Analysis of Polyphenolic Compounds in Water-Based Extracts of *Vicia faba* L.: A Potential Innovative Source of Nutraceutical Ingredients. *Antioxidants* 2022, 11, 2453. https://doi.org/10.3390/antiox11122453.

Caubel, F. & J. Le Guen, 1992. In: *1st European conference on grain legumes*, pp. 337–338. 1–3 June 1992. Angers, France.

Chapman, G. P. (1981). Genetic variation within *Vicia faba*. *Fabis Newsletter*, *3*, 1–12.

Conner, R. L., & Bernier, C. C. (1982). Host range of Uromyces viciae-fabae. *Phytopathology,* 72(6), 687-689.

Cubero, J. 1., A. Pieterse, A. R. Saghir & S. Borg, 1988. In: R. J. Summerfield (Ed.), *World crops: cool season food legumes*, pp. 549- 563. Kluwer Academic Publishers, Dordrecht.

Cubero, J. I. (1973). The problem of the taxonomy of *Vicia faba* L. Anales del Instituto Nacional de Investigaciones Agrarias, *Serie: Producción Vegetal*, 3, 107-118.

Dawkins, T. C. K., & Brereton, J. C. (1984). The Effects of Poor Soil Physical Conditions on the Growth and Yield of *Vicia faba*. In *Vicia faba: Agronomy, Physiology and Breeding: Proceedings of a Seminar in the CEC Programme of Coordination of Research on Plant Protein Improvement,* (pp. 113-126). Springer Netherlands.

De Angelis, D., Pasqualone, A., Costantini, M., Ricciardi, L., Lotti, C., Pavan, S., Summo, C. Data on the proximate composition, bioactive compounds, physicochemical and

functional properties of a collection of faba beans (*Vicia faba* L.) and lentils (Lens culinaris Medik.). *Data Brief* 2021, 34, 106660.

Dieckmann, S., and Link, W. (2010). Quantitative genetic analysis of embryo heterosis in faba bean (*Vicia faba* L.). *Theor. Appl. Genet.* 120, 261–270. doi: https://doi.org/10.1007/s00122-009-1057-7.

Duc, G. (1997). Faba bean (*Vicia faba* L.). Field Crops Research, 53(1-3), 99-109.

Duc, G., and Stoddard, F. (2018). David Bond and Jean Picard: two pivotal breeders of faba bean in the 20th century. *Plant Genet. Resour.* 16, 483–487. doi: https://doi.org/10.1017/S147926211800031X.

Duc, G., Le Guen, J., Picard, J., & Berthelem, P. (1985). Proposed use of a new discovered dominant male sterile allele for breeding purposes in *Vicia faba*. *Fabis Newsletter*, *12*, 8–10.

El Karouri, M. O. H. (1979). Effects of soil salinity on broad bean (*Vicia faba*) in the Sudan. Experimental Agriculture, 15(1), 59-63.

Falconer, D. S. (1996). *Introduction to Quantitative Genetics*. Pearson Education India.

Fernández-Aparicio, M., Flores, F., & Rubiales, D. (2016). The Effect of Orobanche crenata Infection Severity in Faba Bean, Field Pea, and Grass Pea Productivity. *Frontiers in Plant Science*, *7*. https://www.frontiersin.org/articles/10.3389/fpls.2016.01409.

Filippetti, A. (1979). Breeding projects and work for the improvement of broad beans (*Vicia faba*) in Puglia. In D. A. Bond, G. T. Scarascia-Mugnozza, & M. H. Poulsen (Eds.), *Some current research on Vicia fabain western Europe*. EN.

Filippetti, A. (1986). Inheritance of determinate growth habit induced in *Vicia faba* major by ethyl methane sulphonate (EMS). *FABIS Newsletter-Faba Bean Information Service*.

Filippetti, A., & De Pace, C. (1986). Improvement of seed yield in *Vicia faba* L. by using experimental mutagenesis. II. Comparison of gamma-radiation and ethyl-methane-sulphonate (EMS) in production of morphological mutants. *Euphytica*, *35*, 49–59.

Filippetti, A., & Pace, C. de. (1982). Further evidence on the amount of out-crossing in *Vicia faba* grown in southern Italian environments. *FABIS Newsletter*, *5*, 14–15.

Filippetti, A., & Pace, D. (1983). *Improvement of grain yield in Vicia faba l. By using experimental mutagenesis. In: frequency and types of mutations induced by gamma-radiation*.

Filippetti, A., & Ricciardi, L. (1988). Development of new determinate growth habit in *Vicia faba* L. major analysis of cross between determinate mutant and autofertile line. *Genética Agraria*, *42*(3), 299–316.

Free, J. B. (1966). The pollination requirements of broad beans and field beans (*Vicia faba*). *The Journal of Agricultural Science*, *66*(3), 395–397. https://doi.org/10.1017/S002185960006370X.

Fyfe, J. L., & Bailey, N. J. T. (1951). Plant breeding studies in leguminous forage crops. *I. Natural Cross-Breeding in Winter Beans. J. Agric. Sci., Camb.*, *41*.

Gasim, S. M., & Mohamed, A. E. (2018). Heterotic Effect, Combining Ability and Significance of Synthetics over Line Cultivars in Faba Bean (*Vicia faba* L.) Grown under Semi-Arid Zones. *American Journal of Plant Sciences*, *9*(13), Article 13. https://doi.org/10.4236/ajps.2018.913195.

Ghaouti, L., & Link, W. (2009). Local vs. formal breeding and inbred line vs. synthetic cultivar for organic farming: Case of *Vicia faba* L. *Field Crops Research*, *110*(2), 167–172. https://doi.org/10.1016/j.fcr.2008.07.013.

Ghaouti, L., Vogt-Kaute, W., & Link, W. (2008). Development of locally-adapted faba bean cultivars for organic conditions in Germany through a participatory breeding approach. *Euphytica*, *162*(2), 257–268. https://doi.org/10.1007/s10681-007-9603-3.

Graff, R. (1988). Factors affecting ovule fertilization frequency and seed set in faba bean (*Vicia faba* L). Ph.D. Thesis, University of Saskatchewan.

Griffiths, D. W., & Lawes, D. A. (1978). Variation in the crude protein content of field beans (*Vicia faba* L.) in relation to the possible improvement of the protein content of the crop. *Euphytica*, 27(2), 487–495. https://doi.org/10.1007/BF00043174.

Gutierrez, N., Avila, C. M., Duc, G., Marget, P., Suso, M. J., Moreno, M. T., & Torres, A. M. (2006). CAPs markers to assist selection for low vicine and convicine contents in faba bean (*Vicia faba* L.). *Theoretical and Applied Genetics*, 114(1), 59–66. https://doi.org/10.1007/s00122-006-0410-3.

Gutierrez, N., Avila, C. M., Moreno, M. T., & Torres, A. M. (2008). Development of SCAR markers linked to zt-2, one of the genes controlling absence of tannins in faba bean. *Australian Journal of Agricultural Research,* 59, 62–68.

Gutiérrez, N., Fernández-Romero, M. D., Atienza, S. G., Ávila, M., & Torres, M. (2017). Identifying positional candidate genes controlling low vicine-convicine in faba bean (*Vicia faba* L.). *Advances in grain legume breeding, cultivation and uses for a more competitive value-chain*. 27–28 september 2017 (Novi Sad, Serbia).

Hanelt, P. (1972). *Kulturpflanze* 20: 75–128.

Hawtin, G. C., Bond, D. A., Lawes, D. A., Saxena, M. C., & Stephens, J. H. (1985). Faba bean (*Vicia faba* L.). In *Grain legume crops*. Collins, London, GB.

Hebblethwaite, P. D., Dawkins, T. C. K., Heath, M. C., & Lockwood, G. (Eds.). (1984). *Vicia faba: Agronomy, Physiology and Breeding: Proceedings of a Seminar in the CEC Programme of Coordination of Research on Plant Protein Improvement,* Springer Netherlands. https://doi.org/10. 1007/978-94-017-3647-3.

Hern΄ Andez-Infante, M., Sousa, V., Montalvo, I., & Tena, E. (1998). Impact of microwave heating on hemagglutinins, trypsin inhibitors and protein quality of selected legume seeds. *Plant Foods for Human Nutrition,* 52(3), 199–208. https://doi.org/10.1023/A:1008033610737.

Jacquier, N. M. A., Gilles, L. M., Pyott, D. E., Martinant, J.-P., Rogowsky, P. M., and Widiez, T. (2020). Puzzling out plant reproduction by haploid induction for innovations in plant breeding. *Nature Plants* 6, 610–619. doi: https://doi.org/10.1038/s41477-020-0664-9.

Jellis, G. J., D. B. Smith, and E. S. Scott. (1990). Identification of Sclerotinia spp. on *Vicia faba*. *Mycological Research*, 94, (3), 407-409

Jellis, G. J., Plumb, A. S., & Clarke, M. H. (1991). Resistance to *Didymella fabae* in faba beans. *Aspects of Applied Biology*, 27, 63-66.

Jiang, S. Y., Ma, Z., & Ramachandran, S. (2010). Evolutionary history and stress regulation of the lectin superfamily in higher plants. *BMC Evolutionary Biology*, 10 (1), 79. https://doi.org/10.1186/1471-2148-10-79.

Kew Science. (2017). *Data and resources*. Retrieved from https://www.kew.org/science.

Khan, A. M., Megahed, H. A., Hussein, M. M., El-Harty, E. H., Magdi, A. O., Muhammad, F., Salem, S. A. Comparative nutritional profiles of various faba bean and chickpea genotypes. *Int. J. Agric. Biol.* 2015, 17, 449–457.

Khazaei, H., & Vandenberg, A. (2020). Seed mineral composition and protein content of faba beans (*Vicia faba* L.) with contrasting tannin contents. *Agronomy*, 10, 511. https://doi.org/10.3390/agronomy10040511.

Khazaei, H., Purves, R. W., Hughes, J., Link, W., O'Sullivan, D. M., Schulman, A. H., Stoddard, F. L. (2019). Eliminating vicine and convicine, the main anti-nutritional factors restricting faba bean usage. *Trends in Food Science & Technology,* 91, 549–556. https://doi.org/10.1016/j.tifs.2019.07.051.

Khazaei, H., Hawkins, G., & Vandenberg, A. (2021). Historical review of faba bean improvement in western Canada. *Legume Science*, *3*(4), e92.

Kinman, M. L., & Sprague, G. (1945). *Relation between number of parental lines and theoretical performance of synthetic varieties of corn.*

Köpke, U., & Nemecek, T. (2010). *Ecological services of faba bean. Field Crops Research*, 115, 217–233. https://doi.org/10.1016/j.fcr.2009.10.012.

Kosinska, A., Karamać, M., Penkacik, K. *et al.,* Interactions between tannins and proteins isolated from broad bean seeds (*Vicia faba* Major) yield soluble and non-soluble complexes. *Eur Food Res Technol* 233, 213–222 (2011). https://doi.org/10.1007/s00217-011-1506-9.

Kumar, A., Nidhi, Prasad N., Sinha S. K. Nutritional and antinutritional attributes of faba bean (*Vicia faba* L.) germplasms growing in Bihar, India. *Physiol Mol Biol Plants.* 2015 Jan;21(1):159-62. doi: https://doi.org/10.1007/s12298-014-0270-2.

Kwok, D., Elsworth J. D., Roth R. H. (1997) Dopamine synthesis, uptake metabolism, and receptors: relevance to gene therapy of Parkinson's disease. *Exp Neurol* 144:4–9.

Labba, M., Frøkiær H., Sandberg A. S. (2021). Nutritional and antinutritional composition of fava bean (*Vicia faba* L., var. minor) cultivars. *Food Res Int.* 2021 140: 110038. doi: https://doi.org/10.1016/j.foodres.2020.110038.

Ladizinsky, G. (1998). *Plant evolution under domestication.* Springer.

Landry, E. J., Coyne, C. J., McGee, R. J., & Hu, J. (2017). A modified mass selection scheme for creating winter-hardy faba bean (*Vicia faba* L.) lines with a broad genetic base. *Euphytica*, *213*(3), 72. https://doi.org/10.1007/s10681-017-1843-2.

Landry, E. J., Fuchs, S. J., & Hu, J. (2016). Carbohydrate composition of mature and immature faba bean seeds. *Journal of Food Composition and Analysis,* 50, 55–60. https://doi.org/10.1016/j.jfca.2016.05.010.

Lazarte, C. E., Carlsson, N. G., Almgren, A., Sandberg, A. S., & Granfeldt, Y. (2015). Phytate, zinc, iron and calcium content of common Bolivian food, and implications for mineral bioavailability. *Journal of Food Composition and Analysis,* 39, 111–119. https://doi.org/10.1016/j.jfca.2014.11.015.

Lin, Z. S., Zhao, Y. Z., Bao, S. Y., & Guang, W. (1987). *Low temperature and faba bean (Vicia faba L.) yield.*

Link, W. (1990). Autofertility and rate of cross-fertilization: Crucial characters for breeding synthetic varieties in faba beans (*Vicia faba* L.). *Theoretical and Applied Genetics*, *79*(5), 713–717. https://doi.org/10.1007/BF00226888.

Link, W. (1998). "Neues über CMS bei Ackerbohnen," in *Arbeitstagung der Arbeitsgemeinschaft der Saatzuchtleiter (Gumpenstein),* 25. bis 27, 95–101.

Link, W. (2009). Züchtungsforschung bei der Ackerbohne: Fakten und Potentiale. *Journal für Kulturpflanzen*, 341-347 Pages. https://doi.org/10.5073/JFK.2009.09.08.

Link, W., & Ederer, W. (1993). The Concept of Varietal Ability for Partially Allogamous Crops. *Plant Breeding*, *110*(1), 1–8. https://doi.org/10.1111/j.1439-0523.1993.tb00561.x.

Link, W., and Stützel, H. (1995). "Faba bean. Genetics," in *Advances in Plant Breeding. Physiological Potentials for Yield Improvement of Annual Oil and Protein Crops,* Vol. 17, eds W. Diepenbrock, and H. C. Becker (Berlin: Blackwell), 239–278.

Link, W., Ederer, W., Metz, P., Buiel, H., & Melchinger, A. E. (1994). Genotypic and Environmental Variation for Degree of Cross-Fertilization in Faba Bean. *Crop Science*, *34*(4), https://doi.org/10.2135/cropsci1994.0011183X003400040024x.

Link, W., Ederer, W., Gumber, R. K., Melchinger, A. E. Detection and characterization of two new CMS systems in faba bean (*Vicia faba*). *Plant Breed.* 1997, 116, 158–162.

Liu, X. X., Wilson K., Charlton C. G. (2000) Effects of L-DOPA treatment on methylation in mouse brain: implications for the side effects of L-DOPA. *Life Sci* 66:2277–2288.

Luo, Y., Xie, W., & Luo, F. (2012). Effect of several germination treatments on phosphatases activities and degradation of phytate in Faba Bean (*Vicia faba* L.) and Azuki Bean (Vigna angularis L.). *Journal of Food Science,* 77(10), C1023–C1029. https://doi.org/10.1111/j.1750-3841.2012.02733.x.

Maalouf, F., Hu, J., O'Sullivan, D. M., Zong, X., Berger, J., Yadav, S. S., & Ryder, J. (2013). Faba bean genomics: results and future opportunities. In *Grain Legumes* (pp. 225-247). Springer, New York, NY.

Maalouf, F., Hu, J., O'Sullivan, D. M., Zong, Z., Hamwieh, A., Kumar, S., & Baum, M. (2019). Breeding and genomics status in faba bean (*Vicia faba*). *Plant Breeding,* 138, 465–473. https://doi.org/10.1111/pbr.12644.

Maalouf, F., Suso, M., & Moreno, M. T. (2002). Comparative performance of faba bean synthetics developed from different parental number. *Journal of Genetics and Breeding*, *56*, 251–258.

Makkar, H. P. S., Becker, K., Abel, H., & Pawelzik, E. (1997). Nutrient contents, rumen protein degradability and antinutritional factors in some color- and white- flowering cultivars of *Vicia faba* beans. *Journal of the Science of Food and Agriculture*, 75, 511–520.

Małgorzata, S., Jerzy, K., Ewa, H. The Effect of Feeding Native Faba Bean Seeds (*Vicia faba* L.) to Sows and Supplemented With Enzymes to Piglets and Growing Pigs. *Ann. Anim. Sci.* 2018, 18, 1007–1027.

Marzinzig, B., Brünjes, L., Biagioni, S., Behling, H., Link, W., and Westphal, C. (2018). Bee pollinators of faba bean (*Vicia faba* L.) differ in their foraging behaviour and pollination efficiency. *Agric. Ecosyst. Environ.* 264, 24–33. doi: https://doi.org/10.1016/j.agee.2018.05.003.

Mejri, S., Mabrouk, Y., Belhadj, O., & Saidi, M. (2018). Orobanche foetida resistance in two new faba bean genotypes produced by radiation mutagenesis. *International Journal of Radiation Biology*, *94*(7), 671–677. https://doi.org/10.1080/09553002.2018.1486517.

Metz, P. L. J., Buiel, A. A. M., Norel, A. V., & Helsper, J. P. F. G. (1994). Genetic factors controlling outcrossing in faba bean (*Vicia faba*): Effects of pollen donor and receptor genotypes and estimates of inter-receptor crossfertilization. *The Journal of Agricultural Science*, *122*(2), 249–254. https://doi.org/10.1017/S0021859600087438.

Millar, K. A., Gallagher, E., Burke, R., McCarthy, S., & Barry-Ryan, C. (2019). Proximate composition and anti-nutritional factors of fava-bean (*Vicia faba*), green-pea and yellow-pea (Pisum sativum) flour. *Journal of Food Composition and Analysis*, 82, Article 103233. https://doi.org/10.1016/j.jfca.2019.103233.

Muller, H. J., 1968. *Entomologica experimentalis et applicata* 11: 355-371.

Multari, S., Stewart, D., & Russell, W. R. (2015). Potential of fava bean as future protein supply to partially replace meat intake in the human diet. *Comprehensive Reviews in Food Science and Food Safety,* 14, 511–522. https://doi.org/10.1111/1541-4337.12146.

Muratova, V. S. (1931). Common beans (*Vicia faba* L.*). Bulletin of Applied Botany, Genetics and Plant Breeding*, 50(2), 1-298.

Picard, J. (1960). Données sur l'amélioration de la féverole de printemps *Vicia faba* L. *Annales de l'Amélioration Des Plantes II*, 121–123.

Poehlman, J. M. (2013). *Breeding Field Crops*. Springer Science & Business Media.

Poulsen, M. H. (1975). *Pollination seed setting, cross-fertilisation and in breeding in Vicia faba* L. *Z. Pfl Zucht.*

Pradhan, A., Shahi V. K., Sinha S. K. (2014) Evaluation of genetic diversity in faba bean (*Vicia faba* L.) using seed protein and isozymes electrophoresis. *Natl Acad Sci Lett* 4:303–309.

Rabey, J. M., Vered Y., Shabtai H., Graff E., Korczyn A. D. (1992) Improvement of Parkinsonian features correlate with high plasma levodopa values after broad bean (*Vicia faba*) consumption. *J Neurol Neurosurg Psychiatry* 55:725–727.

Randhir, R., Shetty K. (2004) Microwave-induced stimulation of L-DOPA, phenolics and antioxidant activity in Fava bean (*Vicia faba*) for Parkinson's diet. *Process Biochem* 39:1775–1784.

Randhir, R., Shetty P., Shetty K. (2002) L-DOPA and total phenolics stimulation in dark germinated fava bean in response to peptide and phytochemical elicitors. *Process Biochem* 37:1247–1256.

Rharrabti, Y., Villegas, D., García del Moral, L., Aparicio, N., Elhani, S., & Royo, C. (2001). Environmental and genetic determination of protein content and grain yield in durum wheat under Mediterranean conditions. *Plant Breeding,* 120, 381–388.

Robertson, L. D. and Saxena, M. C. (1993). Problems and prospects of stress resistance breeding in faba bean. In: Breeding for Stress Tolerance in Cool-Season Food Legumes, pp. 37–50 (eds. K. B. Singh and M. C. Saxena), Chichester, UK, John Wiley & Sons.

Rubiales, D., & Fernández-Aparicio, M. (2012). Innovations in parasitic weeds management in legume crops. A review. *Agronomy for Sustainable Development*, *32*(2), 433–449. https://doi.org/10.1007/s13593-011-0045-x.

Rubiales, D., Rojas-Molina, M. M., & Sillero, J. C. (2016). Characterization of Resistance Mechanisms in Faba Bean (*Vicia faba*) against Broomrape Species (Orobanche and

Phelipanche spp.). *Frontiers in Plant Science*, *7*. https://www.frontiersin.org/articles/10.3389/fpls.2016.01747.

Sahabuddin, A. (1984). *Mass selection of Fababean (Vicia faba L.)*. https://digital.library.adelaide.edu.au/dspace/handle/2440/109050.

Saleh, H. M., Hassan, A. A., Mansour, E. H., Fahmy, H. A., & El-Bedawey, A. E.-F. A. (2019). Melatonin, phenolics content and antioxidant activity of germinated selected legumes and their fractions. *Journal of the Saudi Society of Agricultural Sciences,* 18(3), 294–301. https://doi.org/10.1016/j.jssas.2017.09.001.

Shi, L., Mu, K., Arntfield, S. D., & Nickerson, M. T. (2017). Changes in levels of enzyme inhibitors during soaking and cooking for pulses available in Canada. *Journal of Food Science and Technology,* 54(4), 1014–1022. https://doi.org/10.1007/s13197-017-2519-6.

Sillero, J. C., Villegas-Fernández, A. M., Thomas, J., Rojas-Molina, M. M., Emeran, A. A., Fernández-Aparicio, M., & Rubiales, D. (2010). Faba bean breeding for disease resistance. *Field Crops Research, 115*(3), 297-307.

Singh, A., Bhardwaj, R., & Singh, I. (2014). Assessment of nutritional quality of developed Faba Bean (*Vicia faba* L.) lines. *Journal of AgriSearch,* 1, 96–101.

Singh, S. P. (2007). Drought resistance in the race durango dry bean landraces and cultivars. *Agronomy Journal,* 99, 1219–1225.

Sjödin, J. (1971). Induced morphological variation in *Vicia faba* L. *Hereditas*, *67*(2), 155–179. https://doi.org/10.1111/j.1601-5223.1971.tb02371.x.

Stoddard, F. L., Nicholas, A. H., Rubiales, D., Thomas, J., & Villegas-Fernández, A. M. (2010). Integrated pest management in faba bean. *Field Crops Research*, 115(3), 308-318.

Suso, M. J., Moreno, M. T., & Melchinger, A. E. (1999). Variation in outcrossing rate and genetic structure on six cultivars of *Vicia faba* L. as affected by geographic location and year. *Plant Breeding*, *118*(4), 347–350. https://doi.org/10.1046/j.1439-0523.1999.00389.x.

Tanno, K. I., & Willcox, G. (2006). The origins of cultivation of Cicer arietinum L. and *Vicia faba* L.: early finds from Tell el-Kerkh, north-west Syria, late 10th millennium BP. *Vegetation History and Archaeobotany,* 15(3), 197-204.

Thiellement, H., Ferault, M., & Blaisoneau, J. (1979). *Cytoplasmic male sterility in Vicia faba l. Iv: a grand-mother effect.*

Torres, A. M., Moreno, M. T., & Cubero, J. I. (1993). Genetics of Six Components of Autofertility in *Vicia faba*. *Plant Breeding*, *110*(3), 220–228. https://doi.org/10.1111/j.1439-0523.1993.tb00581.x.

Van Loon, J. J. A., A. van Norel & L. M. W. Dellaert, 1989. In: J. Huisman, T. FB van der Poel & I. E. Liener (Eds.), *Recent advances of research in antinutritional factors in legume seeds*, pp. 23-25. Pudoc, Lanham, Maryland, USA.

Vidal-Valverde, C., Frias, J., Sotomayor, C., Diaz-Pollan, C., Fernandez, M., Urbano, G. Nutrients and anti-nutritional factors in faba beans as affected by processing. *Eur. Food Res. Technol.* 1998, 207, 140–145.

Villalobos, M. P., & Jellis, G. J. (1990). Factors influencing establishment in *Vicia faba*. *The Journal of Agricultural Science*, *115*(1), 57-62.

Warsame, A., O'Sullivan, D. M., & Tosi, P. (2018). Seed storage proteins of faba bean (*Vicia faba* L): Current status and prospects for genetic improvement. *Journal of Agricultural and Food Chemistry,* 66, 12617–12626. https://doi.org/10.1021/acs.jafc.8b04992.

Zeid, M., Schön, C. C., and Link, W. (2004). Hybrid performance and AFLP-based genetic similarity in faba bean. *Euphytica* 139, 207–216. doi: https://doi.org/10.1007/s10681-004-3156-5.

Chapter 7

From Field to Fork: A Bird's Eye View of the Genetic Diversity, Obstacles, and Advances Made in Faba Bean Cultivation and Breeding

Aitezaz Ali Asad Shahani[1,*]
Sana Ullah[2,†]
and Muhammad Afnan Rabbi[3,‡]
[1]Key Laboratory of Crop Sciences and Plant Breeding Genetics, College of Agriculture, Yanbian University, Yanji, Jilin, China
[2]Department of Plant Breeding and Genetics, University of Agriculture Faisalabad, Faisalabad, Pakistan
[3]Faculty of Agriculture, Gomal University, Dera Ismail Khan, Pakistan

Abstract

This chapter discusses an overview of faba bean (*Vicia faba* L.) genetic diversity and challenges and recent advancements in faba bean breeding and cultivation. Faba beans, an ancient crop with nutritional and versatility benefits, are being renewed. The chapter traces their historical significance and global cultivation, noting their high protein, fiber, and mineral content. This chapter also highlights allelic variation and gene flow in breeding programs. Pest and disease pressures, soil fertility, and water management are addressed as major challenges. The use of advances in traditional breeding, genetic engineering, and tissue culture to develop disease-resistant and high-yield varieties is discussed. In

* Corresponding Author's Email: atezazshahani@gmail.com, 2193609022@ybu.edu.cn
† Corresponding Author's Email: sanaullahpbg1@gmail.com
‡ Corresponding Author's Email: afnanrabi@yahoo.com

In: Faba Beans
Editors: Aamir Raina and Samiullah Khan
ISBN: 979-8-89530-447-1

conclusion, prospects are discussed, emphasizing the need for increased genetic diversity and enhanced agronomic practices to achieve sustainable production.

1. Introduction

Faba bean (*Vicia faba* L.), or broad bean, is an important grain crop grown for its edible seeds. It is an ancient crop, and its cultivation has been documented as far back as 5,000 years ago (Jensen et al., 2010). Faba beans are highly nutritious, providing a good source of dietary protein, dietary fiber, and essential minerals and vitamins. They can be used for human consumption and livestock feed (Merga et al., 2019; Rahate et al., 2021).

Today, faba bean cultivation and breeding is an important area of research, focusing on improving its genetic diversity and yield (Alghamdi et al., 2012). Faba bean crop improvement aims to increase the crop's genetic diversity while increasing its yield. This is done using traditional plant breeding techniques and the development of new genetic technologies (Maalouf et al., 2019).

Faba bean crops are highly susceptible to diseases and pests, which can lead to significant yield losses. For this reason, breeders focus on developing disease and pest-resistant varieties of faba beans. This is accomplished by traditional plant breeding techniques such as selection, hybridization, and mutation breeding and the development of new genetic technologies such as marker-assisted selection and transgenics (Torres et al., 2010; Alghamdi et al., 2012).

The development of new genetic technologies has allowed for the rapid identification of genes associated with important traits in faba bean crops, such as disease and pest resistance. This makes it possible to identify and select desirable traits more quickly and accurately than ever (Gnanasambandam et al., 2012). In addition to the development of new genetic technologies, advances in faba bean breeding have also been made by molecular markers. Molecular markers are useful for identifying and selecting desirable traits and for identifying genetic diversity in the crop (Torres et al., 2010; Warsame et al., 2018).

Advances in faba bean breeding have also been made using traditional plant breeding techniques, such as hybridization and selection. Hybridization is the process of crossing two varieties of faba beans to create a new variety with improved traits. Selection is the process of selecting desirable traits in a

population of faba beans, such as disease and pest resistance (Sillero et al., 2010).

Finally, advances in faba bean breeding have also been made using advanced agronomic practices, such as irrigation and fertilization (Siddique et al., 2000). Improved irrigation and fertilization can help to improve the yield of faba beans and make them more resistant to drought and disease. These advances in faba bean cultivation and breeding have increased the genetic diversity and yield of this important crop. This has allowed for higher quality and more nutritious faba beans, which can be used for both human consumption and livestock feed (Chinthapalli et al., 2015; Karkanis et al., 2018).

Faba bean cultivation and breeding is an important area of research, focusing on improving the genetic diversity and yield of this important crop. Advances in this area of research have been made using traditional plant breeding techniques, the development of new genetic technologies, and the use of advanced agronomic practices. These advances have increased the genetic diversity and yield of this important crop, which can be used for both human consumption and livestock feed (Robinson et al., 2019).

This chapter provides a bird's eye view of genetic diversity, obstacles, and advances in faba bean cultivation and breeding. This chapter will begin by providing an overview of the crop's genetic diversity and how it is maintained. It will then discuss the various obstacles that have hampered progress in faba bean breeding, followed by a review of recent advances in faba bean cultivation and breeding, including developing new varieties and using genomic and biotechnological tools. Finally, the chapter will conclude with a discussion of prospects for faba bean cultivation and breeding.

2. History of Cultivation

Faba beans are an ancient and important crop in the history of agriculture. They are believed to have originated in the Near East and have been cultivated for thousands of years. Faba beans are a high-protein staple crop and have been an important part of diets in many cultures worldwide (Kosterin, 2015). Faba beans spread westward from the Near East to Europe and North Africa around 5000 BC and eventually made their way to the Americas in the 16th century. They have been cultivated in the Mediterranean region since the Bronze Age and were a major crop in Ancient Egypt. The Romans grew faba

beans as a food crop and for animal feed, and they were a major part of the diet of the medieval peasantry (Caracuta et al., 2016).

Faba beans were also an important crop in India and China. In India, faba beans have been grown for food and feed since at least 2000 BC. In China, faba beans were grown for food and medicinal purposes and were believed to have aphrodisiac properties (Li-Juan et al., 1993). In the 19th century, faba beans were introduced to the United States and Canada. They were grown mainly for animal feed but also for human consumption (Khazaei et al., 2021). In the early 20th century, faba beans became an important crop in the United States and Canada and were grown in many states and provinces (Minguez & Rubiales, 2021).

Today, faba beans are grown in over 70 countries around the world. They are a major crop in the Mediterranean, North Africa, the Middle East, South Asia, and China. The United States and Canada are also major producers of faba beans, and in recent years, other countries such as Australia and Argentina have increased their production (Singh et al., 2013; Maalouf et al., 2021).

3. Nutritional Value and Health Benefits

They are believed to have been cultivated for thousands of years and are a popular ingredient in many cuisines. Faba beans are a rich source of vitamins and minerals and contain substantial dietary fiber and protein. As such, they provide various health benefits and can be used in various dishes (Turco et al., 2016; Etemadi et al., 2019).

3.1. Nutritional Value

Faba beans are an excellent source of plant-based protein, fiber, and minerals such as iron, zinc, magnesium, and phosphorus. They are also a good source of vitamins B1, B2, B3, B6, and folate (Crépon et al., 2010). A single 100-gram serving of cooked faba beans contains approximately:

Calories: 110
Protein: 7.6 grams
Carbohydrates: 19.6 grams

Fiber: 5.4 grams
Fat: 0.4 grams
Vitamin C: 0.36 milligrams
Vitamin B-6: 0.072 milligrams
Vitamin E: 0.02 milligrams
Vitamin K: 2.9 micrograms
Folate: 104 micrograms
Iron: 1.5 milligrams
Magnesium: 43 milligrams
Phosphorus: 125 milligrams
Potassium: 268 milligrams
Zinc: 1.01 milligrams
(https://fdc.nal.usda.gov).

3.2. Health Benefits

The protein in faba beans is of high quality and contains all nine essential amino acids. These amino acids are important for building muscle, maintaining healthy bones and skin, and producing hormones and enzymes (Turco et al., 2016). Faba beans are also an excellent source of vitamins and minerals. One cup of cooked faba beans contains significant amounts of iron, potassium, magnesium, phosphorus, and zinc. Faba beans are also a good source of B-group vitamins, including thiamin, riboflavin, niacin, vitamin B-6, and folate (Johnson et al., 2020a; Labba et al., 2021).

Studies have shown that consuming faba beans can help reduce cholesterol levels, lower blood pressure, and reduce the risk of heart disease. Faba beans are also a good source of polyphenols, which are natural compounds with antioxidant and anti-inflammatory properties (Siah et al., 2012; Kwon et al., 2018; Johnson et al., 2020b).

3.3. Uses of Faba Bean

Faba beans can be used as a replacement for other legumes like lentils and chickpeas in recipes. They can be cooked in various ways, including boiling, steaming, and baking. Faba beans can be used in soups, stews, salads, casseroles, side dishes, and desserts. They can also be used as a substitute for

meat in vegetarian and vegan dishes (Njoumi et al., 2019; Schmelter et al., 2021).

Faba beans can also be ground into flour, which can be used as a substitute for wheat flour in baking. Faba bean flour has a nutty flavor and is a good source of protein and dietary fiber. It is also gluten-free, making it a good option for people with gluten sensitivities or allergies (Petitot et al., 2010; Kaya et al., 2018). Faba beans have been found to have several health benefits, including reducing cholesterol levels, lowering blood pressure, and reducing the risk of heart disease (Poonia et al., 2022). Faba beans can also be substituted for other legumes and grains in recipes and as flour for baking. With their versatile uses, faba beans are a great way to get more nutrients and flavor into the diet (Tazrart et al., 2016; Verni et al., 2019).

4. Genetic Diversity of Faba Beans

4.1. Characterization of Faba Bean Genetic Diversity

Faba bean genetic diversity is characterized by a wide range of genetic variability in the form of allelic diversity, genetic structure, and gene flow (Duc et al., 2010). Allelic diversity is the variation in the genetic code of an organism, which is determined by the number of different alleles present in the population. Alleles can be different forms of a gene, which may result in different traits expressed by the organism. For faba beans, different alleles have been found to control various traits such as seed size, seed color, and disease resistance (Gnanasambandam et al., 2012; O'Sullivan and Angra, 2016).

Genetic structure refers to the distribution of genetic variation within and between populations. In faba beans, the genetic structure is determined by the geographic distance between populations and their relative genetic similarity (Karchani-Balma et al., 2008; Oliveira et al., 2016). Studies have shown a high degree of genetic differentiation between populations of faba beans that are separated by large distances, suggesting that gene flow is limited between these populations (Kaur et al., 2014).

Gene flow is the movement of genetic material between populations. In faba beans, gene flow is primarily driven by pollen movement between individuals and populations. Studies have found that gene flow is higher between closely located populations than between those that are farther apart.

This suggests that faba bean populations are locally adapted to their local environment.

The genetic diversity of faba beans can be divided into two broad categories: genetic variation within a single species and genetic diversity across multiple species (Göl et al., 2017). Within a single species, genetic diversity refers to the range of alleles (variant forms of a gene) within a population. This type of genetic diversity is important for breeding programs, as it allows breeders to select the best combination of traits for a particular crop (Zong et al., 2010; Zhang et al., 2023).

The genetic diversity of faba beans across multiple species includes the variation in genetic makeup between two or more species (Kwak & Gepts, 2009). This type of diversity is known as intraspecific diversity and is essential for further crop development (Godbout et al., 2020). Intraspecific diversity can be measured in terms of the number of alleles present in the population and the frequency of those alleles (Moran et al., 2016).

The genetic diversity of faba beans is important for developing new varieties that are well-adapted to different growing conditions and climates and for preserving their genetic resources (Maxted and Bennett, 2001). The genetic diversity of faba beans can be characterized by analyzing the genetic markers of the different varieties. Genetic markers are small regions of DNA used to identify and compare the genetic variations of different plants (Rebaa et al., 2017; Avramidou et al., 2023).

The first studies of faba bean genetic diversity used traditional morphological and physiological markers, such as seed size and color, to characterize accessions from different geographical regions (Abid et al., 2015). These studies revealed considerable genetic variation within the species and that different accessions exhibit varying levels of adaptation to different environments (Karaköy et al., 2014; Malek et al., 2021).

More recent studies have used DNA-based methods to characterize faba bean genetic diversity. These methods include Random Amplified Polymorphic DNA (RAPD), Amplified Fragment Length Polymorphism (AFLP), and Simple Sequence Repeat (SSR) markers. These markers have been used to identify genetic differences between accessions and determine their phylogenetic relationships (Duc et al., 2010) (Maalouf et al., 2019).

RAPD and AFLP markers have been used to investigate the genetic diversity of faba bean accessions from different countries and regions. These studies have revealed that faba bean accessions from different geographical regions exhibit different levels of genetic diversity. For example, a study using AFLP markers showed that accessions from Ethiopia had higher genetic

diversity than those from the United Kingdom (Zeid et al., 2003) (Alghamdi et al., 2020).

Studies using SSR markers have also been conducted to characterize faba bean genetic diversity. These studies have revealed considerable genetic variation within the species, and accessions from different regions have different levels of genetic variation (Qahtan et al., 2021). For example, accessions from Ethiopia have been shown to have higher levels of genetic diversity than those from Turkey (Terzopoulos & Babeli, 2008).

In addition to these studies, research has been conducted to identify genes associated with important traits, such as disease resistance and yield. Studies that have used molecular markers to identify genes associated with these traits have revealed that the faba bean has considerable genetic diversity that can be used to improve existing cultivars and develop new cultivars with improved traits (Zong et al., 2010; Abid et al., 2015).

4.2. Distribution of Faba Bean Genetic Diversity

The genetic diversity of a species can be studied at different levels, including chromosomal, morphological, biochemical, and molecular. Molecular markers, such as simple sequence repeats (SSRs), have been used to study the genetic variation of the faba bean. SSRs are short tandem repeats of nucleotides located in non-coding genome regions, which can differentiate between individuals and varieties (Kaur et al., 2014; Elshafei et al., 2019).

Faba bean genetic diversity is distributed worldwide, with different populations exhibiting varying levels of variability (Ammar et al., 2015). Studies have found that faba bean genetic diversity is higher in regions where the crop has been cultivated for longer periods, such as the Mediterranean, Middle East, and North Africa. These regions have a long history of faba bean cultivation, which has led to the development of diverse landraces adapted to the local environment (Sýmkal et al., 2015).

In contrast, faba bean genetic diversity is lower in newly cultivated regions, such as South America and Asia. This is likely because these regions have only recently begun cultivating faba beans, and the crop has not had enough time to develop diverse landraces (Saxena, 1991).

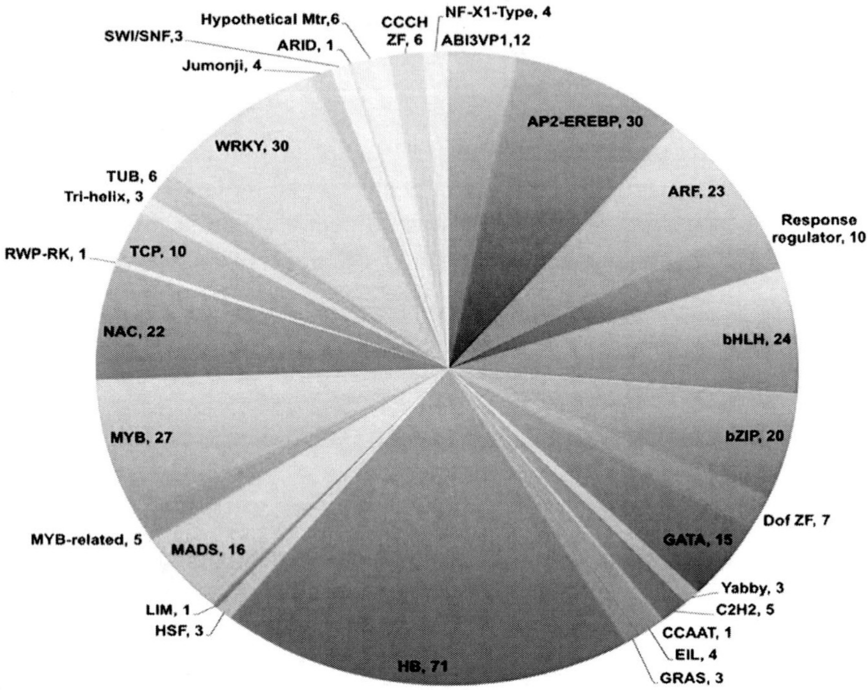

Source: Arun-Chinnappa & McCurdy, 2015.

Figure 1. Distribution of transcripts of faba bean into different transcriptional factors.

The section also highlighted the importance of conserving faba bean genetic resources. Studies have shown that populations from different regions have different genetic compositions, suggesting they have adapted to their local conditions. This variation should be preserved to maintain the genetic diversity of the species and ensure its long-term sustainability (Elshafei et al., 2019).

4.3. Examples of Faba Bean Genetic Diversity

Faba bean genetic diversity is exemplified by the wide range of landraces developed over time. Landraces are crops adapted to local environments and are usually the result of traditional farming techniques. One example of a faba bean landrace is Egypt's "Giza 3" landrace. This landrace is characterized by small, round seeds and is resistant to fusarium wilt, a fungal disease that can

cause significant crop losses (Elshaer et al., 2020). Another example of a faba bean landrace is the "*Omani*" landrace, which is found in Oman. This landrace is characterized by medium-sized, oval seeds and is resistant to several common diseases, including root rot, stem canker, and ascochyta blight (Al-Saady et al., 2018). Finally, the "*Tuscan*" landrace is an example of a faba bean landrace in Italy. This landrace is characterized by large, oblong seeds and is resistant to several common diseases, including anthracnose, botrytis, and ascochyta blight (Giambanelli, 2016).

5. Major Challenges in Cultivating Faba Beans

5.1. Pest and Disease Pressure

Insect pests are one of the major biotic constraints of faba bean production. Commonly reported insect pests include the bean aphid, *Aphis fabae*, the pea aphid, *Acyrthosiphon pisum*, the bean fly, *Ophiomyia phaseoli*, the bean weevil, *Acanthoscelides obtectus*, the bean leaf beetle, *Cerotoma trifurcata* and the bean thrips, *Caliothrips fasciatus* (Pergande) (Kumari & Mukkok, 2007). These pests cause direct damage to the plants by feeding on the foliage and stem and developing pods, reducing yield and quality (Hailu et al., 2014).

Fungal diseases are also major constraints to faba bean production. Common fungal diseases include rust (*Uromyces viciae-fabae*), powdery mildew (*Erysiphe polygoni*), and downy mildew (*Peronospora viciae-fabae*) (Akem & Bellar, 1999). Rust is the most serious disease, causing severe leaf damage, yellowing, and defoliation of the plants. Powdery mildew causes white spots on the leaves, while downy mildew leads to leaf yellowing and early defoliation. These diseases cause significant yield losses in faba bean production (Abebe et al., 2014; Ademe et al., 2018).

Viral diseases are another major constraint to faba bean production. Commonly reported viral diseases include bean common mosaic virus (BCMV) and cowpea aphid-borne mosaic virus (CABMV). BCMV is the most serious viral disease, causing severe stunting, leaf yellowing, and mottling. It also leads to reduced seed size and yield losses of up to 50%. CABMV causes a mosaic pattern on the leaves and stunting of the plants (Link et al., 2008) (Ademe et al., 2018).

Bacterial diseases are also major constraints to faba bean production. Commonly reported bacterial diseases include common bacterial blight

(*Xanthomonas axonopodis* pv. phaseoli) and bean anthracnose (*Colletotrichum lindemuthianum*) (Belete & Bastas, 2017) (Girma et al., 2022). Common bacterial blight causes dark-brown lesions on the leaves and pods, while bean anthracnose causes black spots on the leaves and pods. These diseases cause significant yield losses in faba bean production (Salau et al., 2023).

5.2. Soil Fertility

Soil fertility is another major challenge in faba bean cultivation. Faba beans are sensitive to low soil fertility and pH and require adequate nitrogen, phosphorus, and potassium levels to achieve optimal yields (Tadele, 2020). Farmers can address this issue by testing soil and adjusting the pH and nutrient levels. Additionally, adding organic matter such as compost or manure to the soil can help improve fertility and increase yields (Jensen et al., 2010).

5.3. Water Management

Water management is another key challenge in faba bean cultivation. Faba beans are sensitive to waterlogging and drought and are best grown in well-drained soils with even moisture (Sivasankar, 2018). Farmers should use effective irrigation methods to address this issue and consider soil amendments such as mulch to help retain moisture. Water conservation techniques such as drip irrigation and rainwater harvesting can reduce water use and maximize yields (Hussain et al., 2010).

5.4. Harvesting

Harvesting is a major challenge in faba bean cultivation. Faba beans are a crop that requires careful timing and handling during harvest, as they can be easily damaged (Huber et al., 2023). Farmers should use appropriate harvesting equipment and techniques to address this issue, such as a combine harvester. Additionally, they should time the harvest carefully, as faba beans are sensitive to frost and should be harvested before the first frost (Uebersax et al., 2022).

6. Breeding Challenges in Cultivating Faba Bean

Despite their potential, faba beans have not been widely adopted as a commercial crop and remain largely confined to subsistence agriculture. This is partly due to several breeding and variety development challenges that have hindered their widespread adoption.

One of the major challenges in breeding and variety development for faba beans is the lack of genetic diversity in the available germplasm (Keneni et al., 2006). Faba beans are believed to have originated in the Near East or Mediterranean region and have a limited geographic distribution. As a result, the available gene pool is narrow and lacks the diversity necessary to develop varieties adapted to different conditions. This can make the process of selecting and breeding varieties for different climates and soil types difficult (Caracuta et al., 2016; Pasqualone et al., 2020).

Another challenge is the low seed yield of faba beans. This is partly due to several factors, including the slow rate of seed maturation, low seed set rate, and susceptibility to certain diseases. All these factors can limit the potential yield of faba beans and make it difficult for farmers to get a good return on their investment. Additionally, faba beans are sensitive to certain environmental conditions, such as heat and drought, which can further reduce yields (Tammeorg et al., 2014; Al-Suhaibani, 2020). Faba beans are vulnerable to various pests and diseases. These include bacterial and fungal diseases and insect pests like aphids, caterpillars, and mites. The susceptibility of faba beans to these pests and diseases can reduce yields and make it difficult to obtain good crop yields (Hailu et al., 2014).

Several potential solutions have been proposed to address the various challenges associated with breeding and variety development for faba beans. The first is the introduction of new germplasm to increase the genetic diversity of the crop (El-Esawi, 2017). This could be done by using wild relatives of the crop, which may have the traits necessary to adapt to different climates and soil types. Additionally, the use of genetic engineering and marker-assisted selection could be used to introduce desirable traits into existing varieties (Wang et al., 2012; Aschi et al., 2017).

Another potential solution is the development of improved varieties through traditional breeding methods. This could include the selection of varieties for traits such as increased seed yield, resistance to pests and diseases, and better adaptation to different environmental conditions. Additionally, tissue culture and mutagenesis technologies could produce new varieties with improved traits (Tester & Langridge, 2010; Bhat et al., 2016).

Finally, agronomic practices such as proper fertilization and pest control could help improve yields and reduce losses due to pests and diseases. These practices could be used in conjunction with the development of improved varieties to maximize the potential of faba beans (Liliane and Charles, 2020).

7. Advances in Faba Bean Breeding

Recently, an increased focus has been on improving faba bean varieties through breeding and selection. This section will discuss the advances in faba bean breeding and some major examples of improvement.

To improve the crop, breeders must first understand the genetic diversity and inheritance of traits in the faba bean germplasm. This involves studying the genetic relationships between the various lines, varieties, and cultivars. This research type helps identify the most suitable lines for crossing and selection. It also provides insights into managing the crop for optimal yield and quality. Once genetic relationships are understood, breeders can focus on selecting lines with desirable traits and crossing them to create new varieties (Richards et al., 2010; Torres et al., 2010).

7.1. Conventional Breeding Methods

Breeding for improved yields is one of the key goals of faba bean breeding. This involves selecting traits such as increased seed size, higher seed number per plant, improved disease resistance, and higher tolerance to environmental stresses (Gnanasambandam et al., 2012). These traits can be achieved through a combination of genetic selection and traditional breeding techniques. For example, backcrossing is a commonly used technique that involves crossing a selected line with the original cultivar and repeating the process until the desired traits are achieved (Acquaah, 2015). Similarly, hybridization is used to combine the best traits of two different lines and create an improved variety (Tolessa et al., 2015).

7.1.1. Examples of Conventionally Bred Varieties of Faba Bean

Another important goal of faba bean breeding is to improve the nutritional quality of the crop. This involves selecting traits such as increased protein content, increased levels of essential amino acids, increased levels of minerals, and higher levels of vitamins. To achieve this, breeders must identify the genes involved in the synthesis and accumulation of these nutrients. Once identified, they can then select lines that have the desired traits (Warsame et al., 2018). For example, the genes responsible for the synthesis of the essential amino acid lysine have been identified in faba beans, allowing breeders to select for lines with higher lysine content (Ray and Georges, 2010; Singh et al., 2012).

Table 1. Major varieties of faba bean and their characteristics

Sr. Number	Variety	Characteristics
1	Ace	This variety is high-yielding, with a medium-sized seed. It is resistant to rust and anthracnose and is well adapted to cooler climates.
2	Aquastar	This variety produces medium-sized seeds and has a high yield. It is resistant to rust and anthracnose, as well as fusarium wilt.
3	Barada	This variety has a high yield and produces large-sized seeds. It is resistant to rust and anthracnose, as well as fusarium wilt and nematodes.
4	Cobra	This variety has an excellent yield and produces medium-sized seeds. It is resistant to rust and anthracnose, as well as fusarium wilt and nematodes.
5	Eland	This variety produces large-sized seeds and has a medium yield. It is resistant to rust and anthracnose, as well as fusarium wilt and nematodes.

7.2. Genetic Engineering for Improving Faba Bean

In addition, modern breeding techniques have been used to improve the resistance of faba bean varieties to various diseases and pests. For example, breeders have developed lines with resistance to rust, bean common mosaic virus, and Fusarium wilt. These lines are then used to create new hybrid varieties with increased disease resistance (Singh and Schwartz, 2010).

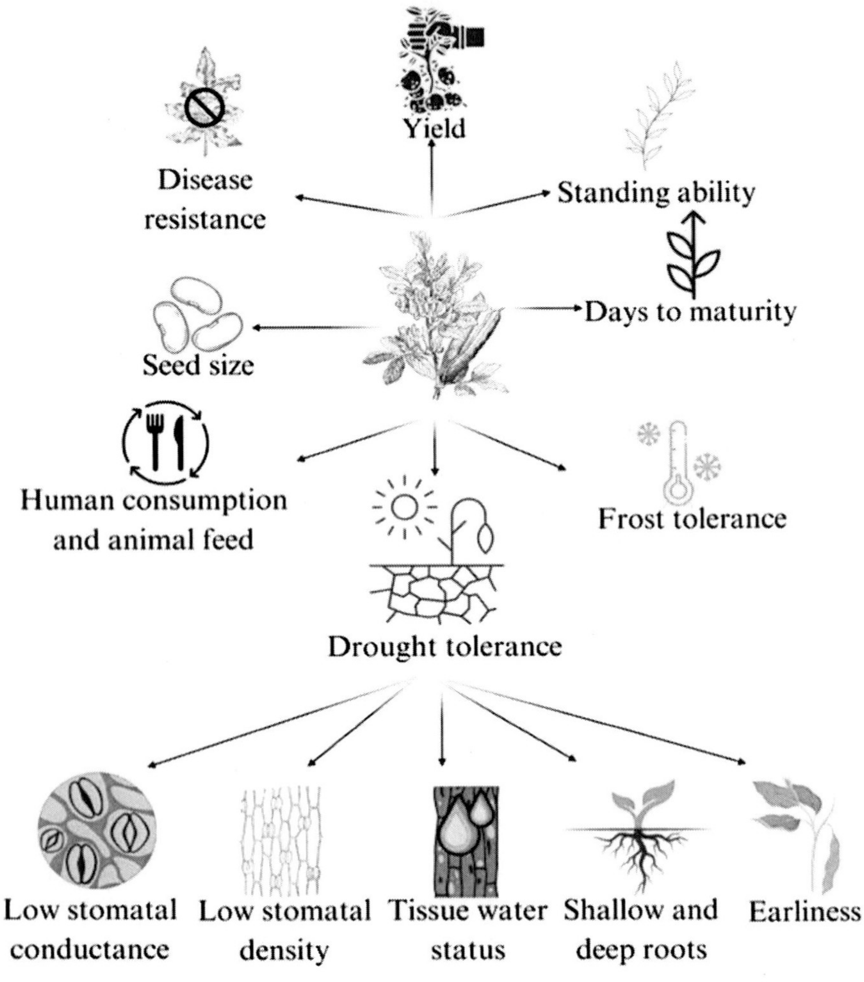

Figure 2. Selection criterion for breeding of faba bean.

Genetic engineering has made it possible to create new varieties of faba beans that are more resistant to disease, more productive, and more nutritious. This has enabled farmers to increase their yields and profits while providing more nutritious food to consumers (Multari et al., 2015) (Van Esse et al., 2020).

7.2.1. List of Genetically Engineered Varieties of Faba Bean

Table 2. Major varieties developed through genetic engineering

Sr. Number	Variety	Characteristics
1	Barbara	This variety was developed in Australia and is known for its drought tolerance, high yields, and resistance to fusarium wilt and Ascochyta blight. It is well-suited to dryland farming and has a short growing season.
2	Lentini	This variety was developed in Italy, has a short growing season, and is well suited to cultivation in cool climates. It is resistant to fusarium wilt, Ascochyta blight, and drought tolerance. It has large, round seeds and produces high yields.
3	Vicia	This variety was developed in Spain and is known for its high yields, resistance to fusarium wilt, Ascochyta blight, and drought tolerance. It is well-suited to dryland farming, has a short growing season, and produces large, round seeds.
4	Era	This variety was developed in France and has a short growing season, drought tolerance, and resistance to fusarium wilt and Ascochyta blight. It produces high yields and has a large, round seed.

Source: Aguilar-Benitez et al., 2020, Carrillo-Perdomo et al., 2020, Khazaei et al., 2021.

One way in which genetic engineering has been used to improve faba beans is through the introduction of disease-resistant genes. Plant diseases can cause a great deal of damage to crops, reducing yields and reducing the quality of the final product (Zaidi et al., 2020). Genetic engineering has allowed scientists to introduce disease-resistance genes into faba bean plants, making them more disease-resistant. This has improved yields and better quality faba bean products (Carrillo-Perdomo et al., 2020; Tyagi et al., 2021).

Another way in which genetic engineering has been used to improve faba beans is through the introduction of genes that increase the nutritional content of the crop. For example, researchers have used genetic engineering to introduce genes that increase essential amino acids, vitamins, and minerals in faba beans. This has resulted in faba bean varieties that are more nutritious and healthier for consumers (Zalabák et al., 2013).

In addition, genetic engineering has been used to improve the productivity of faba beans. This has been done by introducing genes that increase the plant's ability to utilize light and water more efficiently. This has resulted in higher yields of faba beans, which has allowed farmers to increase their profits (Abou-Khater et al., 2022).

7.3. Tissue Culture Technique for Improvement of Faba Bean

Tissue culture is a technique used to reproduce plants and animal cells in an artificial environment. This technique is used to quickly produce genetically identical plants with desirable traits and new varieties of plants resistant to diseases and other factors. It is used in the breeding and improvement of many plants, including the faba bean (Gantait and Mukherjee, 2021).

Tissue culture is used to produce genetically identical plants, which helps to eliminate the need for multiple generations of breeding to achieve the desired traits. This is done by taking a small piece of tissue from a donor plant, called an explant, and culturing it in a nutrient medium. This allows for the rapid multiplication of the explant, producing hundreds of genetically identical plants in weeks. This process is known as micropropagation and is used to produce many plants for commercial sale (Ahmar et al., 2020; Zhu et al., 2020; Gao, 2021).

Tissue culture has also been used to aid in improving faba bean. The technique has been used to produce disease-resistant varieties of faba bean, as well as to produce varieties that are more tolerant of drought and other environmental stresses. It has also been used to produce varieties with improved yields and higher nutritional value (Adhikari et al., 2022).

Conclusion

Faba bean is one of the most important grain legumes grown in the world. It is an important source of nutrients, particularly in areas where other protein sources are limited. Its importance has been increasing in recent years due to its potential to contribute to food security and its ability to fix nitrogen, thus reducing the number of chemical fertilizers used in crop production.

The genetic diversity of the faba bean is vast and complex. Various morphological and physiological traits have been identified and used in breeding programs. However, most of these traits have not been studied properly, leaving many potential improvements untapped. Furthermore, the faba bean has a relatively low level of genetic diversity due to its domestication process that mainly relied on self-pollination and selection. This has narrowed the genetic base, making it more difficult to select the best genotypes for certain traits.

In recent years, there have been significant advances in faba bean breeding and cultivation. New varieties have been developed with improved traits such as higher yields, better disease resistance, and improved nutritional content. Molecular markers, as well as genomic and transcriptomic studies, have provided additional insights into the genetic diversity of the faba bean and allowed for a more precise selection of desirable traits.

Despite these advances, there are still many challenges facing faba bean cultivation and breeding. This includes the limited availability of high-quality seeds, the lack of resources to support breeding programs, and limited farmer access to new varieties and technologies. Furthermore, the effects of climate change on faba bean production are still not fully understood.

In conclusion, faba bean cultivation and breeding have made great strides in recent years. However, there is still much to be done to ensure the sustainable and profitable production of this valuable crop. Future efforts should focus on increasing genetic diversity, developing improved varieties, and increasing access to new technologies and varieties for farmers. Such efforts will ensure that the faba bean remains an important source of nutrition and a valuable contributor to global food security.

Prospects

Despite the various obstacles to faba bean breeding, there is still much potential for further improvement. Recent advances in conventional and biotechnological techniques have enabled the development of new varieties with improved yields, disease and pest resistance, and nutritional content. Additionally, the use of genomic tools, such as molecular markers and genomic selection, has enabled the efficient tracking and selection of desirable traits, further increasing the potential for successful breeding. In the future, there is potential for further advances in faba bean breeding, such as developing cultivars with improved trait combinations and increased yields. Additionally, using genomic tools, such as gene editing, could enable the development of new varieties with even greater yield potential and disease and pest resistance.

References

Abebe, T., Birhane, T., Nega, Y., & Workineh, A. (2014). The prevalence and importance of faba bean diseases with special consideration to the newly emerging faba bean gall in Tigray, Ethiopia. *Discourse Journal of Agriculture and Food Sciences*, *2*(2), 33-38.

Abid, G., Mingeot, D., Udupa, S. M., Muhovski, Y., Watillon, B., Sassi, K., ... & Jebara, M. (2015). Genetic relationship and diversity analysis of faba bean (*Vicia faba* L. var. Minor) genetic resources u1sing morphological and microsatellite molecular markers. *Plant molecular biology reporter*, *33*, 1755-1767.

Abou-Khater, L., Maalouf, F., Jighly, A., Alsamman, A. M., Rubiales, D., Rispail, N., ... & Kumar, S. (2022). Genomic regions associated with herbicide tolerance in a worldwide faba bean (*Vicia faba* L.) collection. *Scientific Reports*, *12*(1), 158.

Acquaah, G. (2015). Conventional plant breeding principles and techniques. *Advances in plant breeding strategies: Breeding, biotechnology and molecular tools*, 115-158.

Ademe, A., Ebabuye, Y., Gelaye, M., Gezachew, S., & Telahun, G. (2018). Survey of faba bean (*Vicia faba* L.) diseases in major faba bean growing districts of North Gondar. *African Journal of Plant Science*, *12*(2), 32-36.

Adhikari, K. N., Khater, L. A., & Maalouf, F. (2022). Application of Genetic, Genomic Strategies to Address the Biotic Stresses in Faba Bean. In *Genomic Designing for Biotic Stress Resistant Pulse Crops* (pp. 353-380). Cham: Springer International Publishing.

Aguilar-Benitez, D., Casimiro-Soriguer, I., & Torres, A. M. (2020). The first approach to pod dehiscence in faba bean: genetic and histological analyses. *Scientific Reports*, *10*(1), 1-14.

Ahmar, S., Gill, R. A., Jung, K. H., Faheem, A., Qasim, M. U., Mubeen, M., & Zhou, W. (2020). Conventional and molecular techniques from simple breeding to speed breeding in crop plants: recent advances and future outlook. *International journal of molecular sciences*, *21*(7), 2590.

Akem, C., & Bellar, M. (1999). Survey of faba bean (*Vicia faba* L.) diseases in the main faba bean-growing regions of Syria. *Arab Journal of plant protection*, *17*(2), 113-116.

Alghamdi, S. S., Migdadi, H. M., Ammar, M. H., Paull, J. G., & Siddique, K. H. M. (2012). Faba bean genomics: current status and future prospects. *Euphytica*, *186*, 609-624.

Alghamdi, S. S., Migdadi, H. M., Khan, M. A., & Afzal, M. (2020). AFLP-based analysis of variation and population structure in mutagenesis induced faba bean. *Diversity*, *12*(8), 303.

Al-Saady, N., Nadaf, S., Al-Lawati, A., & Al-Hinai, S. (2018). Faba bean (*Vicia faba* L.) germplasm collection and its diversity in Oman.

Al-Suhaibani, N. A. (2009). Influence of early water deficit on seed yield and quality of faba bean under arid environment of Saudi Arabia. *Am-Eurasian J Agric Environ Sci*, *5*(5), 649-654.

Ammar, M. H., Alghamdi, S. S., Migdadi, H. M., Khan, M. A., El-Harty, E. H., & Al-Faifi, S. A. (2015). Assessment of genetic diversity among faba bean genotypes using agro-morphological and molecular markers. *Saudi Journal of Biological Sciences*, *22*(3), 340-350.

Arun-Chinnappa, K. S., & McCurdy, D. W. (2015). De novo assembly of a genome-wide transcriptome map of *Vicia faba* (L.) for transfer cell research. *Frontiers in plant science*, *6*, 217.

Aschi, A., Aubert, M., Riah-Anglet, W., Nélieu, S., Dubois, C., Akpa-Vinceslas, M., & Trinsoutrot-Gattin, I. (2017). Introduction of Faba bean in crop rotation: Impacts on soil chemical and biological characteristics. *Applied Soil Ecology*, *120*, 219-228.

Avramidou, E., Ganopoulos, I., Mylona, P., Abraham, E. M., Nianiou-Obeidat, I., Osathanunkul, M., & Madesis, P. (2023). Comparative Analysis of the Genetic Diversity of Faba Bean (*Vicia faba* L.). *Sustainability*, *15*(2), 1016.

Belete, T., & Bastas, K. K. (2017). Common bacterial blight (Xanthomonas axonopodis pv. phaseoli) of beans with special focus on Ethiopian condition. *J Plant Pathol Microbiol*, *8*(2).

Bhat, J. A., Ali, S., Salgotra, R. K., Mir, Z. A., Dutta, S., Jadon, V., ... & Prabhu, K. (2016). Genomic selection in the era of next generation sequencing for complex traits in plant breeding. *Frontiers in genetics*, *7*, 221.

Caracuta, V., Weinstein-Evron, M., Kaufman, D., Yeshurun, R., Silvent, J., & Boaretto, E. (2016). 14,000-year-old seeds indicate the Levantine origin of the lost progenitor of faba bean. *Scientific reports*, *6*(1), 37399.

Carrillo-Perdomo, E., Vidal, A., Kreplak, J., Duborjal, H., Leveugle, M., Duarte, J., ... & Aubert, G. (2020). Development of new genetic resources for faba bean (*Vicia faba* L.) breeding through the discovery of gene-based SNP markers and the construction of a high-density consensus map. *Scientific Reports*, *10*(1), 6790.

Chinthapalli, B., Dibar, D. T., Chitra, D. V., & Leta, M. B. (2015). A comparative study on the effect of organic and inorganic fertilizers on agronomic performance of faba bean (*Vicia faba* L.) and pea (Pisum sativum L.). *Agriculture, Forestry and Fisheries*, *4*(6), 263-268.

Crépon, K., Marget, P., Peyronnet, C., Carrouée, B., Arese, P., & Duc, G. (2010). Nutritional value of faba bean (*Vicia faba* L.) seeds for feed and food. *Field crops research*, *115*(3), 329-339.

Duc, G., Bao, S., Baum, M., Redden, B., Sadiki, M., Suso, M. J., ... & Zong, X. (2010). Diversity maintenance and use of *Vicia faba* L. genetic resources. *Field Crops Research*, *115*(3), 270-278.

El-Esawi, M. A. (2017). SSR analysis of genetic diversity and structure of the germplasm of faba bean (*Vicia faba* L.). *Comptes Rendus Biologies*, *340*(11-12), 474-480.

Elshaer, H., & Hellal, A. G. (2020). Genetic Diversity Based on SSR Markers and Morphological Analysis of faba bean (*Vicia faba* L.) in Egypt. *Taeckholmia*, *40*(1), 130-142.

Elshafei, A. A. M., Amer, M. A. E., Elenany, M. A. M., & Helal, A. G. A. E. (2019). Evaluation of the genetic variability of faba bean (*Vicia faba* L.) genotypes using agronomic traits and molecular markers. *Bulletin of the National Research Centre*, *43*(1), 1-10.

Etemadi, F., Hashemi, M., Barker, A. V., Zandvakili, O. R., & Liu, X. (2019). Agronomy, nutritional value, and medicinal application of faba bean (*Vicia faba* L.). *Horticultural Plant Journal*, *5*(4), 170-182.

Gantait, S., & Mukherjee, E. (2021). Tissue culture-based genetic improvement of fava bean (*Vicia faba* L.): analysis on previous achievements and future perspectives. *Applied Microbiology and Biotechnology*, *105*(18), 6531-6546.

Gao, C. (2021). Genome engineering for crop improvement and future agriculture. *Cell*, *184*(6), 1621-1635.

Giambanelli, E. (2016). *Traditional crops and foods: documentation, analytical characterization, retention factors of bioactive compounds*.

Girma, F., Fininsa, C., Terefe, H., & Amsalu, B. (2022). Distribution of common bacterial blight and anthracnose diseases and factors influencing epidemic development in major common bean growing areas in Ethiopia. *Acta Agriculturae Scandinavica, Section B—Soil & Plant Science*, *72*(1), 685-699.

Gnanasambandam, A., Paull, J., Torres, A., Kaur, S., Leonforte, T., Li, H., ... & Materne, M. (2012). Impact of molecular technologies on faba bean (*Vicia faba* L.) breeding strategies. *Agronomy*, *2*(3), 132-166.

Godbout, J., Gros-Louis, M. C., Lamothe, M., & Isabel, N. (2020). Going with the flow: Intraspecific variation may act as a natural ally to counterbalance the impacts of global change for the riparian species Populus deltoides. *Evolutionary applications*, *13*(1), 176-194.

Göl, Ş., Doğanlar, S., & Frary, A. (2017). Relationship between geographical origin, seed size and genetic diversity in faba bean (*Vicia faba* L.) as revealed by SSR markers. *Molecular Genetics and Genomics*, *292*, 991-999.

Hailu, E., Getaneh, G., Sefera, T., Tadesse, N., Bitew, B., Boydom, A., ... & Temesgen, T. (2014). Faba bean gall; a new threat for faba bean (*Vicia faba*) production in Ethiopia. *Adv Crop Sci Tech*, *2*(144), 2.

Huber, J., Chaluppa, N., Voit, B., Steinkellner, S., & Killermann, B. (2023). Damage potential of the broad bean beetle (Bruchus rufimanus Boh.) on seed quality and yield of faba beans (*Vicia faba* L.). *Crop Protection*, 106227.

Hussain, N., Sarwar, G., Schmeisky, H., Al-Rawahy, S., & Ahmad, M. (2010). Salinity and drought management in legume crops. *Climate change and management of cool season grain legume crops*, 171-191.

Jensen, E. S., Peoples, M. B., & Hauggaard-Nielsen, H. (2010). Faba bean in cropping systems. *Field crops research*, *115*(3), 203-216.

Johnson, J. B., Collins, T., Skylas, D., Quail, K., Blanchard, C., & Naiker, M. (2020a). Profiling the varietal antioxidative contents and macrochemical composition in Australian faba beans (*Vicia faba* L.). *Legume Science*, *2*(2), e28.

Johnson, J. B., Walsh, K., & Naiker, M. (2020b). Application of infrared spectroscopy for the prediction of nutritional content and quality assessment of faba bean (*Vicia faba* L.). *Legume Science*, *2*(3), e40.

Karaköy, T., Baloch, F. S., Toklu, F., & Özkan, H. (2014). Variation for selected morphological and quality-related traits among 178 faba bean landraces collected from Turkey. *Plant Genetic Resources*, *12*(1), 5-13.

Karchani-Balma, S., Gautier, A., Raies, A., & Fournier, E. (2008). Geography, plants, and growing systems shape the genetic structure of Tunisian Botrytis cinerea populations. *Phytopathology*, *98*(12), 1271-1279.

Karkanis, A., Ntatsi, G., Lepse, L., Fernández, J. A., Vågen, I. M., Rewald, B., ... & Savvas, D. (2018). Faba bean cultivation–revealing novel managing practices for more sustainable and competitive European cropping systems. *Frontiers in plant science*, 1115.

Kaur, S., Cogan, N. O., Forster, J. W., & Paull, J. G. (2014). Assessment of genetic diversity in faba bean based on single nucleotide polymorphism. *Diversity*, *6*(1), 88-101.

Kaya, E., Yılmaz Tuncel, N., & Tuncel, N. B. (2018). Utilization of lentil, pea, and faba bean hulls in Turkish noodle production. *Journal of food science and technology*, *55*, 1734-1745.

Keneni, G., Jarso, M., & Wolabu, T. (2006). Faba bean (*Vicia faba* L.) genetics and breeding research in Ethiopia: A Review. *Food and Forage Legumes of Ethiopia: Progress and Prospects*, *42*, 52.

Khazaei, H., Hawkins, G., & Vandenberg, A. (2021). Historical review of faba bean improvement in western Canada. *Legume Science*, *3*(4), e92.

Kosterin, O. E. (2014). The lost ancestor of the broad bean (*Vicia faba* L.) and the origin of plant cultivation in the Near East. *Вавиловский журнал генетики и селекции*, *18*(4-1), 831-840.

Kumari, S. G., & Makkouk, K. M. (2007). Virus diseases of faba bean (*Vicia faba* L.) in Asia and Africa. *Plant viruses*, *1*(1), 93-105.

Kwak, M., & Gepts, P. (2009). Structure of genetic diversity in the two major gene pools of common bean (Phaseolus vulgaris L., Fabaceae). *Theoretical and Applied Genetics*, *118*, 979-992.

Kwon, S. J., Kim, D. G., Kim, J. M., Kang, K. Y., Lee, M. K., Hong, M. J., ... & Ryu, J. (2018). Phytochemical compounds and antioxidant activity in the grain of selected faba bean (*Vicia faba*) genotypes. *Plant breeding and biotechnology*, *6*(1), 65-73.

Labba, I. C. M., Frøkiær, H., & Sandberg, A. S. (2021). Nutritional and antinutritional composition of fava bean (*Vicia faba* L., var. minor) cultivars. *Food Research International*, *140*, 110038.

Li-Juan, L., Zhao-hai, Y., Zhao-jie, Z., Ming-shi, X., & Han-qing, Y. (1993). Faba bean in China: state-of-the-art review. *International Center for Agricultural Research in the Dry Areas (ICARDA). Aleppo, Syria*, 127-128.

Liliane, T. N., & Charles, M. S. (2020). Factors affecting yield of crops. *Agronomy-climate change & food security*, 9.

Link, W., Hanafy, M., Malenica, N., Jacobsen, H. J., & Jelenic, S. (2008). Faba bean. *Compendium of Transgenic Crop Plants: Part*, *3*, 71-88.

Maalouf, F., Hu, J., O'Sullivan, D. M., Zong, X., Hamwieh, A., Kumar, S., & Baum, M. (2019). Breeding and genomics status in faba bean (*Vicia faba*). *Plant Breeding*, *138*(4), 465-473.

Maalouf, F., Ahmed, S. and Bishaw, Z., (2021). Faba bean. In the beans and the peas (pp. 105-131). Woodhead Publishing.

Malek, N., Aci, M. M., Khamassi, K., Lupini, A., Rouissi, M., & Hanifi-Mekliche, L. (2021). Agro-Morphological and Molecular Variability among Algerian Faba Bean (*Vicia faba* L.) Accessions. *Agronomy*, *11*(8), 1456.

Maxted, N., & Bennett, S. J. (Eds.). (2001). *Plant genetic resources of legumes in the Mediterranean* (Vol. 39). Springer Science & Business Media.

Merga, B., Egigu, M. C., & Wakgari, M. (2019). Reconsidering the economic and nutritional importance of faba bean in Ethiopian context. *Cogent Food & Agriculture*, *5*(1), 1683938.

Minguez, M. I., & Rubiales, D. (2021). Faba bean. In *Crop physiology case histories for major crops* (pp. 452-481). Academic Press.

Moran, E. V., Hartig, F., & Bell, D. M. (2016). Intraspecific trait variation across scales: implications for understanding global change responses. *Global Change Biology*, *22*(1), 137-150.

Multari, S., Stewart, D., & Russell, W. R. (2015). Potential of fava bean as future protein supply to partially replace meat intake in the human diet. *Comprehensive Reviews in Food Science and Food Safety*, *14*(5), 511-522.

Njoumi, S., Josephe Amiot, M., Rochette, I., Bellagha, S., & Mouquet-Rivier, C. (2019). Soaking and cooking modify the alpha-galacto-oligosaccharide and dietary fibre content in five Mediterranean legumes. *International Journal of Food Sciences and Nutrition*, *70*(5), 551-561.

O'Sullivan, D. M., & Angra, D. (2016). Advances in faba bean genetics and genomics. *Frontiers in genetics*, *7*, 150.

Oliveira, H. R., Tomás, D., Silva, M., Lopes, S., Viegas, W., & Veloso, M. M. (2016). Genetic diversity and population structure in *Vicia faba* L. landraces and wild related species assessed by nuclear SSRs. *PloS one*, *11*(5), e0154801.

Pasqualone, A., Abdallah, A., & Summo, C. (2020). Symbolic meaning and use of broad beans in traditional foods of the Mediterranean Basin and the Middle East. *Journal of Ethnic Foods*, *7*(1), 39.

Petitot, M., Boyer, L., Minier, C., & Micard, V. (2010). Fortification of pasta with split pea and faba bean flours: Pasta processing and quality evaluation. *Food Research International*, *43*(2), 634-641.

Poonia, A., Vikranta, U., Chaudhary, N., & Dangi, P. (2022). Current and Potential Health Claims of Faba Beans (*Vicia faba*, L.) and Its Components. In *Faba Bean: Chemistry, Properties and Functionality* (pp. 331-355). Cham: Springer International Publishing.

Qahtan, A. A., Al-Atar, A., Abdel-Salam, E. M., El-Sheikh, M. A., Gaafar, A. R. Z., & Faisal, M. (2021). Genetic diversity and structure analysis of a worldwide collection of faba bean (*Vicia faba*) genotypes using ISSR markers. *Int. J. Agric. Biol*, *25*, 683-691.

Rahate, K. A., Madhumita, M., & Prabhakar, P. K. (2021). Nutritional composition, anti-nutritional factors, pretreatments-cum-processing impact and food formulation potential of faba bean (*Vicia faba* L.): A comprehensive review. *Lwt*, *138*, 110796.

Ray, H., & Georges, F. (2010). A genomic approach to nutritional, pharmacological and genetic issues of faba bean (*Vicia faba*): prospects for genetic modifications. *GM crops*, *1*(2), 99-106.

Rebaa, F., Abid, G., Aouida, M., Abdelkarim, S., Aroua, I., Muhovski, Y., ... & Jebara, M. (2017). Genetic variability in Tunisian populations of faba bean (*Vicia faba* L. var. major) assessed by morphological and SSR markers. *Physiology and Molecular Biology of Plants*, *23*, 397-409.

Richards, R. A., Rebetzke, G. J., Watt, M., Condon, A. T., Spielmeyer, W., & Dolferus, R. (2010). Breeding for improved water productivity in temperate cereals: phenotyping,

quantitative trait loci, markers and the selection environment. *Functional Plant Biology*, *37*(2), 85-97.

Robinson, G. H. J., Balk, J., & Domoney, C. (2019). Improving pulse crops as a source of protein, starch and micronutrients. *Nutrition bulletin*, *44*(3), 202-215.

Salau, A. O., Abeje, B. T., Faisal, A. N., & Asfaw, T. T. (2023, January). Faba Bean Disease Detection Using Deep Learning Techniques. In *2023 International Conference On Cyber Management And Engineering (CyMaEn)* (pp. 344-349). IEEE.

Saxena, M. C. (1991). Status and scope for production of faba bean in the Mediterranean countries. *Options Méditerranéennes*, *10*(1), 5-20.

Schmelter, L., Rohm, H., & Struck, S. (2021). Gluten-free bakery products: Cookies made from different *Vicia faba* bean varieties. *Future Foods*, *4*, 100038.

Siah, S. D., Konczak, I., Agboola, S., Wood, J. A., & Blanchard, C. L. (2012). *In vitro* investigations of the potential health benefits of Australian-grown faba beans (*Vicia faba* L.): chemopreventative capacity and inhibitory effects on the angiotensin-converting enzyme, α-glucosidase and lipase. *British Journal of Nutrition*, *108*(S1), S123-S134.

Siddique, K. H. M., Brinsmead, R. B., Knight, R., Knights, E. J., Paull, J. G., & Rose, I. A. (2000). Adaptation of chickpea (Cicer arietinum L.) and faba bean (*Vicia faba* L.) to Australia. In *Linking Research and Marketing Opportunities for Pulses in the 21st Century: Proceedings of the Third International Food Legumes Research Conference* (pp. 289-303). Springer Netherlands.

Sillero, J. C., Villegas-Fernández, A. M., Thomas, J., Rojas-Molina, M. M., Emeran, A. A., Fernández-Aparicio, M., & Rubiales, D. (2010). Faba bean breeding for disease resistance. *Field Crops Research*, *115*(3), 297-307.

Singh, A. K., Bharati, R. C., Manibhushan, N. C., & Pedpati, A. (2013). An assessment of faba bean (*Vicia faba* L.) current status and future prospect. *African Journal of Agricultural Research*, *8*(50), 6634-6641.

Singh, A. K., Bhatt, B. P., Upadhyaya, A., Kumar, S., Sundaram, P. K., Singh, B. K., ... & Bharati, R. C. (2012). Improvement of faba bean (*Vicia faba* L.) yield and quality through biotechnological approach: A review. *African Journal of Biotechnology*, *11*(87), 15264-15271.

Singh, S. P., & Schwartz, H. F. (2010). Breeding common bean for resistance to diseases: a review. *Crop Science*, *50*(6), 2199-2223.

Sivasankar, S. (2018). Efficient and sustainable production of faba bean. In *Achieving sustainable cultivation of grain legumes Volume 2* (pp. 291-318). Burleigh Dodds Science Publishing.

Smýkal, P., Coyne, C. J., Ambrose, M. J., Maxted, N., Schaefer, H., Blair, M. W., ... & Varshney, R. K. (2015). Legume crops phylogeny and genetic diversity for science and breeding. *Critical Reviews in Plant Sciences*, *34*(1-3), 43-104.

Tadele, M. (2020). Impact assessment of abiotic production constrains of Faba Bean (*Vicia faba* L.) and breeding mechanisms for acid soil, drought and waterlogging affected environments. *J Biol Agric Healthc*, *10*(11), 49-56.

Tammeorg, P., Simojoki, A., Mäkelä, P., Stoddard, F. L., Alakukku, L., & Helenius, J. (2014). Biochar application to a fertile sandy clay loam in boreal conditions: effects

on soil properties and yield formation of wheat, turnip rape and faba bean. *Plant and soil*, *374*, 89-107.

Tazrart, K., Zaidi, F., Lamacchia, C., & Haros, M. (2016). Effect of durum wheat semolina substitution with broad bean flour (*Vicia faba*) on the Maccheronccini pasta quality. *European Food Research and Technology*, *242*, 477-485.

Terzopoulos, P. J., & Bebeli, P. J. (2008). Genetic diversity analysis of Mediterranean faba bean (*Vicia faba* L.) with ISSR markers. *Field Crops Research*, *108*(1), 39-44.

Tester, M., & Langridge, P. (2010). Breeding technologies to increase crop production in a changing world. *Science*, *327*(5967), 818-822.

Tolessa, T. T., Keneni, G., & Mohammad, H. (2015). Genetic progress from over three decades of faba bean ('*Vicia faba*'L.) breeding in Ethiopia. *Australian Journal of Crop Science*, *9*(1), 41-48.

Torres, A. M., Avila, C. M., Gutierrez, N., Palomino, C., Moreno, M. T., & Cubero, J. I. (2010). Marker-assisted selection in faba bean (*Vicia faba* L.). *Field crops research*, *115*(3), 243-252.

Turco, I., Ferretti, G., & Bacchetti, T. (2016). Review of the health benefits of Faba bean (*Vicia faba* L.) polyphenols. *Journal of Food & Nutrition Research*, *55*(4).

Tyagi, S., Kumar, R., Kumar, V., Won, S. Y., & Shukla, P. (2021). Engineering disease-resistant plants through CRISPR-Cas9 technology. *GM Crops & Food*, *12*(1), 125-144.

Uebersax, M. A., Siddiq, M., Cramer, J., & Bales, S. (2022). Harvesting, postharvest handling, distribution, and marketing of dry beans. *Dry beans and pulses: Production, processing, and nutrition*, 81-104.

Van Esse, H. P., Reuber, T. L., & van der Does, D. (2020). Genetic modification to improve disease resistance in crops. *New Phytologist*, *225*(1), 70-86.

Verni, M., Coda, R., & Rizzello, C. G. (2019). The use of faba bean flour to improve the nutritional and functional features of cereal-based foods: Perspectives and future strategies. In *Flour and bread and their fortification in health and disease prevention* (pp. 465-475). Academic Press.

Wang, H. F., Zong, X. X., Guan, J. P., Yang, T., Sun, X. L., Ma, Y., & Redden, R. (2012). Genetic diversity and relationship of global faba bean (*Vicia faba* L.) germplasm revealed by ISSR markers. *Theoretical and applied genetics*, *124*, 789-797.

Warsame, A. O., O'Sullivan, D. M., & Tosi, P. (2018). Seed storage proteins of faba bean (*Vicia faba* L): Current status and prospects for genetic improvement. *Journal of Agricultural and Food Chemistry*, *66*(48), 12617-12626.

Zaidi, S. S. E. A., Mahas, A., Vanderschuren, H., & Mahfouz, M. M. (2020). Engineering crops of the future: CRISPR approaches to develop climate-resilient and disease-resistant plants. *Genome biology*, *21*(1), 1-19.

Zalabák, D., Pospíšilová, H., Šmehilová, M., Mrízová, K., Frébort, I., & Galuszka, P. (2013). Genetic engineering of cytokinin metabolism: a prospective way to improve agricultural traits of crop plants. *Biotechnology advances*, *31*(1), 97-117.

Zeid, M., Schön, C. C., & Link, W. (2003). Genetic diversity in recent elite faba bean lines using AFLP markers. *Theoretical and Applied Genetics*, *107*, 1304-1314.

Zhang, H., Liu, Y., Zong, X., Teng, C., Hou, W., Li, P., & Du, D. (2023). Genetic Diversity of Global Faba Bean Germplasm Resources Based on the 130K TNGS Genotyping Platform. *Agronomy*, *13*(3), 811.

Zhu, H., Li, C., & Gao, C. (2020). Applications of CRISPR–Cas in agriculture and plant biotechnology. *Nature Reviews Molecular Cell Biology*, *21*(11), 661-677.

Zong, X., Ren, J., Guan, J., Wang, S., Liu, Q., Paull, J. G., & Redden, R. (2010). Molecular variation among Chinese and global germplasm in spring faba bean areas. *Plant breeding*, *129*(5), 508-513.

Chapter 8

Agronomic Practices in Faba Beans

Anurag Tripathi[1]
Anupam Tiwari[2,*]
Akankhya Guru[1]
Abhay Vikram Singh[2]
Ashutosh Mishra[3]
and Amit Nagar[1]
[1]Assistant Professor, Department of Genetics and Plant Breeding, JV College, Baraut, Baghpat, India
[2]Assistant Professor, Department of Horticulture, JV College, Baraut, Baghpat. India
[3]Research Scholar, Banaras Hindu University, Varanasi, India

Abstract

An important legume crop with both nutritional and environmental advantages, the faba bean (*Vicia faba* L.) is an essential part of sustainable farming systems. Although improving its yield and quality requires effective agronomic techniques, faba bean cultivation is complicated by problems such as soil nutrient constraints, disease risks, and fluctuating climate conditions. This study looks at how different agronomic techniques, such as crop rotation, tillage, sowing rates, irrigation, and fertilizer management, affect the crop's capacity to fix nitrogen, grow, and produce. Because of their nitrogen-fixing symbiosis, faba beans can increase soil fertility when included in various cropping systems, reducing the need for synthetic fertilizers. The results show that in order to maximize yields and increase resistance to environmental

* Corresponding Author's Email: uniquetiwari77@gmail.com

In: Faba Beans
Editors: Aamir Raina and Samiullah Khan
ISBN: 979-8-89530-447-1

challenges, a balanced nutrient application, an adequate water supply, and an appropriate plant density are essential. By improving soil health and overall production in a variety of farming scenarios, optimizing these techniques promotes the sustainability of faba bean cultivation. This study emphasizes how crucial agronomic strategies are to attaining the best yields and promoting ecologically friendly faba bean farming.

Introduction

The faba bean (*Vicia faba* L.), typically produced in India, is a substantial cool-season legume vegetable. Broad bean, Faba bean, Horse bean, Windsor bean, and Tick beans (small forms) are only a few of the many names for this plant that carry significant cultural and agricultural value around the world. While it is known as Boby kurmouvje in Russia, it is known as Bakela in Ethiopia, Faveira in Portugal, Feve in France, Ful Masri in Sudan, and Yeshil Bakla in Turkey. This ancient crop, native to the Middle East, has been cultivated for centuries due to its tender pods and dried seeds. The tender pods are a popular vegetable, and the dried seeds are used for making pulses (Fageria et al., 2022). Faba bean is crucial in Indian agriculture; it is the only bean cultivated during winter. It ranks third in area and production among feed grain legumes in India, following *Glycine max* and *Pisum* sativum (Singh et al., 2013). As a legume crop, faba bean plays a vital role in crop rotation. Its symbiotic relationship with Rhizobium allows it to fix atmospheric nitrogen, enhancing soil fertility sustainably (Mekkei, 2014). Depending on factors such as cultivar, cultural practices, and environmental conditions, faba bean can fix nitrogen ranging from 40 to 330 kg per hectare in a crop cycle (Galloway et al., 2004; Etemadi et al., 2018a). Furthermore, its cultivation promotes soil microbial activity, improves soil physical properties, and enhances the solubilization of insoluble phosphorus (Rashid et al., 2016).

Botany

Faba bean, an annual legume crop, exhibits an upright growth pattern, reaching a height of 0.3 to 1.0 m, which varies based on the genotype, either dwarf or giant. It boasts a robust taproot system and features a singular, square, hollow, and erect stem accompanied by compound leaves. The plant produces clusters of four flowers, white with black blotches, nestled in the leaf axils

(Rana and Vilas, 2018). Faba beans typically take 45-60 days to flower after seed sowing, with the crop maturing between 110-130 days post-sowing. Faba bean plants are pollinated through entomophily. These plants develop large, fleshy green pods, each measuring up to 10 cm in length and 1-2 cm in width, arranged in clusters of five or more along the stem within the leaf axils. Within each pod, three to four oblong or oval seeds are found (Rana and Vilas, 2018).

Nutritional Composition and Uses

Pods of faba beans are a rich source of lysine-rich proteins, carbohydrates, vitamins, minerals, and numerous bioactive compounds (Dhull et al., 2021) (Table 1). Mature seeds of faba bean contain proteins (26.1%), fat (1.3%), carbohydrates (58.3%), calcium (1s03 mg/100 g), iron (6.7 mg/100 g), magnesium (192 mg/100 g), phosphorous (421 mg/100 g).

Area and Production

Faba bean holds significant importance as a legume crop worldwide, although its cultivation is primarily concentrated in select countries. In 2020, the global cultivation of faba bean covered an area of 2.66 million hectares, yielding a production of 56.8 million tons (FAOSTAT, 2022). China emerged as the leading grower and producer of faba bean that year, with Ethiopia, Australia, and the UK following suit (FAOSTAT, 2022). Australia stands out as the primary global exporter of faba beans (FAOSTAT, 2022). Within India, major faba bean cultivation centres include Madhya Pradesh, Odisha, Jharkhand, and Uttar Pradesh, with peri-urban cultivation being a common practice in the country (Singh, 2012).

Climatic Requirement

Faba bean, a cool-season vegetable crop, thrives in moist environments and performs exceptionally well during winter under cool growing conditions. It is crucial to avoid regions with significant temperature fluctuations. In subtropical and warm temperate areas, it is cultivated as a winter annual, while in tropical regions, it is grown at higher altitudes. The optimal temperature

range for flowering and pod formation is between 15-20°C. When temperatures exceed 27°C, the flowers tend to abort, and hot, dry conditions adversely impact seed setting during pod formation. Faba bean is considered to be a legume vegetable with low drought resistance (Subash et al., 2012). The crop's moisture requirement is highest between 9-12 weeks after establishment. Faba beans can be cultivated in dry areas where the total annual rainfall exceeds 400 mm, although well-distributed rainfall ranging from 650-1000 mm annually is ideal for its cultivation.

Table 1. Nutritional composition of immature and mature seeds of faba bean (per 100 g)

Nutrient	Quantity	
	Immature Seeds	Mature seeds
Water (%)	81	10.98
Protein (%)	5.6	26.12
Total lipid (%)	0.6	1.53
Ash (%)	1.1	3.08
Carbohydrates (%)	11.7	58.29
Total dietary fiber (%)	4.2	25
Calcium (mg)	22	103
Iron (mg)	1.9	6.7
Magnesium (mg)	38	192
Phosphorus (mg)	95	421
Potassium (mg)	250	1062
Sodium (mg)	50	13
Zinc (mg)	0.58	3.14
Total ascorbic acid(mg)	33	1.4
Niacin (mg)	1.5	2.832
Folate (µg)	96	423
Vitamin A (IU)	350	53
Phylloquinone (µg)	-	9

Source: USDA (2021).

Soil

Faba beans can thrive in a wide range of soils and flourish in fine-textured soils (Jensen et al., 2010). The optimal pH is 6.5-7.5 for successful cultivation of faba bean. In regions with high precipitation, acidic soils tend to form, necessitating liming when the soil pH drops below 6. Although faba beans can

grow in sandy loam soils, these require more frequent irrigation. Faba beans have shallow roots, making them vulnerable to drought stress in rapidly drying soils. However, these plants can tolerate brief periods of water-logging due to their shallow root system (Tekalign et al., 2016). It is crucial not to cultivate faba beans repeatedly in the same field. Crop rotation, particularly with cereal crops, is essential to minimize insect infestation and the occurrence of soil-borne diseases.

Cultivars

For a particular region to produce high-quality faba bean yields, the appropriate variety selection is crucial. Climate, cultural customs, and intended application are all considered throughout the selection process (Zandvakili et al., 2018). Based on the size of the seeds, faba bean varieties can be broadly divided into two groups: those with little seeds (*Vicia faba* L. subsp. minor) and those with medium-to-large seeds (*Vicia faba* L. subsp. minor). Medium- to large-seeded faba beans, also known as broad beans, are used for both green and dried beans, whereas small-seeded varieties are frequently used for animal feed and cover crops (Crépon et al., 2010). These cultivars can withstand temperatures as high as 35°C. As a result of their introduction, faba bean yields in Sudan increased by a significant 40%, and local farmers began to favour them. In 2011, Egypt released the drought-tolerant cultivar Nubaria 3 (ICARDA, 2018). The Windsor cultivar predominates in the faba bean cultivation area in the New England region (Etemadi et al., 2017).

Propagation

Faba beans are propagated commercially through seeds and are sown twice a year, depending on the specific climatic conditions of the region. Planting is typically done once between February and March in hilly areas and again between September and October in plains. It is crucial to avoid late sowing as it significantly reduces the yield potential of faba beans. The quantity of seeds required for cultivation typically ranges from 70-100 kg/ha and varies according to their size.

Seed Treatment

Faba bean, a nitrogen-fixing leguminous crop, can fulfil all its nitrogen needs through biological nitrogen fixation. It is advisable to inoculate the seeds before sowing to enhance nitrogen fixation efficiency. The recommended inoculant for faba bean seeds is the bacteria *Rhizobium leguminosarum* cv. Viciae (Abd-Alla et al., 2014). Seed inoculation will be carried out before planting, as suggested by Etemadi et al. (2015). It is important to note that fungicides should not be used after treating the seeds with *Rhizobium* to preserve the effectiveness of the inoculation process.

Spacing

In India, faba bean seeds are typically sown in shallow furrows measuring 15 cm in width, with a spacing of 75 cm between furrows. Within each furrow, two rows of seeds are sown in a zigzag pattern, with a spacing of 25 cm between seeds. Alternatively, a single-row system with a spacing of 45 x 15 cm can be employed. In regions with shorter growing seasons, like the Northeastern US, transplanting faba beans instead of direct seeding can be advantageous. Transplanting allows for early sowing and double cropping opportunities and helps prevent diseases like chocolate spot (*Botrytis fabae*) (Etemadi et al., 2018b). This method offers benefits such as higher survival rates, early maturity, and increased yield (Lee et al., 2018).

To calculate the seed rate, the following formula is used:

Seed rate (kg/ha) = (Plant Density (plants per square meter) × 100×seed weight (g)×10)/Percent germination.

Faba bean seeds typically germinate 10-15 days after sowing. To expedite germination, it is recommended to soak the seeds in water overnight before planting, as soaked seeds tend to germinate faster than untreated ones. Dry seeds are fragile and can split easily during handling, making them highly susceptible to mechanical damage during transportation. Therefore, it's crucial to handle the seeds with care. Using mulch during sowing enhances germination rates.

Nutritional Management

Faba bean, being a legume crop, requires a lesser amount of nitrogen compared to other crops. The application of exogenous fertilizer depends on factors such as soil fertility, genotype, environmental conditions, and cultural practices. It is advisable to conduct soil testing and analysis before determining the specific fertilizer dose for faba beans in a particular growing condition. During field preparation, incorporating Farmyard Manure at a rate of 10 quintals per hectare between the 2nd and 3rd ploughing is recommended. At the time of sowing, it is advisable to apply 15–20 kg of nitrogen, 40–50 kg of phosphorus, and 30–40 kg of potassium per hectare. An additional dose of 10-20 kg of nitrogen per hectare may be top-dressed after 30-40 days of sowing (Shukla, 2023).

Water Management

Faba bean is recognized as a legume vegetable with low drought resistance (Subash et al., 2012). Its highest moisture requirement occurs between the 9th and 12th weeks after crop establishment. Although it can be cultivated in regions with annual rainfall exceeding 400 mm, ample moisture is crucial, especially during seed germination. The initial irrigation should be administered immediately after sowing, followed by a light irrigation on the third day. Subsequent light irrigations should be carried out every 12-15 days. The water requirement varies based on weather conditions, genotype, and cultural practices. To conserve water, mulches have gained popularity in arid regions. It is advisable to test the quality of irrigation water before application.

Intercultural Operations

Weed Management

Regular weeding is necessary to keep the field free from weeds. Weeds compete with the main crop for nutrients and water. Poor growth and yield of the crop are associated with severe weed infestation. Weeding should be carried out either manually or mechanically, i.e., hoeing.

Stacking and Pinching

Faba beans, particularly tall varieties, need support, which may be given with wooden sticks or twigs. Staking protects against high-speed winds. Bamboo stakes are driven into the soil at an interval of 1 meter above the plants on either side of the double rows, and then, the plants are tied around the stakes with plastic string at 30 to 60 cm above the ground. Side shoots are removed at regular intervals to keep the plants upright. When five or six flowers are obtained, the growth points are pinched off at the top of the stems. This operation encourages pod formation and discourages aphid infestation.

Harvesting

Harvesting faba beans at the right time ensures a high-quality yield. Manual harvesting is a common practice for obtaining beans. Any remaining pods on the plant should be left to dry after reaching maturity. These dried pods are then processed to extract the beans. The maturity period of faba beans varies based on the specific genotype, environmental conditions, and climate. Generally, for spring sowing, pods are ready for harvest within 90-120 days, while for autumn sowing, it takes 180-210 days. The ideal time to harvest is when approximately two-thirds of the pods on the plant have turned brown. At this stage, the water content in the plant is around 77.4%, and in the pod, it's about 61.8%, signifying the late "milky-maturity" stage, as described by Lia-Juan et al. in 1993. During maturity, the lower leaves darken and fall off, and the pods at the bottom of the plant dry up first, gradually progressing to the top. After harvesting, it's recommended to incorporate the plant biomass back into the soil. This process helps capture fixed nitrogen, benefiting future plant growth.

Yield

Factors like genotype, cultural habits, and the growing environment all affect the production of faba beans. Well-managed production techniques can produce harvests between 70 and 100 quintals of green pods and 18 to 30 quintals of dry shell beans per hectare, depending on the specific growing season.

Frost Injury

Frost occurring during the early pod-filling stage in faba beans can lead to discolouration and deformation of pods and seeds. Subsequently, these affected pods and seeds have a water-soaked, rubbery, and wilted appearance. Pods at the top of the plant are more vulnerable to frost damage than those developing closer to the ground. Therefore, it's advisable to harvest the top green pods sooner, especially for the fresh market, to avoid frost-related issues.

Diseases

Chocolate Spot

This is the most common disease in faba beans and is caused by fungi like *Botrytis fabae* and *Botrytis cinerea*. According to Torres et al., in 2004, this disease primarily affects crops that were sown late and has a considerable negative influence on the foliage of the plant. On the leaves, the disease appears as tiny, rounded, and reddish-brown dots. These patches rapidly spread, turn a deeper brown, and cause leaf shrivelling under favourable circumstances, such as extended periods of high humidity and high temperatures. Either contaminated seeds or plant waste mixed into the soil are the two main ways for disease spread.

Integrated Disease Management of Chocolate Spot

- Crop rotation practice with non-host crops and avoid buckwheat in the rotation.
- Sowing of the crop at the recommended time.
- Seeds should be free from the spores of this disease.
- Follow optimum spacing to ensure adequate airflow around the plants.
- Avoid its cultivation in damp and humid sites.
- Destroy infected plant debris at the end of the crop season.
- Spray the crop with carbendazim, benomyl, or mancozeb 0.2% at 10-day intervals.

Broad Bean Rust (Uromyces Viciae Fabae)

It is a fungal disease that affects both leaves and stems. It has symptoms like minute, dusty, dark brown spots frequently encircled by a pale-yellow halo. Defoliation may happen in extreme circumstances. To guarantee optimal air circulation around the plants, proper control procedures include maintaining adequate spacing between crops. This disease can be prevented by using disease-free sowing materials, rotating crops, and treating the soil.

Powdery Mildew

The fungus *Erysiphe cichoracearum* causes powdery mildew and feeds on leftover plant material or other hosts. It first appears as yellow spots on the upper leaf surface before spreading to cover the entire plant in powdery grey-white patches. In severe circumstances, the plant may appear light blue or grey. The disease, especially harmful to faba bean plants, flourishes in warm, dry conditions with cool nights.

Integrated Disease Management of Powdery Mildew

The steps followed in the management of Chocolate Spot are same, with the following specific measures.

1. Properly dispose of the infected plant debris at the end of the crop season to prevent disease spread.
2. In severe infestations, consider the application of sulfur-based fungicides for effective control (Van Emden et al., 1988).

Fusarium Root Rot

The fungus *Fusarium solani* is responsible for causing fusarium root rot, which is particularly problematic in warm, densely populated soils with little moisture and few nutrients. Stunted plant growth, yellowing, and lower leaf necrosis are a few symptoms. In the study by Abdel-Kader et al. in 2011, it was discovered that these lesions may even spread above the soil surface.

To control *Fusarium root* rot:

1. Implement crop rotation with non-leguminous plants.
2. Well-draining soils during cultivation is effective in preventing excess moisture.
3. Treat seeds with Emison, captan, foltaf, captafol, or bavistin at a rate of 2.5–3.0 g/kg of seeds.

Seed Leaf Blight

According to Belete and Bastas (2017), the bacteria *Xanthomonas campestris*, also called *Xanthomonas axonopodis*, which causes seed leaf blight, can persist in agricultural waste during the winter and spread through contaminated seeds. This disease can develop more easily in warm, muggy climates. Water-soaked leaf regions gradually turn necrotic, and they frequently have yellow discolouration around them. Defoliation is rare, but if enough of these spots come together, the plant could seem to look burned. Additionally, in humid conditions, pods can get rounded, reddish-brown lesions that leak moisture also appear. Effective management strategies include using uncontaminated seeds, growing resistant varieties, and treating seeds before sowing with the appropriate antibiotics. Plants should be sprayed with some fungicides, which are copper-based before symptoms manifest (Tar'an et al., 2001; Gillard et al., 2009; Buruchara et al., 2010; Omafra, 2009).

Bacterial brown spot: Caused by *Pseudomonas syringae*, spreads through contaminated crop waste. It has symptoms like small, dark-brown necrotic patches that are frequently encircled by yellow tissue that appear on leaves. This disease also impacts pods; water-soaked patches develop on them before turning brown. According to Schwartz et al. (2005), these patches get necrotic and cause twisting and deformation in the surrounding area. According to suggestions from Harveson and Schwartz (2007), it is best to use uncontaminated seeds, carry out correct crop rotation, and remove crop leftovers from the field after harvest to stop the disease's development.

Conclusion

Improved soil health, increasing yields, and promoting sustainable farming depend on improved agronomic techniques in faba bean agriculture. This analysis highlights the importance of personalized nutrient management,

strategic crop rotation, and ideal seeding rates to maximise nitrogen fixation and encourage robust development. Stable yields under various circumstances are maintained through efficient irrigation, disease and pest management, and ideal planting density, all enhancing resistance to environmental challenges. Adding faba beans to crop rotations improves soil quality by fixing nitrogen and reduces the need for artificial fertilizers, which promotes environmental sustainability. Even though faba beans have a lot of potential as a core legume crop, more study and flexible approaches are required to optimize these methods in various ecological zones.

References

Abd-Alla, M. H., El-Enany, A. W. E., Nafady, N. A., Khalaf, D. M. and Morsy, F. M., 2014. Synergistic interaction of Rhizobium leguminosarum bv. viciae and arbuscular mycorrhizal fungi as plant growth promoting biofertilizers for faba bean (*Vicia faba* L.) in alkaline soil. *Microbiol Res,* 169: 49–58.

Abdel-Kader, M., El-Mougy, N. and Lashin, S., 2011. Essential oils and Trichoderma harzianum are an integrated control measure against faba bean root rot pathogens. *J Plant Protect Res,* 51: 306–313.

Anonymous, 2023. https://ecoursesonline.iasri.res.in/mod/page/view.php?id=101588.

Belete, T. and Bastas, K. K., 2017. Common Bacterial Blight (Xanthomonas axonopodis pv. phaseoli) of beans with a special focus on Ethiopian condition. *J Plant Pathol Microbiol*, 8: 2.

Buruchara, R, Mukankusi, C. and Ampofo, K., 2010. Bean disease and pest identification and management. *Handbooks For Small-Scale Seed Producers International Center For Tropical Agriculture.* (CIAT), Pan-Africa Bean Research Alliance (PABRA), Uganda, Kampala.

Crépon, K., Marget P, Peyronnet C, Carrouee B, Arese P, Duc G (2010). Nutritional value of faba bean (*Vicia faba* L.) seeds for feed and food. *Field Crops Res* 115:329–339.

Etemadi, F., Hashemi, M., Mangan F. and Weis, S., 2015. *Faba beans; Growers guide in New England.* http://ag.umass.edu/sites/ag.umass.edu/files/researchreports/faba_bean_guide_2.pdf .

Etemadi, F., Hashemi, M., Shureshjani, R. A., Autio and W. R., 2017. Application of data envelopment analysis to assess performance efficiency of eight faba bean varieties. *Agron J*, 109: 1225–1231.

Etemadi, F., Hashemi, M., Zandvakili, O., Dolatabadian, A. and Sadeghpour, A., 2018a. Nitrogen contribution from winter-killed faba bean. Agronomy, Nutritional Value, and Medicinal Application of Faba Bean (*Vicia faba* L.) cover crop to spring-sown sweet corn in conventional and no-till systems. *Agronomy Journal*, 110: 455–462.

Etemadi, F., Hashemi, M., Zandvakili, O. and Mangan, F., 2018b. Phenology yield and growth pattern of faba bean varieties. *Int J Plant Prod*, 1–8.

Fageria, M. S., Dhaka, R. S. and Choudhary, B. S., 2022. Beans. In book *Vegetable Crops Production Technology* (Vol.-2). Kalyani Publisher.

FAOSTAT (2022) Statistics database of the food and agriculture organization of the United Nations. Available at: *http://www.fao.org/statistics/databases/en/.*

Galloway, J. N., Dentener, F. J., Caone, D. G., Boyer, E. W., Howarth, R. W., Seitzinger, S. P., Asner, G. P., Cleveland, C. C., Green, P. A., Holland, E. A., Karl, D. M., Michaels, A. F., Porter, J. H., Townsend, A. R. and Voosmarty, C. J., 2004. Nitrogen cycles: past, present, and future. *Biogeochemistry*, 70: 153–226.

Gillard, C. L., Conner, R. L., Howard, R. J., Pauls, K. P., Shaw, L. and Taran, B., 2009. The performance of dry bean cultivars with and without common bacterial blight resistance in field studies across Canada. *Can J Plant Sci,* 89: 405–416.

Harveson, R. M. and Schwartz, H. F., 2007. Bacterial diseases of dry edible beans in the central high plains. *Plant Health Prog* doi:10.1094/PHP-2007-0125-01-DG.

ICARDA. https://www.icarda.org/crop/fabaBeans. July/10/2018.

Jensen, E. S., Peoples, M. B. and Hauggaard-Nielsen, H., 2010. Faba bean in cropping systems. *Field Crops Research*, 115: 203–216.

Lee, M. K., Kim, D. G., Kim, J. M., Ryu, J., Eom, S. H., Hong, M. J. and Kwon, S. J. (2018). Selection of the elite lines with high adaptability to autumn sowing of South Korea in faba bean. *Plant breeding and biotechnology*, *6*(1), 57-64.

Mekkei, M. E., 2014. Effect of intra-row spacing and seed size on yield and seed quality of faba bean (*Vicia faba* L.). *International Journal of Agriculture Crop Sciences*, 7: 665–6.

OMAFRA, 2009. Diseases of Field crops. Edible beans. *Agronomy Guide for Field Crops* Publication: 811.

Rana, M. K. and Vilas, C. A. 2017. Broad Bean. In book: *Vegetable Crop Science* edited by Rana, M. K., CRC Press. Tylor and Francis Group. doi: 10.1201/9781315116204-75

Rashid, M. I., Mujawar, L. H., Shahzad, T., Almeelbi, T., Ismail, I. M. and Oves, M., 2016. Bacteria and fungi can contribute to nutrient bioavailability and aggregate formation in degraded soils. *Microbiol Research*, 183: 26–41.

Schwartz, H. F., Steadman, J. R., Hall, R. and Forster, R. L., 2005. Compendium of Bean Diseases, Second ed APS Press, CO.

Shukla, Y R (2023). https://ecoursesonline.iasri.res.in/mod/page/view.php?id=101588.

Singh, A. K., Bharti, R. C., Chandra, N. and Dimree, S., 2012. Effect of Seed Size and Seeding depth on Fava Bean (*Vicia fava* L.) productivity. *Environment and Ecology*, 28(3A): 1722-1727.

Singh, A. K., Bharti, R. C., Manibhushan, N. C. and Pedpati, Anitha., 2013. An assessment of faba bean (*Vicia faba* L.) current status and future prospect. *African Journal of Agricultural Research*, 8(50): 6634-6641.

Subash, N., Priya, N. and Harish, Abdul., 2012. Climatic Requirements for Optimizing Faba Bean (*Vicia faba*.L) Production.n book: Faba Bean(*Vicia faba* L.) *A Potential Leguminous Crop of India* by A. K. Singh and B. P. Bhatt. (Ed) Publisher: ICAR Research Complex for Eastern Region, Patna.

Tar'an, B., Michaels, T. E. and Pauls, K. P., 2001. Mapping genetic factors affecting the reaction to Xanthomonas axonopodis pv. phaseoli in Phaseolus vulgaris L. under field conditions. *Genome,* 44: 1046– 1056.

Tekalign, A., Derera, J., Sibiya, J., Fikre and A., 2016. Participatory assessment of production threats, farmers' desired traits and selection criteria of faba bean (*Vicia faba* L.) varieties: opportunities for faba bean breeding in Ethiopia. *Indian Journal of Agricultural Research,* 50: 295–302.

Torres, A. M., Roman, B., Avila, C. M., Satovic, Z., Rubiales, D., Sillero, J. C., Cubero, J. I. and Moreno, M. T., 2004. Faba bean breeding for resistance against biotic stresses: towards application of marker technology. *Euphytica*, 147: 67–80.

USDA (US Dept of Agriculture). (2021). *Food Data Central* (Nutrient Database). https://fdc.nal.usda.gov.

Van Emden, H. F., Ball, S. L. and Rao, M. R., 1988. Pest, disease and weed problems in pea, lentil, faba bean and chickpea, World crops: Cool season Food Legumes. Springer, *Dordrecht:* 519–534.

Zandvakili, O. R., Barker, A. V., Hashemi, M., Etemadi, F. and Autio, W. R., 2018. Comparisons of commercial organic and chemical fertilizer solutions on growth and composition of lettuce. *Journal of Plant Nutrition* doi: 10.1080/01904167.2019. 1589505.

Chapter 9

Insect Pest and Disease Management in Faba Beans

Showket Ahmad Dar[1,*]
Mohamed A. M. Iesa[2,†]
Kounser Javeed[3,‡]
Mohamed Morsi M. Ahmed[4,§]
Mohammed Tarique[5,‖]
Rizwan Rashid[6]
Wajid Hasan[7,¶]
and Yendrembam K. Devi[8,#]

[1]Faculty of Forestry -Social and Basic Sciences, (Entomology)-SKUAST, Kashmir, India
[2]Department of Physiology, Al Qunfudah Medical College, Umm Al Qura University, Mecca, Saudi Arabia
[3]Faculty of Horticulture FOA Wadura Sopore, Sher-e-Kashmir University of Agricultural Sciences and Technology of Kashmir, Srinagar, India
[4]Department of Biological Sciences, Faculty of Science, King Abdulaziz University, Jeddah, Saudi Arabia
[5]Almanac Life Science India, Okhla Village, Jamia Nagar, New Delhi, India
[6]Department of Vegetable Sciences, SKAUST, Kashmir, India

---

[*] Corresponding Author's Email: drshowketshameen@gmail.com
[†] Corresponding Author's Email: mmmahmed6@yahoo.ca
[‡] Corresponding Author's Email: kousarjavaid11@gmail.com
[§] Corresponding Author's Email: mmmahmed6@yahoo.ca
[‖] Corresponding Author's Email: tariqueaiims@gmail.com
[¶] Corresponding Author's Email: entowajid@gmail.com
[#] Corresponding Author's Email: krishna.23363@lpu.co.in

In: Faba Beans
Editors: Aamir Raina and Samiullah Khan
ISBN: 979-8-89530-447-1

[7]Subject Matter Specialist Entomology, Krishi-Vigyan Kendra, Bihar 16 Agricultural University, Bihar, India
[8]Department of Entomology, Lovely Professional University, Punjab India

Abbreviations

IPM	Integrated Pest Management
DL	Doppler Lidar {An active remote-sensing instrument which provides long range and time resolved information using radial velocity and aerosol backscatter}.
FAO	Food and Agriculture Organization (Roma Lazio Italy)
ETL	Economic Thresholds Level
UK	United Kingdom
ICARDA	International Centre for Agricultural Research in the Dry Areas

Abstract

Integrated pest and disease management (IPM) comprises diverse approaches to reduce crop losses caused by insect pests and diseases in agriculture. IPM is a universal approach but unlikely to be very efficient as every agro-climatic region growing faba bean crops has a unique and different complex of insect pests and diseases. Faba bean is commonly attacked by arthropod, nematode, fungal, and broomrape pests. The most damaging pests of faba beans are efficiently prevented by applying various agronomic, chemical, host plant resistance, resistance breeding and cultural techniques in a sustainable manner. The most recommended cultural practices are sanitation, clean cultivation and crop rotation. In addition to above mentioned management practices the disease management in faba bean is also controlled by using synthetic fungicides applied at prescribed dosages to avoid environmental degradation. Further, based on data obtained from forecasting and monitoring techniques, millimetre-wavelength radar and Doppler Lidar (DL), the management of insect pests become more efficient, quick and profitable. World agriculturists have designed several models for faba bean insect and disease management, but all are in their infancy, and therefore, the problem of crop loss persists. The elevated prices of synthetic chemicals, labour charges and environmental hazards have further restricted effective management methods. The biological method of insect pests and disease control has attained some success, but due to quick and

overnight crop damage caused by many agents, the efficiency of biological control has become less. The other techniques employed in pest management include breeding for genetic resistance, introduction of transgenic crops and biorational control measures.

Keywords: faba bean, insects, diseases, doppler lidar, millimetre wavelength, Fabaceae

Introduction

Integrated pest management (IPM) is a broad approach introduced by R. F. Smith, and R. van den Bosch (1967) and aims at the integration of various control strategies for economic control of insect pests, reducing the losses in yield as well as preventing unnecessary expenditure on crop production inputs. The faba bean is dominantly used as a vegetable worldwide and, in a culinary sense, is also known as the fava bean. It is a vetch species, and like other species, it is also a flowering plant from the pea and bean family. The faba bean is commonly used for human consumption and works as a cover and shade crop in kitchen gardens Abdelwahd et al. 2008). The profitable crop production of faba beans depends on various biotic and abiotic factors that hinder crop yield quality and quantity. Despite various biotic factors such as fungi, weeds, nematodes, insects, mites, and vertebrate pests, crop production depends on integration and cumulative effects of genetic resistances, hygienic environment, monitoring of pests, and scheduled and timely application of various chemicals as well as biological natural enemies (Akpinar et al., 2001).

Among various biotic stresses, the impact of arthropods on faba beans is highly significant, and their management is of high importance. The management of insect pests in faba beans is still overly reliant on broad-spectrum insecticides. Among various insect pests attacking faba beans, the aphids were found dominant, and various cultural practices such as field assessment, tillage, date of sowing, plant population, application of NPK fertilizers and weed control were found very useful to control as well as minimize the insect pest damage (Almaghrabi 2014).

Normally, insect pest monitoring, population dynamics and prediction are done based on crop assessment using feeding baits, pheromone traps and economic thresholds. Further, various biological measures such as natural enemy conservation, bio-insecticides and selective synthetic chemicals are very useful in reducing the damage (Aly and Hattori 2007). The damage

assessment is mostly done from leaf materials, including succulent tissues, veins, midribs, per cent defoliation, specific portions damaged such as pods and fruits, notches, as well as cut areas from leaf margin. Under field and storage conditions, the maintenance of crop hygiene and seed demise is of huge significance in the management of broomrape since maximum damage is done before the parasite is visible on the soil surface (Anwar et al. 2011).

Herbicide treatment, either as spraying or dipping, prevents the ripening of broomrape seeds (English broom, Flex leaf broom and Montpellier broom) with the least damage to the faba bean crop. Most diseases spread rapidly when moisture conditions are conducive for the perpetuation of spores, especially rust, Ascochyta blight, Chocolate spot, Downy Mildews, Alternaria leaf spot, stem rot and many virus diseases. Each of the above diseases has specific conducive weather conditions suitable for pathogen growth and proliferation. For each disease attacking *Vicia faba,* there is a strategic management option using diverse but recommended fungicides at the time of disease threshold and possible risk (Bahgat et al., 2008).

Crop breeders and entomologists have found several varieties showing significant degrees of resistance to various key diseases under field conditions. Most of the cultivars were found resistant to more than one disease. In most of the susceptible varieties, root diseases, especially root lesions, stem rot and root rot by nematodes, have caused considerable yield loss; however, many resistant varieties have shown a notable level of resistance that has increased the total harvestable yield (Binding and Nehls (1978a; 1978b).

Several varieties had revealed good degrees of resistance to several root nematodes (root-lesion nematodes and root-knot nematodes) and stem nematodes and had been brought into breeding programmes to develop resistant varieties. The above-mentioned nematodes had a wide host range; therefore, the rotation of crops for their control is most recommended. For nematodes, chemical control is very tedious, inefficient and has various associated environmental concerns, while biological control was thought better and is under experimental trials (Borowska et al., 2003). For effective chemical control of various pathogens and nematodes, the inputs, especially the seeds, planting material, as well as fields and implements, must be neat and clean so that new crops must not get contaminated by debris, and volunteer and alternate hosts growing inside or outside the main field. In parallel to the continuous selection pressure on pests, the requirements of breeding for novel resistance genes, newer chemicals (neonicotinoid), biocontrol agents, as well as new designs and experimental methodologies will be important to save the yield loss (Böttinger et al., 2001).

Faba Bean Origin and Pest Management

The wild progenitors, relatives, and the origin of the faba bean are unknown. Faba bean hosts a diverse number of insect pests and diseases (fungal, bacterial, and viral) from regional, native to exotic areas (Caracuta et al., 2015) (Table 1). Further, the crop is weakened by various annual to perennial weed species (Canada thistle, perennial sow thistle, cow cockle, blue burr, cleavers and wild tomato). It is inherently associated with many other weeds from all geographical levels. Generally, there are three major types of faba bean, namely Mediterranean-adapted, cool temperate winter-adapted, and spring-adapted, all exposed to a range of biotic and abiotic production constraints. In tropical areas of the Mediterranean region and several cold places, the sowing of *Vicia faba* is performed in autumn conditions when pest activity and damage are least as compared to the spring season when the temperature is increasing and the losses are higher (Caracuta et al., 2016). *Vicia faba* is by nature a cool weather annual crop generally planted in early spring to fall when insect pests and disease chances are least. The sowing of faba beans from autumn till spring optimizes the use of various resources and inputs. However, the crop is exposed to multiple diseases and pests over a long period of time. Late autumn to early spring sown varieties of faba bean are directly subjected to insect pests and diseases from emergence, and very often, a different category of parasites and pathogens are involved in crop loss (Crépon et al., 2010).

The term Integrated Pest Management was put forward by Smith (1972), while FAO employed this as an inter-country program for developing and applying the IPM concept in south and southeast Asia in 1980. The United States of America in 1996 used a widely accepted definition as a sustainable approach to managing pests by combining biological, cultural, physical and chemical tools in a manner that minimizes economic, health and environmental risks (Croser et al., 2006).

Pest refers to any pathogen, insect or animal that harms the properties and accessories of economic interest to human beings. The basic and key aspect of pest management is the understanding of biology and the most susceptible stage of pests (Desouky et al., 2021). Understanding the biological stage helps to predict the outbreak threshold of damage and the time of application of control measures. The use of synthetic chemicals has played a significant role in the management of *Vicia faba* insect pests. Among all insect pests attacking the faba bean, the damage done by *Helicoverpa armigera* is most significant and chemical control is only a successful strategy to curb this pest below the

economic threshold damage level. The continued use of synthetic chemicals leads to the development of pesticide resistance and other associated environmental, health, biodiversity and pollinator issues, which demand a strong concern (Etemadi et al., 2018). Cultural practices are a regular routine in the management of insect pests. However, complete control is never achieved. Therefore, the integration of other control practices is essential. The clubbing of genetic resistance, crop hygiene and monitoring data helps in the determination of threshold levels of infestation; in other words, it gives the final decision on the effectiveness and economic feasibility of any control programme for the enhancement of overall crop growth (Etemadi et al., 2019). This publication, in the form of a book chapter, is aimed at an integrated management of various insect pests and major diseases caused by fungi and nematodes in faba bean.

Arthropods as Pests of Faba Bean

The *Vicia faba* is damaged by many insect pests from almost 70 species. The damage to the plant is caused at every growth and developmental stage (Bachmann et al., 2020). Among the most important insect pests attacking *Vicia faba*, aphids are considered an insect pest causing maximum damage worldwide (Barrett et al., 1976). Aphids having piercing and sucking mouth parts feed either by direct attack on the phloem of *Vicia faba* or work as vectors of several plant diseases especially caused by viruses, and do indirect damage. Blackman and Eastop (2000, 2007) reported 14 species of aphids that colonize the faba bean, and among them, three species are the most important, which cause huge damage to foliage. Besides damage by sucking and piercing of tissue, more than 50 plus diseases are vectored by aphids among plants causing significant damage. The management of aphids dominates all other pest management programmes in terms of expenditure and yield loss. Similarly, 50 and 21 insect pest species were found by Bardner (1983) and Van Emden et al. (1988), respectively. Nuessly et al. (2004) from Southern Florida recorded 31 families of insects belonging to 61 species comprised of 4 each stem and flower feeder, 1 root feeder, 15 foliage feeder and 13 pod feeders on *Vicia faba*. Among natural enemies' complex, 39 species of predators and parasites were found, while among flower feeders, nine insect species were also found feeding on pollen and nectar. Based on secondary sources of information about insect species interacting with faba bean, it is concluded that species complex

showed a good variation in their composition, unlike their groups and categories (Bartomeus et al., 2015; Fakhrai et al., 1989).

Table 1. The taxonomic position and status of insect pests identified from Faba beans field and storage

The taxonomic position and status of insect pests identified from stored faba beans in the study area					
Insect pests recorded	Common name	Order	Families	Status	Frequency*
Callosobruchus sp	Cowpea beetle	Coleopteran	Bruchidae	Major	92.59
Callosobruchus chinensis	Adzuki bean weevil	Coleopteran	Bruchidae	Major	86.42
Callosobruchus maculatus	Cowpea weevil	Coleopteran	Bruchidae	Major	80.25
Zabrotes subfasciatus	Mexican bean weevil	Coleopteran	Bruchidae	Major	67.90
Acanthoscelides obtectus	Common bean weevil	Coleopteran	Bruchidae	Major	74.07
Carpophilus sp	Sap beetle	Coleopteran	Bruchidae	Major	49.38
Carpophilus dimidatus	Corn sap weevil	Coleopteran	Nitidulidae	Major	43.21
Liposcelis sp	Book lice	Psocoptera	Nitidulidae	Major	24.69
Bruchus sp	Pea weevil	Coleopteran	Liposcelididae	Major	18.52
Bruchidius sp	Bruchid weevil	Coleopteran	Bruchidae	Major	12.35
Ants	Ants	Formicidae	Formica spp.	Major	15
Aphids (*Megoura spp.)*	Plant lice	Hemipteran	*Megoura crassicauda*	Major	78
Liriomyza sativae	Leaf minors	Dipteran	Liriomyza spp.	Major	56

Source: Wondale E and Berhanu 2020, *(% of samples containing each species).

Integrated Pest Management of *Vicia faba*

Cultural Control

The cultural control methods are used to make crops less suitable for insect pests, and in most cases, it is used as a preventive measure. In cultural control, various things are considered. The major factors are site and aspect selection, crop rotation, cultivar and seed selection, ploughing, tilling, seed depth, row-to-row spacing, plant-to-plant distance, plant density, soil condition, date of sowing and harvesting, irrigation and clearance of crop residues (Bates et al., 2020; Bishop et al., 2020). The geophysical locations and the physical environment surrounding the faba bean have a direct influence on its insect pest status, emergence and possible outbreak. Cultural control methods are very important in the management of aphids; for instance, sloping sites and boarder hedgerows lessen the wind speed and flow, resulting in higher aphid landing as well as distribution in the field (Bishop et al., 2016).

Faba bean showed poor growth response in stony calcareous soils, which are weaker in major nutrients (NPK). Therefore, the plant is stressed by poor-quality soil and demands nutrient-rich soil. Generally, plants growing in nutrient-poor soils are infested by field thrips, slugs and snails. It has been found that snails and slugs are dominant in nutrient-rich soils. From an insect pest management point of view, the various practices such as crop rotation, volunteer plant removal and weed management activities lessen the availability of alternate host plants that, in return, prevent pest damage (Cote et al., 2016).

The members of the order Hemiptera mostly suck in their feeding habits and feed on almost all tender plant parts. Aphids generally have long, thin legs and transparent wings. Aphids possess cornicles, which are abdominal tubes that assist in releasing cornical wax. In aphids, this fluid released works as a defence mechanism against various adverse conditions (Cuny et al., 2018). The dominant insect pest that attacks faba beans is *Aphis fabae* (Black bean aphid). It is predisposed to enter fields from the margins, with bigger fields seeing higher aphid colonisation rates than smaller ones. No cultivars are advantageously resistant to any insect pests that are widely marketed, even though it has been determined that 36 breeding lines are resistant to aphid feeding and damage. Seed Quality, yield, and disease resistance are the three main criteria for choosing cultivars (Epperlein, 1992). Due to the potent interactions between row spacing and either aphicides or broad-spectrum insecticides to control insect pests and insect-transmitted plant diseases that reduce production, various cultivars respond to plant density in different ways.

Normally, it is recommended that for getting pest and disease-free yields, even susceptible cultivars require fewer chemical inputs than cultivars of high resistance, such as Minden, which are better suitable under high pest incidence situations (Fisher et al., 2021). Pest populations, particularly aphid infestation and subsequent growth rate, can be significantly impacted by the planting dates. Crops sown in temperate regions too late in the spring or too early in the winter are typically more vulnerable to viral infections and aphid infestations than those sown in the more favourable summer periods. Newly emerged juvenile seedlings are vulnerable to harm from the feeding of aphids, which cause wilting and mortality. Early seedling maximises the potential yield in subtropical winter cropping systems, but in the case of summers in the temperate and tropical regions, if seedlings are exposed to early virus infection or late aphid attack, there would be enough yield loss. Late-planted crops may yield less due to considerably more vulnerability to early emerging insect pests and diseases. Pea pod borer *Helicoverpa* species generally attack beans at the pod stage, while aphids attack plants with bare ground surrounding them, besides crops being succulent and fields irrigated. Plants with higher densities in areas with less canopy are related to less aphid populations. It was shown that increasing plant density in coastal Syria from 22 plants/m^2 to 33 plants/m^2 decreased viral prevalence by 10–20%. Beans were planted on borders and ridges close to Faba bean fields; however, they had little effect on the propagation of continually transmitted viruses. Mulching with straw and mustard (*Sinapis alb*a L.) can reduce the number of aphids to 80% and 75%, respectively. While plants are young, escaping the earlier colonization phase is easy, and the crop is saved from yield damage by the mechanism that works as per the principle of Host Plant Resistance (HPR). Direct planting of beans into leftover crop stubble also lessens the aphid colonisation in lupines and the wireworm infection in sunflowers (Gagic et al., 2019; Gagic et al., 2016). Although, this consequence has not been demonstrated in faba beans. It has been shown that utilising press wheels to compress the soil surrounding the seed during sowing reduces the destruction and loss of seed from pests like wireworms and weevils (Garcia and Eubanks 2019). In Australia, where kitchen farming is normal, a technique known as "pupae busting" was performed by superficial soil disturbance, a technique widely used to control the proliferation of *Helicoverpa* species. Pupae busting is suggested when more delicate crops, such as Faba bean, are alternated with cotton, and the results showed less susceptibility of the crop to *Helicoverpa armigera* attack. However, it is resistant to several pesticides.

Techniques of Chemical and Biological Control

Integrated pest management in faba beans still significantly relies on broad-spectrum insecticides, which are frequently used preventatively, especially to control aphids. Due to their low cost, necessity and cost of pest monitoring, there are many methods for reducing the usage of broad-spectrum insecticides (Garibaldi et al., 2018; Heinze, 2020). The core of IPM systems is the economic threshold level (ETL), which lets growers determine whether a control measure is necessary or not. Prevention is needless under heavy infestation, while the blind spraying of chemicals is harmful, and the goal of IPM is to enable the protection of pollinators, predators, and parasites in the environment (Jaquiery and Keller 1978). The timing and choice of insecticide treatments must be carefully chosen because the faba bean depends on bee pollination. In some faba bean cultivars, the economic threshold for *Aphis Fabae* was 5%, green vegetable bugs (1-2 m^2), *Helicoverpa* spp. (2-4 larvae/m^2 or 1 m^2 for human use), thrips (4-6 flowers), and mites >50/100 cm^2. Insect traps that employ insect sex attractants, pheromones and feeding baits, such as chicken mash to attract snails and seed baits for wireworms, are used to monitor pest frequency in the field (Järemo et al.1999). Aphid flights are monitored in the UK using a suction sampling network. Vegetable growers used this knowledge to aid in control decision-making. Strong aphicides like pirimicarb and pyrethroids control pests like cutworms and silvery moths. However, some other insect pests of faba bean were previously controlled by insecticides like carbamates and pyrethroids (Jensen et al., 2010; Karkanis et al., 2018)

Nevertheless, a new generation of more specific pesticides, such as indoxacarb and spinosad, as well as a number of biocontrol agents, have been created in accordance with the global shift towards IPM (Kendall and Smith 1975). These new, more targeted pesticides assist in minimising secondary pest outbreaks and cause less harm to beneficial insects. Indoxacarb and spinosad mode of attack is based on chemical poisoning, while biocontrol products such as Nucleo-polyhedroviruses (NPVs) utilising polyhedral inclusion bodies derived from *Helicoverpa armigera* in the laboratory, which are later employed for the control of *Helicoverpa* species in Australia (Krupnick and Weis 1999). In order to protect faba beans from Lepidopteran larvae such as *Helicoverpa* species, *Spodoptera armyworms*, *diamondback moth*, *Plutella xylostella* and *Chrysodeixis loopers*, faba beans are treated with *Bacillus thuringiensis* ssp. Kurstaki. Insecticides such as imidacloprid are relatively new and have a lot of potential for faba bean IPM programmes. Its

efficiency against broad bean weevil *Bruchus rufimanus* Boh, wireworms, thrips, and aphids is still being investigated (Lenth et al., 2021). Bean leaf roll virus, faba bean necrotic yellows virus, and soybean dwarf virus infection are successfully reduced by imidacloprid. Imidacloprid dramatically decreased leaf miner *Liriomyza huidobrensis* (Blanchard) populations when used at the pod stage as a foliar spray. However, it has adverse impacts in the form of lowering *Diglyphus isaea* (Walker) populations, a parasite of the leaf miner. Based on experiences, most faba bean growers employ cultural approaches, such as changing the planting date or plant density for the sustainable control of insect pests (López-Bellido et al., 2005). Nevertheless, broad-spectrum pesticides continue to be overused in faba bean IPM. Even though many of the cultural practices have been examined and applied in isolation and have been proven to be successful, their potential has rarely been shown. There are several ways to improve IPM in faba beans to reduce the need for pesticides, including contemporary selective insecticides and bio-insecticides (Lundin and Raderschall 2021).

Broomrapes

Faba beans suffer from reductions in total biomass due to broomrape parasitism (Fernandez-Aparicio 2016). The broomrape greatly impacts plant reproductive biology, thereby hampering the overall yield. The three most important broomrape species that have the ability to parasitize with plants are crenate broomrape, foetid broomrape, and Egyptian broomrape. Crenate broomrape has been known to harm agricultural yields of legumes since antiquity (Punia et al., 2018). The faba bean is commercially significant in the Mediterranean Basin and the Middle East, along with other grain and forage legumes (lentil, pea, vetches, grass pea), as well as plants from the Asteraceae and Apiaceae families, such as safflower and carrot. Large, upright plants that only branch from their laterals and tubercles are major characteristics of the faba bean (Rathore et al., 2014). The spikes may grow up to 1 m tall and are covered in numerous blooms with different floral colours, ranging from yellow to white to pink and violet. One of the important plant parasites of faba bean from Tunisia is *Orobanche foetid, which attacks the plant when it is widespread in its natural habitats in North Africa, Spain,* and France. Unbranched stalks of the plant are smothered in ugly red or purple blossoms. *Phelipanche aegyptiaca* parasitizes the faba bean, chickpea, and lentil. The attack of broomrape is widespread and invades several other plants from

various families, such as Asteraceae, Brassicaceae, Cucurbitaceae, Fabaceae, and Solanaceae (Sheoran et al., 2014).

Faba bean is broadly dispersed in the Middle East, eastern Mediterranean, and Asia. A broomrape plant in good health has the capacity to generate 200,000 seeds and, in rare instances, up to 500,000 seeds. These seeds are dormant in the soil until they detect a host root and begin to grow. The seedling adheres to the host root's surface and punctures the cortex, endodermis, and central cylinder with the help of mechanical pressure and enzyme activity (Alarcón et al., 2018). Over weeks, the parasite forms a tubercle on the surface of the host root, which eventually causes one (*Orobanche crenata or Orobanche foetida*) or numerous (*Phelipanche aegyptiaca*) blooming shoots to emerge from the ground and generate seeds. The tiny seeds are easily disseminated by farming practices, as well as by water, wind, animals, and particularly by cars and agricultural machinery. According to one study, there is no single approach that can be used to manage broomrape. Identifying the parasites before the crop is permanently damaged is cumbersome due to small and long-lasting seeds and seed complexity (Amanuel et al., 2000). The various conventional methods adopted have the least efficacy in the management of broomrape in field conditions. Herbicides can't effectively control parasites because of the close relationship between the host and parasite. Controlling these weedy root parasites necessitates an integrated strategy that includes sanitation, containment, direct and indirect defences against parasite harm, and, finally, eliminating parasite seed banks from the soil.

Agronomic Management Techniques

Though difficult, the faba bean broomrape resistance breeding has made significant progress (Ahmed et al., 2017). The selection of the *Orobanche crenata*-resistant line F402 was the first notable resistance finding (Nassib et al., 1982) in agricultural sciences. Tunisian Faba bean germplasm has been found to exhibit *Orobanche foetida* resistance. It has been proven that it is feasible to combine resistance in faba bean against *Orobanche crenata* by selecting *Orobanche foetida* resistance traits from germplasm resistant to *Orobanche crenata*. However, this genetic background of the resistance appears to be complex, but it is one feasible method by Orobanche to control yield loss. Resistance management of herbicide is being investigated in several crops as a potential method of broomrape management, especially in the recent

release of non-transgenic resistant sunflowers, which showed resistance to imidazolinone (Ashworth et al., 2014). Preventative measures are crucial, but it is possible to curtail infestation spread by limiting the use of infected seed lots, broomrape-fed cattle's farmyard dung, or even by simply eliminating badly afflicted crops. While other methods like soil solarization or fumigation effectively destroy seed banks, they are not economically appropriate for lower-value and lower-intake crops like faba beans. Despite the potential of several biological control agents in managing broomrape at the commercial scale, the technology is not yet appropriate to manage it below an economic threshold level (ETL) (Ashworth et al., 2018). Even though delayed sowing reduces yield potential, it is perhaps the most well-established conventional technique for *Orobanche crenata* management in Mediterranean nations (Avižienytė et al., 2013). Faba beans are generally extremely affected by weather factors, and according to some researchers, the crop sowing from October to November and December, or January, reduces the quantity and dry weight of faba beans. Even though the faba bean has a long development cycle, delayed and late planting implies a shorter grain-filling phase, which is unfavourable for overall crop production, especially in Mediterranean climates (Badagliacca et al., 2018). The best sowing times depend on the weather, cultivar, and pest density to produce the maximum yields. When an infection first appears in a field, manual weeding is only necessary to prevent plants from spreading and seed banks from growing further, but it is not cost-effective in industrialised agriculture (Buczek et al., 2021). Crop rotation has little significance because of the prolonged seed life and the vast range of hosts. Use of imidazolinone herbicide-resistant host oilseed rape (*Brassica napus* L.) and rotation with non-host cereals are two effective methods for reducing branching broomrape (*Orobanche ramosa* L.) in southern Australia (Confalone et al., 2010).

Allelopathy is generally an interspecific interaction and an important and effective strategy. Allelopathy is an economical and eco-friendly strategy for the control of weeds in various crops. The release of allelopathic compounds from leaves, seeds, flowers and bark influences weed growth and density. Allelopathy is a key factor in decreasing *Orobanche crenata* infection on faba beans in intercrops with cereals or fenugreek. Techniques of chemical and biological control have been well-known for a long for the management of weeds; however, in recent times, instead of high resistance development in weeds, glyphosate can still effectively control broomrape on faba beans (Daoui and Fatemi 2019).

Monitoring of underground vegetative stages of broomrape development is necessary because too-large parasitic attachments render management ineffective and result in high yield loss. Conversely, too early application of the herbicide results in inadequate control of the parasitic attachments (Dhull et al., 2021). Normal foliar herbicide treatment doses for controlling broomrape are lower than those for controlling autotrophic weeds. The imazapic or imazethapyr herbicides can successfully suppress the *Orobanche crenata* in faba beans when applied three times in succession to the leaves; however, these herbicides are currently restricted for use (Fatemi 2019). *Orobanche crenata* in faba bean has been successfully controlled by seed treatments with imidazolinones, resulting in the significant benefit of reduced application costs because the herbicide is built for the seed treatment (Du et al., 2019).

Seed treatment is environmentally friendly because it alters the pre-emergent germination of weed seeds, saves money on mechanical application expenses, and reduces the amount of herbicide needed at the frequency of two to three times per season (Hiywotu et al., 2023). The seed treatment therapy is supplemented in areas where the environment is particularly conducive to broomrape outbreaks to achieve effective parasite control. Although several microorganisms have been discovered and much literature has been published in the past, all efforts have failed to employ biocontrol methods extensively against broomrape species management. However, the biopesticides control technique is not yet ready for application on a broader scale and demands more research to be done to develop an appropriate method to handle, store, and use fungal propagules effectively (Giambalvo et al., 2012; Hou et al., 2013).

In the area where broomrape is present, the fly *Phytomyza orobanchia* is common and consumes a significant amount of seeds. However, in regions where broomrape infestations are severe, this natural infestation is insufficient (Janusauskaite et al., 2022). This effect of broom rape infestation can be managed by tremendous reproduction and inundative release of *Phytomyza orobanchia*. (Kimbirauskienė et al., 2023) The simulation of short- and long-term control methods and the consequences of many control measures in combination with the seed bank model supported the superiority of integrated pest control in faba bean (Mo et al., 2019). However, no design has a large impact on agricultural practices as there exist huge communication gaps and models that were simple for farmers to adopt but were found least effective. Phenological models that forecast host and parasite development in response to temperature seem more likely to be used in practical settings. A similar

approach may be created for faba beans after recently being used for Octopus minor that infects red clover (Lal 2019; Mínguez and Rubiales 2021).

Faba Bean Diseases and Their Management

Several pathogens attack the faba bean, but the most important among them are fungi (Rubiales and Khazaei, 2022; You et al., 2021). The faba bean is attacked by several diseases such as stem rot, root rot, wilt and Alternaria leaf spot. However, research does not include the extensive host ranges of stem rot, root rot, wilt, or Alternaria leaf spot (Yitayih and Azmeraw, 2017; Wang et al., 2021). Contrarily, in addition to faba bean, only a small number of host species are affected by Cercospora leaf spot, rust, chocolate spot, downy mildew, and ascochyta blight. One of the most detrimental ailments that affect faba beans is chocolate stain (Webb et al., 2016). *Botrytis Fabae* is a necrotrophic fungus in almost all faba bean farming locations. *Botrytis Fabae* is characterised by symptoms of dark-brown specks with orange-brown rings on leaves, petals, and stems (Watson et al., 2018; Warsame, 2021).

Under temperatures of 15-25°C and relative humidity above 80%, the limited-lesion stage appears, and it is followed by an aggressive phase in which necrosis spreads rapidly, defoliating the plant and finally killing it. Since environmental factors are frequently more favourable for disease development during flowering, while the decaying flower petals provide food for the fungus to grow, yield reductions may be significant as a result of the pathogen spreading from the blooms into the growing pods. The damage at the bloom stage is the most significant, and subsequent yield loss is the maximum (Villegas-Fernández et al., 2014; Varshney et al., 2021).

Conidiophores causing *Botrytis Fabae* create the spores, which are visible on the senescent leaves of the affected plants (Varshney et al., 2019). Due to the airborne transmission of spores, the disease spread swiftly. The infection spreads to fresh tissue, producing more conidiophores and beginning a new infection cycle. Whenever the humidity and temperature are favourable, conidiophores begin to proliferate. In agricultural waste, the fungus lives between growing seasons as mycelium or sclerotia. Additional host plant species include lentils, beans, and common vetch. Rarely *Botrytis fabae* infection to faba bean seeds is thought to be a significant factor in the onset of the disease. The disease, known as chocolate spot, sometimes shows symptoms similar to *Botrytis cinerea* (Tiwari et al., 2021). In moderate oceanic to Mediterranean climates where faba beans are produced year-round,

fungal diseases are most common and spread quickly in the early spring before the plant begins to bloom. The elongated lesions appear on stems and round lesions on leaves; however, leaves commonly show distinct bands of black pycnidia and grey colour. The elongated, somewhat curved, and splash-dispersed septate pycnidiospores are on infected parts. *Aphis fabae* populations might vary greatly, and research indicates that the presence of resistant races in populations is important for developing disease management systems (Ghanbari et al., 2020). The subsequent breeding programmes for disease resistance make it necessary to look for additional sources of resistance in other crops to incorporate it in faba bean against various adverse factors (Sudheesh et al., 2019).

The pathogens are subjected to and eventually adjust to the selective pressure brought on by the activation of host resistance genes. The sexually reproducing teleomorph *Didymella Fabae* may have a substantial impact on the disease cycle by making it easier for the fungus to thrive on infected agricultural waste, creating ascospores that may travel great distances by the wind, and fostering more genetic diversity in the pathogen. These factors could negatively affect chemical control methods and improve the yield by using resistance breeding programmes (Sillero et al., 2017; Shtaya et al., 2021).

Jellis and Punithalingam (1991) from the UK made the initial discovery of bacteria infecting faba bean. They later found it overwintering in straw, and similarly, several other nations have since then identified it. The bacteria Uromyces is responsible for the rust disease in faba bean, which affects crop yield and leads to huge economic losses worldwide. In Australia, some of North Africa, and the Middle East, rust is very dangerous and causes huge losses. Rust is hosted and brought on by the same species of common vetch and lentil. Because *Uromyces viciae-Fabae* is monoecious, it completes its life cycle on several host plants. Rust disease organisms such as *Viciae-Fabae, viciae-sativae,* and *lentil-culinaris* have been classified as "host specialised isolates" based on differences in host specialisation, spore size, and the form and size of the substomal vesicle (Seidenglanz and Huňady 2016; Satovic et al., 2013; O'Sullivan et al., 2019).

It has been proven twice that faba bean isolates have a racial gene structure exhibiting high virulence. Conner and Bernier (1982) classified six races of *Uromyces viciae-Fabae* isolates based on changes in colony size collected from nine faba bean lines in North America. The presence or absence of necrosis on six faba bean lines served as the determining factor for resistance and susceptibility of genotype (Mínguez and Rubiales 2020). In another

experiment, 23 isolates of Uromyces viciae-Fabae belonging to 16 races were identified based on the infection type. Generally, the above-ground portion of faba beans is impacted more by rust invasion. During the filling stage in late of the late growth season, the infection frequently starts and spreads from the lower parts of the plants to other parts (Mokhtar et al., 2020; Mobini et al., 2020). However, the mean infection has little effect on yield components, and losses generally range from 5 to 20%. Yet severe early-season outbreaks have been found to cause crop losses of up to 70%.

Uredial pustules look oval, brown in colour, and up to 1 mm in diameter and develop on both sides of leaflets, branches, stems, and pods. *Echinulate uredospores* have 3–4 germination holes or germ pores, measure 22–28 mm by 19–22 mm, and are carried singly on pedicels. On stems, the epidermis always covers the elongated, dark brown to black telia, even though spores emerge on leaves in the late season from beneath the epidermis (Mahmoud and Ab El-Fatah 2020). Globose to subglobose unicellular teliospores measure 25–40 mm and 18–26 mm, contain a single germination pore and are carried individually on pedicels. The main spore mass and pathogenic structure that survives between crop seasons is the teliospore, which is resistant to harsh environmental conditions (Maalouf et al., 2019; Lee et al., 2021; Lee et al., 2020).

In agricultural areas of most European and African countries under extended hot and dry summers, the predominant inoculum originates from infected crop debris. The teliospores are dispersed through seeds or mostly left in the field in crop debris (Laloum et al., 2021). Teliospores can remain dormant for up to two years and swiftly germinate between temperature ranges of 17 - 22 degrees Celsius. The uredospore may also be crucial to intercrop survival in cooler production zones. The inoculum was also found to be present in other hosts, especially on vetch. Aeciospores and uredospores are employed for secondary spreading because they both germinate in humid conditions on plant surfaces and are quickly disseminated by wind (Kumari et al., 2018).

The humid weather of temperatures around 17-22 degrees Celsius favours the development of the downy mildew disease. Although many diseases have been documented in most areas where faba beans are produced, the most destructive epidemics of the downy mildew caused by *Peronospora viciae* on faba beans occur in northwest Europe. According to several researchers, the primary cause of the infection is soil-borne oospores that germinate and infect the hypocotyls and upper regions of the faba bean roots. This leads to systemic colonisation of infectious spots on plant leaves and stems (Khazaei et al.,

2021). Humidity encourages sporulation and results in conidia-based cycles of secondary infection. On seedlings growing on bunds and borders of the main field, the downy mildew persists well and, in later stages, infects the main field. Further, in the areas where crops are sown in both winter and spring seasons, the occurrence of downy mildew is most frequent, resulting in high survival and spreading of disease in the later stages of the crop. Attention has been paid to the Cercospora leaf spot caused by Cercospora zonata to faba beans from late summer to winter in Australia, China, and Poland. Besides faba bean, it also infects vetch, beans and lentils. The disease is characterised by small circular spots and brown to grey lesions on leaves. The Ascochyta blight is characterised by black pycnidia and chocolate spots due to spore zonation. When the lesion is wet, clusters of conidiophores give it a hairy look, and the infection spreads to all fields, resulting in crop loss. Cercospora prefers the same growing circumstances as Ascochyta and is, therefore, more common throughout the vegetative stage of the crop cycle.

Agronomic Management Techniques

The agronomic practices are not always sufficient for disease control; besides this, the fungicides are expensive, and the only option left is through the application of host plant resistance. Researchers have used many assessment techniques, and the way in which faba bean collections respond to chocolate spots has been investigated. Due to the resistance being transferred into hybrids from local and wild cultivars, the International Centre for Agriculture Research in Dry Areas (ICARDA) has published new cultivars from several countries, including Ethiopia, Egypt, and Australia. However, the types of resistance that are now in place are inadequate, and more research is needed so that crop loss can be minimized. Instead of regions where faba beans have been grown for thousands of years, most resistant cultivars come from Colombia and Ecuador. According to many researchers, resistance to Ascochyta blight is conferred by both major and minor genes. Some of the genes have been localised to specific chromosomal regions. In many different countries, there are several cultivars with varied levels of disease resistance (Mahapatra et al., 2022).

Rust resistance is frequently and partially effective, and there is only limited resistance to the several infection types caused by many pathogens since the necrotic response proceeds slowly. After a longer latent period, a smaller colony and a lower incidence of infections are connected to a low level

of resistance, normally called partial resistance. The only difference between necrotic and partial resistance is whether or not there is macroscopically evident necrosis (Maurya et al., 2018).

According to Thomas and Kenyon (2004), downy mildew resistance levels are significant among faba bean cultivars, and from the UK, the bean cultivar Betty is regarded as resistant. In southern Australia, breeding for resistance to Cercospora leaf spot was conducted, and many cultivars were found resistant. Further, in addition to resistant cultivars, disease control through crop rotation was also found to be significant; however, the crop rotation period must be taken into account with respect to other host species growing in the surroundings (Papan et al., 2021). All of these diseases have been linked to sources of infection that include crop residue, crop stubble, and volunteer seedlings. For various diseases such as Cercospora leaf spot, rust, Ascochyta blight, and chocolate spot, the minimum gap is thought to be almost four years (Wettberg and Guerra-Garcia 2023). Given that oospores may persist in substantial numbers for at least 10 years and that this Oospore count may be extracted from soil using microbial extraction techniques, fields can still be indexed to forecast the risk of infection from the soil. However, to define sample schedules and possible infection levels, an in-depth study would be necessary. It is recommended that at least 500 meters apart from other fields is important, especially when some off-type plants and stubble are present on field bunds (Tantasawat et al., 2020).

The spread of Ascochyta blight is prevented using clean, perfect, and ideally certified seed since the disease inoculums are present beneath the seed coat. The spread of *Cercospora zonata, Uromyces viciae-Fabae, or Botrytis Fabae* through seeds is less likely to be caused by related plant material transported on the exterior. According to some views, the downy mildew infection may have suddenly developed in regions where bean production had never occurred. It may have been spread by tiny particles of infected plant debris sticking to the seed.

Chemical seed washing tests against downy mildew pathogens help to create disease-free seeds. The infection process can be reduced by lowering the relative humidity of seeds close to a minimum to promote aeration throughout the crop. This is achieved by decreasing the sowing rate (thereby reducing plant density but decreasing prospective yield in a disease-free situation) and excellent weed management. Bean Rust and chocolate spot are two important diseases that have been linked to the intercropping of faba beans with cereals (Zeleke and Legesse, 2023). This approach combines two types of benefits: first, because the grain does not act as a host for the pathogen, it

acts as a hurdle to the inoculi transmission; second, grain cover encourages aeration and prevents faba bean seedlings from developing an infection by thick mycelium mass. Each of these diseases, irrespective of the causal organism, prefers a certain range of temperatures (20-25 degrees Celsius), as well as high relative humidity and wet leaf surfaces. Ascochyta and Cercospora like low temperatures (5–18°C), and the rain splash can successfully spread Ascochyta blight. Leaf Rust and chocolate spot favour higher temperatures of range 15 to 25 and >28 degrees Celsius, respectively.

The time and duration of flowering play a role in downy mildew spread (Denekew et al., 2018). In faba beans seven days before the commencement of blooming, the disease was more likely to swiftly develop on leaves that had been damp for more than 12 hours at temperatures below 10.8 degrees Celsius. Therefore, the UK offers a service for monitoring bean downy mildew during the growing season, especially at flowering stages. To provide farmers with an early warning of disease development and enable successful planning of spraying on sensitive crops, monitoring is based on untreated observation crops in disease-prone areas. From Western Australia, a thorough guide outlining the spraying criteria and forecasting of Ascochyta blight, chocolate spot, and rust was created. Finally, any factor that reduces plant vigour encourages the growth of the disease (Firdu et al., 2022; Merga et al., 2019). Therefore, major and micronutrient deficits such as phosphorus, nitrogen, potassium, iron, zinc, sulphur and silicon should be applied for relevant deficiency symptoms. Nutrient deficiencies lead to increased susceptibility of the faba bean plants to various infestations. It is also important to maintain proper soil drainage to minimize waterlogging. The disease chocolate spots are more susceptible when there has been frost damage. Hence, it is important to choose the sowing date wisely (Gela et al., 2022).

Techniques for Chemical and Biological Control

Fungicides, which are available in the market, are used to manage faba bean diseases. However, some of them may not be legal to use in many countries due to various adverse reasons. Overall, the use of synthetic chemical fungicides for disease control in faba bean is small in overall quantity as the least number of diseases attack the faba bean crop. Of course, the Rhizobium inoculants associated with leguminous seeds or roots must be proper to facilitate nutrient fixation (Zhang et al., 2023). According to the literature, the

seed treatment with P-Pickel T1 (Thiram and Thiabendazole) is advised in Australia to prevent Pythium and Fusarium in faba beans as well as Ascochyta and Botrytis in other grain legumes. Wakil XL (cymoxanil, fludioxonil, and metalaxyl-M) is an efficient product used against downy mildew of peas in Europe. However, it is rarely used in faba bean disease control programmes because the maximum sowing of faba bean is often grown without any seed treatment. Spraying fungicides on leaves is the major method used to control disease. To prevent the spread of the disease chocolate spot, fungicides are often used in fields across the world, and they have been the focus of several researchers. Benzimidazoles, dicarboximides, dithiocarbamates, aromatics, conazoles and strobilurins are just a few of the substances that have been reported to be successful in preventing various foliar diseases in faba bean.

Most pathologists suggest the use of chlorothalonil and mancozeb for Ascochyta blight, while Cercospora leaf spot can be controlled with chlorothalonil and carbendazim, even though these chemicals were found to be limited in use due to adverse impacts on the environment. Mancozeb, chlorothalonil, copper, triadimefon and propineb are among the most important chemicals used for the treatment of rust. The repeated sprays of metalaxyl and chlorothalonil were found to be efficient against pathogens causing downy mildew (Zhou et al., 2022).

Fungicidal sprays are applied during the critical crop development period of early blooming. During early blooming in faba beans, the disease first became established and caused the highest production losses. Under suitable and favourable environmental conditions for disease development in faba beans, the second dose of fungicide is applied during mid-to-late flowering periods. The first spray is recommended seven days after the beginning of flowering, which prevents the losses and enhances the yield to 0.5t/ha. Even though most of the fungicides are very expensive, Metalaxyl-M is very efficient against downy mildew and is frequently required for cultivars vulnerable to it. However, due to the high cost and negative effects on the environment, it is recommended that fungicides should only be used under specific conditions and only in conjunction with other integrated disease management strategies (Wang et al., 2020).

Regional research experience from various agro-climatic zones determined that the early and late fungicidal application and the number of applications depend on weather characteristics. Each disease develops differently depending on the weather; thus, forecasts must be taken into consideration while applying pesticidal sprays (Snowdon et al., 2021). A direct monitoring process should be followed strictly to take into account the

disease symptoms seen on the crop in the field. The local and regional forecasts of the prospective crop production and protection measures and other agro-advisories provide valuable early inputs for effective disease management keys at an appropriate time. To effectively control disease without any harmful impact on the environment, a minimal concentration of fungicide and a lesser number of applications shall be used. This makes sense from an economic and environmental standpoint (Zhang et al., 2020). The waiting period for each herbicide between application and harvest must also be followed by producers, applicators and consumers. One of the most crucial factors considered while deciding on fungicide spraying under any circumstance is the incidence and severity of disease, the resistance development ability of pathogen and alternate hosts in surroundings. The resistances to several chemical fungicides, including benzimidazoles and dicarboximides, have already been documented by many researchers. As per chemical spray recommendations, a proper quantity is prescribed either singly or together with the proper dosage of active ingredients from various groups. As a potential substitute for traditional fungicides, various oils, especially basil oil (*Ocimum basilicum*), have been demonstrated to have antifungal activity against *Botrytis fabae* infected plants under both *in vitro* and *in vivo* conditions. *In vitro* experiments showed that chemical secretions from three cyanobacteria species impeded the development of various fungi, including *Botrytis fabae*. The medium dosage of pyrimidine glycosides @500 ng/ml was effective against *Botrytis cinerea* infecting faba bean *in vitro*. The greater concentrations of the same chemical also worked effectively against several other pathogenic fungi, causing many diseases in faba bean. In the future, biological control will have huge significance in disease and pest control, and several successful stories have already been noted and discovered; thus, research in pathology using potential microbes that are antagonistic to the plant pathogen has been carried out. However, the shift from the lab to the field is usually time-consuming, and it frequently fails to give satisfactory results. Efforts are going on to shift agriculture from synthetic chemical management to organic management. Plants infected with *Botrytis fabae* have increased resistance due to repeated use of ribavirin, citric, oxalic, benzoic, and salicylic acids. The creation of preventive medicines was suggested because all these substances are thought to be inducers of resistance. It has also been suggested that *Rhizobium leguminosarum* and vesicular-arbuscular mycorrhizal fungi can be used in tandem to induce resistance to *Botrytis fabae*, but it is still too early to predict which of these methods will finally find their way onto the agricultural market.

Plant Parasitic Nematodes of Faba Bean

Nematodes, also called roundworms, belong to phylum Nematoda (Nemathelminthes) and constitute a big complex of plant parasitic agents. Nematodes are diverse, inhibiting a wider range of ecosystems. Generally, nematodes feed on almost all parts of plants, including roots, stems, leaves, flowers and seeds. The nematodes damage plants of the family Fabaceae in several ways. More specifically, they use morphologically a specialized spear-like structure called a stylet. Different species vary in their stylet length and shape. The stylet morphological features, such as its size and shape, are used to classify nematodes into several genera and further categorise them according to their mode of damage. The most dominant ectoparasitic nematodes, namely Belonolaimus and Longidorus, use stylet for feeding inside of roots, while Helicotylenchus feeds on the outside of roots and later burrows into the deep of root to continue feeding. Using stylet, the nematodes suck out the cell content and result in cell death. Once this style of feeding occurs, the subsequent symptoms which are visible in plants include large lesions formed on the plant surface. Few species of nematodes don't kill plants immediately but trick the plant cells into voluminous growth, producing nutrient-rich tissue masses for fresh nematode proliferation. It is a fact that nematodes feed on almost all plant parts and, therefore, produce several feeding cell types. Nematodes feed on more than one cell, and the number varies as per nematode species.

The genera Ditylenchus, Meloidogyne, and Pratylenchus contain nematodes most frequently observed to harm faba beans. About 30 taxa make up the Ditylenchus genus complex, which affects more than 500 host species. Two of these, traditionally known as the oat race and the gigantic race, impact faba beans. *Ditylenchus dipsaci* does attack the faba bean occasionally, but reports of the former "Gall nematode race" are considerably more frequent and wider in the host range. The plant stems may enlarge and deform, and plant components may get discoloured. The nematodes may penetrate the growing seed coat, causing killing, wilting, weakening and discolouring seedlings. Cultivars resistant to the "Gall race" have been discovered at ICARDA and in France, but Morocco is home to an especially virulent genotype of the nematode that has surpassed the highest resistance in French line 29H. The maximum area of India, Iraq, and Egypt is hot, dry, tropical and subtropical locations where Meloidogyne species are most dominant. Massive multinucleated syncytia shaped in the roots of the host plants result from a sessile female nematode, leading to the destruction of cell walls and tissues.

All Pratylenchus species have a wide variety of hosts and attack and feed on cortical parenchyma cells of the roots of different species. Root damage reduces a plant's capacity to absorb nutrients and water but promotes fresh bacterial and fungal infections. A survey was done in faba bean fields of Ethiopia, and the frequency and population of plant parasitic nematode genera were determined. The nematodes associated with the rhizosphere of faba bean were determined, and 120 soil samples from faba bean fields were analyzed using the Baermann technique to extract plant parasitic nematodes. The six dominant nematode species were Pratylenchus, Rotyenchulus, Tylenchoryhnchus, Xiphinema, Ditylenchus and Tylenchus. The Xiphinema were most dominant, contributing 12%, followed by Ditylenchus shearing 7% of total populations associated with faba bean.

Agronomic Management of Nematodes

The best method of plant protection is using certified seeds. Meanwhile, the chemically treated seeds that have been tainted by nematicides rarely eradicate the nematode population (El-Nagdi et al., 2019; Yu et al., 2023). The contaminated seeds survive passage *via* a monogastric or ruminant mammal, infected straw and plant waste left in the storage or in the field, which works as a source of inoculum. Therefore, it is still believed that sanitation (burning) of diseased material is the most efficient way to eradicate the inoculums (Ali et al., 2019). It is recommended to wait 3-8 years between successive faba bean plantings in areas where the enormous Ditylenchus species are present. Therefore, it is recommended that management is effective while developing effective rotation strategies. It has been demonstrated that *Meloidogyne javanica* (Treub) is resistant to the Egyptian faba bean cultivar Romy, unlike *Meloidogyne incognita.* Faba bean variety called Fiord was first time introduced in Australia, which worked as a well-known "cleaner" crop that significantly reduced Pratylenchus populations, *Pratylenchus neglectus* in the fields (Chu et al., 2019). In their long-term rotation study, it was discovered that wheat root-knot nematode populations were eight times lower following a Fiord faba bean variety. Researchers found that the faba bean is moderately resistant to *Pratylenchus thornei* but highly resistant to *Pratylenchus neglectus*. Even though some faba bean germplasm from North Africa is resistant to the above two species (Dhull et al., 2022). Research in the Mediterranean basin has shown that *Pratylenchus neglectus, Pasteuria thornei, Pratylenchus penetrans* (Cobb) Filipjev and *Schuurmans Stekhoven*,

and *Pratylenchus pinguicaudatus* Corbett can survive on the faba bean plant. Due to its great genetic variation, the genus *Pratylenchus* represents a huge species complex.

Techniques of Chemical and Biological Control of Nematodes

Chemically, nematodes are hard to manage (Abdel-Razik et al., 2016). In the past, it was believed that the strong pesticide aldicarb was effective against the *Pratylenchus*, *Meloidogyne*, and *Ditylenchus* species (Hammam et al., 2021). However, it has been banned in the European Union and Asia (Abd-Elgawad 2020a; 2020b). It is thought that oxyamyl effectively combats root-knot nematodes. When the temperature under the mulch reached 55-58 degrees Celsius, soil solarization was found to be effective against several species, including *Pratylenchus thornei*. Once the tillage practice was reduced, the populations of Pratylenchus species showed a significant increase. Studies on fungus have been tested using *Monacrosporium lysipagum* (Drechsler) and *Paecilomyces lilacinus* (Thom) Samson as biological controls against different nematodes (Abd-Elgawad, 2021). It was found that fungi Streptomyces isolates dramatically decreased *Pratylenchus penetrans* populations in potted alfalfa plants, whereas *M. lysipagum* showed only limited efficiency against *Pratylenchus neglectus* and *Ditylenchus dipsac*i. In another experiment, the faba bean seeds were treated with *Serratia marcescens*, and a substantial decrease in the *Meloidogyne incognita* population was recorded (Bakr et al., 2020).

Conclusion

A universal IPM technique is unlikely to be successful because each country or region's faba bean crops have a unique complex of arthropod, nematode, fungal, and broomrape pests. Broomrapes, nematodes, and several fungi must be prevented from spreading by using uncontaminated or treated seed and various synthetic pesticides for most pests. The crop must be kept far away from sources of garbage or stray seedlings to prevent contamination. Most fungi and arthropods can be controlled chemically; however, bees must pollinate the crop and resistance in the target organisms must be avoided. Arthropod pest populations can be managed using forecasting techniques.

However, there aren't many models for pest control that attack faba beans, which is still a problem. The chemical price, yield per hectare, and cost of harvesting per tonne are a few elements that affect the economic threshold but are not included in this study. Though many biological control techniques are being researched, very few have been commercialised.

References

Abd-Elgawad, M. M. M (2020a). Biological control agents in the integrated nematode management of pepper in Egypt. *Egypt. J. Biol. Pest Cont.,* 30: 70. https://doi.org/10.1186/s41938-020-00273-6.

Abd-Elgawad, M. M. M (2021). Biological control of nematodes infecting eggplant in *Egypt. Bull. NRC*, 45: 6. https://doi.org/10.1186/s42269-020-00463-0.

Abd-Elgawad, M. M. M. (2020b). Optimizing biological control agents for controlling nematodes of tomato in Egypt. *Egypt. J. Biol. Pest Cont*. 30: 58. https://doi.org/10.1186/s41938-020-00252-x.

Abdel-Razik, A. B., Shoaib, R. M., Taha, E. H., and Ibrahim, S. A. (2016). The effect of Serratia marcescens and genetically improved Pseudomonas fluorescens on Meloidogyne incognita. *Egypt. Acad. J. Biol. Sci. (B Zool.,)* 8(1): 49-59. https://doi.org/10.21608/eajbsz.2016.13463.

Abdelwahd, R., Hakam N., Labhilili M., Udupa S. M. (2008) Use of an adsorbent and antioxidants to reduce the effects of leached phenolics in *in vitro* plantlet regeneration of faba bean. *Afr J Biotechnol* 7:997–1002.

Ahmed, M., Qadir G., Shaheen F. A., Aslam M. A. (2017) Response of proline accumulation in bread wheat (Triticum aestivum L.) under rainfed conditions. *J Agricultural Meteorol* 73:147–155. https://doi.org/10.2480/agrmet.D-14-00047.

Akpinar, N., Akpinar M. A., Türkoğlu S. (2001) Total lipid content and fatty acid composition of the seeds of some *Vicia* L. species. *Food Chem* 74:449–453.

Alarcón, R., Hernández-Plaza E., Navarrete L., Sánchez M. J., Escudero A., Hernanz J. L., Sánchez-Giron V., Sánchez A. M. (2018) Effects of no-tillage and non-inversion tillage on weed community diversity and crop yield over nine years in a Mediterranean cereal-legume cropland. *Soil Tillage Res* 179:54–62.

Ali, M. B., Haridy A. G., Mahmoud A. F. (2019). Evaluation of faba bean genotypes for yield and resistance to fusarium root rot under greenhouse and field conditions. *Int. J. Biosciences.* 14 (2), 374–385. doi: 10.12692/ijb/14.2.374-385.

Almaghrabi, O. A. (2014) Effect of growth hormone 2, 4-D on some callus traits of different Faba bean (*Vicia faba* L.) cultivars. *Life Sci J* 11:98–102.

Aly, H. M. A., Hattori K. (2007) Histological observation on plant regeneration in faba bean cotyledon (*Vicia faba* L.) cultured *in vitro*. *Asian J Plant Sci* 6:723–731.

Amanuel, G., Kühne R. F., Tanner D. G., Vlek P. L. G. (2000) Biological nitrogen fixation in faba bean (*Vicia faba* L.) in the Ethiopian highlands as affected by P fertilization and inoculation. *Biol Fertil Soils* 32:353–359.

Anwar, F., Alghamdi S. S., Ammar M. H., Siddique K. H. M. (2011) An efficient *in vitro* regeneration protocol for faba bean (*Vicia faba* L.). *J Med Plants Res* 5:6460–6467.

Ashworth, A. J., Allen F. L., DeBruyn J. M., Owens P. R., Sams C. (2018) Crop rotations and poultry litter affect dynamic soil chemical properties and soil biota long term. *J Environ Qual* 47:1327–1338. https://doi.org/10.2134/jeq2017.12.0465.

Ashworth, A. J., Allen F. L., Wight J. P., Saxton A. M., Tyler D. D., Sams C. E. (2014) Soil organic carbon sequestration rates under crop sequence diversity, bio-covers, and no-tillage. *Soil Sci Soc Am J* 78:1726–1733. https://doi.org/10.2136/sssaj2013.09.0422.

Avižienytė, D., Romaneckas K., Pališkytė R., Bogužas V., Pilipavičius V., Šarauskis E., Adamavičienė A., Vaiciukevičius E. (2013) The impact of long-term reduced primary soil tillage on maize (Zea mays L.) productivity. *Zemdirbyste-Agriculture* 100:377–382. https://doi.org/10.13080/z-a.2013.100.048.

Bachmann, M., Kuhnitzsch, C., Martens, S. D., Steinhöfel, O., & Zeyner, A. (2020). Control of bean seed beetle reproduction through cultivar selection and harvesting time. *Agriculture, Ecosystems & Environment*, 300, 107005. https://doi.org/10.1016/j.agee.2020.107005.

Badagliacca, G., Benítez E., Amato G., Badalucco L., Giambalvo D., Laudicina V. A., Ruisi P. (2018) Long-term no-tillage application increases soil organic carbon, nitrous oxide emissions and faba bean (*Vicia faba* L.) yields under rain-fed Mediterranean conditions. *Sci Total Environ* 639:350–359.

Bahgat, S., Shabban OA, El-Shihy O, Lightfoot DA, El-Shemy HA (2008) Establishment of the regeneration system for *Vicia faba* L. *Curr Issue Mol Biol* 11(Suppl):47–54.

Bakr, R. A., Mahdy, M.E., and Mousa, E.M., 2020. Survey of root-knot nematodes Meloidogyne spp. associated with different economic crops and weeds in Egypt. *Egypt J. Crop Prot.*, 15(2): 1-14. https://doi.org/10.21608/ejcp.2020.205589.

Bardner, R., 1983. Pests of *Vicia faba* other than aphids and nematodes. In: Hebblethwaite, P.D. (Ed.), The Faba Bean (*Vicia faba* L.) Butterworths, London, pp. 371–390.

Barrett, G. W., van Dyne, G. M., & Odum, E. P. (1976). Stress ecology. *BioScience*, 26(3), 192–194. https://doi.org/10.2307/1297248.

Bartomeus, I., Gagic, V., & Bommarco, R. (2015). Pollinators, pests and soil properties interactively shape oilseed rape yield. *Basic and Applied Ecology*, 16(8), 737–745. https://doi.org/10.1016/j.baae.2015.07.004.

Bates, D., Maechler, M., Bolkeraut, B., Walker, S., Christensen, R. H. B., Singmann, H., Dai, B., Scheipl, F., Grothendieck, G., Green, P., Fox, J., & Bauer, A. (2020). *lme4: Linear Mixed-Effects Models* (1.1-26).

Binding, H., Nehls R (1978a) Regeneration of isolated protoplast of *Vicia faba* L. Z Pflanzenphysiol 88:327–332.

Binding, H., Nehls R (1978b) Somatic cell hybridization of *Vicia faba* and Petunia hybrida. Mol Gen Genet 164:137–143.

Bishop, J., Garratt, M. P. D., & Breeze, T. D. (2020). Yield benefits of additional pollination to faba bean vary with cultivar, scale, yield parameter and experimental method. *Scientific Reports*, 10(1), 2102. https://doi.org/10.1038/s41598-020-58518-1.

Bishop, J., Jones, H. E., Lukac, M., & Potts, S. G. (2016). Insect pollination reduces yield loss following heat stress in faba bean (*Vicia faba* L.). *Agriculture, Ecosystems & Environment*, 220, 89–96. https://doi.org/10.1016/j.agee.2015.12.007.

Blackman, R. L. and V. F. Eastop. (2000). *Aphids on the World's Crops*, Second Edition. John Wiley & Sons with the Natural History Museum, London.

Blackman, R. L. and V. F. Eastop. (2006). Aphids on the World's Herbaceous Plants and Shrubs. Volume 2 *The Aphids.* John Wiley & Sons with the Natural History Museum, London. viii + pages 1025-1439.

Borowska, J., Giczewska A, Zadernowski R (2003) Nutritional value of broad bean seeds. Part 2: Selected biologically active components. *Food/Nahrung* 47:98–101.

Böttinger, P., Steinmetz A, Schieder O, Pickardt T (2001) *Agrobacterium*-mediated transformation of *Vicia faba*. Mol Breed 8:243–254.

Buczek, J., Migut D, Jańczak-Pieniążek M (2021) Effect of Soil Tillage practice on Photosynthesis, Grain Yield and Quality of Hybrid Winter Wheat. *Agriculture* 11:479. https://doi.org/10.3390/agriculture11060479.

Caracuta, V., Barzilai O, Khalaily H, Milevski I, Paz Y, Vardi J, Regev L, Boaretto E (2015). The onset of faba bean farming in the Southern Levant. *Sci Rep* 5:1–9.

Caracuta, V., Weinstein-Evron, M., Kaufman, D., Yeshurun, R., Silvent, J., & Boaretto, E. (2016). 14,000-year-old seeds indicate the Levantine origin of the lost progenitor of faba bean. *Scientific reports*, *6*(1), 37399.

Chu, M., Gu M. Y., Tang Q. Y., Zhu J., Hao X. Y., Zhang Z. D. (2019). Identification of pathogenic fungi causing the root rot disease of dabancheng faba bean and analysis of antagonistic effect by the biocontrol bacteria. *Xinjiang Agric. Sci.* 56 (10), 1904–1911. doi: 10.6048/j.issn.001-4330.2019.010.016.

Confalone, A., Lizaso JI, Ruiz-Nogueira B, López-Cedrón F-X, Sau F (2010) Growth, PAR use efficiency, and yield components of field-grown *Vicia faba* L. under different temperature and photoperiod regimes. *Field Crops Res* 115: 140–148.

Conner, R.L., Bernier, C.C., 1982. Inheritance of rust resistance in inbred lines of *Vicia faba*. Phytopathology 72, 1555–1557.

Cote, I. M., Darling, E. S., & Brown, C. J. (2016). Interactions among ecosystem stressors and their importance in conservation. *Proceedings of the Royal Society B-Biological Sciences*, 283(1824), 20152592. https://doi.org/10.1098/rspb.2015.2592.

Crépon, K., Marget P, Peyronnet C, Carrouee B, Arese P, Duc G (2010) Nutritional value of faba bean (*Vicia faba* L.) seeds for feed and food. *Field Crops Res* 115:329–339.

Croser, J. S., Lülsdorf MM, Davies PA, Clarke HJ, Bayliss KL, Mallikarjuna N, Siddique KHM (2006) Toward doubled haploid production in the Fabaceae: progress, constraints, and opportunities. *Crit Rev Plant Sci* 25:139–157.

Cuny, M. A. C., Gendry, J., Hernández-Cumplido, J., & Benrey, B. (2018). Changes in plant growth and seed production in wild lima bean in response to herbivory are attenuated by parasitoids. *Oecologia*, 187(2), 447–457. doi.org/10.1007/s00442-018-4119-1.

Daoui, K., Fatemi ZA (2019) Augmentation de la production des légumineuses alimentaires entre opportunités et défis. *INRA Meknès Magazine.*

Denekea, S., Shiferawa D, Fininsa C, Yesuf M (2018) Integrated management of faba bean chocolate spot (Botrytis fabae Sard) through host resistance, intercropping and fungicide applications in arsi. Ethiopia. *Inter J Sci Res Publ* 8(5):563–573.

Desouky, A. F., Ahmed AHH, Stützel H, Jacobsen HJ, Pao YC, Hanafy MS (2021) Enhanced abiotic stress tolerance of *Vicia faba* L. plants heterologously expressing the *PR10a* gene from potato. *Plants* 10:173.

Dhull, S. B., Kidwai M. K., Noor R., Chawla P., Rose P. K. (2022). A review of nutritional profile and processing of faba bean (*Vicia faba* l.). *Legume Science.* 4 (3), e129. doi: 10.1002/leg3.129.

Dhull, S. B., Kidwai MK, Noor R, Chawla P, Rose PK (2021) *A review of nutritional profile and processing of faba bean. Vicia faba.*

Du, P., Luo H, He J, Mao T, Du B, Hu L (2019) Different tillage induces regulation in 2-acetyl-1-pyrroline biosynthesis in direct-seeded fragrant rice. *BMC Plant Biol* 19:1–10.

El-Nagdi, W. M. A., Youssef, M.M.A., AbdEl-Khair, H., and Abd-Elgawad, M.M.M. (2019). Effect of certain organic amendments and Trichoderma species on the root-knot nematode, Meloidogyne incognita infecting pea (Pisum sativum L.) plants. *Egypt J. Biol. Pest Cont.,* 29: 75. https://doi.org/10.1186/s41938- 019-0182-0.

Epperlein, K. (1992). Investigation of the damage of broad bean weevil Bruchus rufimanus Bohem. (Col., Bruchidae) on broad bean seed (*Vicia faba* L.). *Anzeiger Für Schaedlingskunde, Pflanzenschutz Umweltschutz*, 65(8), 147–150.

Etemadi, F., Hashemi M, Barker AV, Zandvakili OR, Liu X (2019) Agronomy, nutritional value, and medicinal application of faba bean (*Vicia faba* L.). *Hortic Plant J* 5:170–182.

Etemadi, F., Hashemi M, Randhir R, ZandVakili O, Ebadi A (2018) Accumulation of L-DOPA in various organs of faba bean and influence of drought, nitrogen stress, and processing methods on L-DOPA yield. *Crop J* 6:426–434.

Fakhrai, H., Fakhrai F, Evans PK (1989) *In vitro* culture and plant regeneration in *Vicia faba* subsp. Equina (var. Spring Blaze). *J Exp Bot* 40:813–817.

Fatemi, Z. A. (2019) Amélioration du potentiel et de la stabilité du rendement de la fève et de la féverole *Vicia faba* L. *INRA Meknès Magazine.* https://mag.inrameknes.info/?p=1954. Accessed 3 Oct 2022.

Fernandez-Aparicio, A. M., Flores, F., Rubiales, D., 2016. The effect of Orobanche crenata infection severity in faba bean, field pea, and grass pea productivity. *Frontiers in Plant Science* 7, 1409. doi:10.3389/fpls.2016.01409.

Firdu, Z., Alemu T, Assefa F (2022) The synergistic effects of Trichoderma harzianum AAUT14 and *Bacillus subtilis* AAUB95 on faba bean (*Vicia faba* L.) growth performance and control of chocolate spot compared to chemical fungicides under greenhouse conditions. *Archives Phytopathol & Plant Protec* 55:129–142.

Fisher, A., DeGrandi-Hoffman, G., Smith, B. H., Johnson, M., Kaftanoglu, O., Cogley, T., Fewell, J. H., & Harrison, J. F. (2021). Colony field test reveals dramatically higher toxicity of a widely-used mito-toxic fungicide on honeybees (Apis mellifera). *Environmental Pollution*, 269, https://doi.org/10.1016/j.envpol.2020.115964.

Gagic, V., Marcora, A., & Howie, L. (2019). Additive and interactive effects of pollination and biological pest control on crop yield. *Journal of Applied Ecology*, 56(11), 2528–2535. https://doi.org/10.1111/1365-2664.13482.

Gagic, V., Riggi, L. G., Ekbom, B., Malsher, G., Rusch, A., & Bommarco, R. (2016). Interactive effects of pests increase seed yield. *Ecology and Evolution*, 6(7), 2149–2157. https://doi.org/10.1002/ece3.2003.

Garcia, L. C., and Eubanks, M. D. (2019). Overcompensation for insect herbivory: A review and meta-analysis of evidence. *Ecology*, 100(3), e02585. https://doi.org/10.1002/ecy.2585.

Garibaldi, L. A., Andersson, G. K. S., Requier, F., Fijen, T. P. M., Hipólito, J., Kleijn, D., Pérez-Méndez, N., & Rollin, O. (2018). Complementarity and synergisms among ecosystem services supporting crop yield. *Global Food Security*, 17, 38–47. https://doi.org/10.1016/j.gfs.2018.03.006.

Gela, T. S., Bruce, M., Chang, W. (2022) Genomic regions associated with chocolate spot (*Botrytis fabae* Sard.) resistance in faba bean (*Vicia faba* L.). *Mol Breeding* 42, 35 https://doi.org/10.1007/s11032-022-01307-7.

Ghanbari, M., Taherpurkolai S, Babaeizad V, Mojerloo S (2020) Evaluation of resistance and enzyme production of faba bean cultivars against Alternaria leaf spot. *Iran J Plant Prot Sci* 51(1):55–66.

Giambalvo, D., Ruisi P, Saia S, Di Miceli G, Frenda AS, Amato G (2012) Faba bean grain yield, N2 fixation, and weed infestation in a long-term tillage experiment under rainfed Mediterranean conditions. *Plant Soil* 360:215–227. https://doi.org/10.1007/s11104-012-1224-5.

Hammam, M. M. A., Mohamed, M. M. M., and Abd-Elgawad, M. M. M. (2023). The damage and gain threshold of Meloidogyne arenaria on faba bean favors the use of a safe biological control over chemical nematicides. *Pakistan Journal of Nematology,* 41(1): 8-17.

Heinze, J. (2020). Herbivory by aboveground insects impacts plant root morphological traits. *Plant Ecology*, 221(8), 725–732. https://doi.org/10.1007/s11258-020-01045-w.

Hiywotu, A. M., Abate A, Worede F (2023) Correlation and path coefficient analysis of yield and yield components of some Ethiopian faba bean (*Vicia faba* L.) accessions. *Acta Agriculturae Slov* 119:1–11.

Hou, X., Li R, Jia Z, Han Q (2013) Rotational tillage improves photosynthesis of Winter Wheat during Reproductive Growth stages in a Semiarid Regio. *Agron J* 105:215–221. https://doi.org/10.2134/agronj2012.0201.

Janusauskaite, D., Feisziene D, Feiza V (2022) The effect of tillage, fertilization and residue management on winter wheat and spring wheat physiological performance. *Acta Physiol Plant* 44:75.

Jaquiery, R., & Keller, E. R. (1978). La chute des fruits chez la feverole (*Vicia faba* l.) En relation avec la disponibilite en assimilats marques au 14c. *Revue Suisse D'agriculture*, 10(4), 123–127. 5893.

Järemo, J., Tuomi, J., Nilsson, P., and Lennartsson, T. (1999). Plant adaptations to herbivory: Mutualistic versus antagonistic coevolution. *Oikos*, 84(2), 313 320. https://doi.org/10.2307/3546728.

Jellis, G.J., Punithalingam, E., 1991. Discovery of Didymella fabae sp. nov., the teleomorph of Ascochyta fabae, on faba bean straw. Plant Pathol. 40, 150–157.

Jensen, E. S., Peoples, M. B., and Hauggaard-Nielsen, H. (2010). Faba bean in cropping systems. *Field Crops Research*, 115(3), 203-216.

Karkanis, A., Ntatsi, G., Lepse, L., Fernández, J. A., Vågen, I. M., Rewald, B., Alsiņa, I., Kronberga, A., Balliu, A., Olle, M., Bodner, G., Dubova, L., Rosa, E., and Savvas, D. (2018). Faba bean cultivation – revealing novel managing practices for more sustainable and competitive european cropping systems. *Frontiers in Plant Science*, 9, 1115. https://doi.org/10.3389/fpls.2018.01115.

Kendall, D. A., & Smith, B. D. (1975). The Pollinating Efficiency of Honeybee and Bumblebee Visits to Field Bean Flowers (*Vicia faba* L.). *Journal of Applied Ecology*, 12(3), 709–717. https://doi.org/10.2307/2402083.

Khazaei, H., O'Sullivan DM, Stoddard FL, Adhikari KN, Paull JG, Schulman AH, Andersen SU, Vandenberg A (2021) Recent advances in faba bean genetic and genomic tools for crop improvement. *Legume Sci* 3:e75.

Kimbirauskienė, R., Sinkevičienė A, Jonaitis R, Romaneckas K (2023) Impact of Tillage Intensity on the development of Faba Bean Cultivation. *Sustainability* 15:8956.

Krupnick, G. A., and Weis, A. E. (1999). The effect of floral herbivory on male and female reproductive success in Isomeris Arborea. *Ecology*, 80(1), 135–149.

Kumari, S. G., van Leur J, Maalouf F (2018) Screening faba bean (*Vicia faba* L.) germplasm for resistance to persistently aphid transmitted viruses. In: *Seventh international food legumes research conference*, Marrakesh, Morocco, p 100.

Lal, K. (2019) Genetic variability and diversity analysis in faba bean (*Vicia faba* L). *Trends Biosci* 12:754–760.

Laloum, Y., Gangneux C, Gügi B, Lanoue A, Munsch T, Blum A, Gauthier A, Trinsoutrot-Gattin I, Boulogne I, Vicré M, Driouich A, Laval K, Follet-Gueye ML (2021) Faba bean root exudates alter pea root colonization by the oomycete *Aphanomyces euteiches* at early stages of infection. *Plant Sci* 312:111032.

Lee, R. C., Farfan-Caceres L, Debler JW, Williams AH, Syme RA, Henares BH (2021) *Reference genome assembly for Australian Ascochyta lentis* isolate Al4. G3 11(2):jkab006. https://doi.org/10.1093/g3journal/jkab006.

Lee, R. C., Farfan-Caceres LM, Debler JW, Syme RA (2020) Characterization of growth morphology and pathology, and draft genome sequencing of *Botrytis fabae*, the causal organism of chocolate spot of faba bean (*Vicia faba* L.). *Front Microbiol* 11:217. https://doi.org/10.3389/fmicb.2020.00217.

Lenth, R. V., Buerkner, P., Herve, M., Love, J., Riebl, H., & Singmann, H. (2021). *Emmeans (1.5.5-1)* [Computer software]. https://CRAN.R-project.org/package=emmeans.

López-Bellido, F. J., López-Bellido, L., & López-Bellido, R. J. (2005). Competition, growth and yield of faba bean (*Vicia faba* L.). *European Journal of Agronomy*, 23(4), 359–378. https://doi.org/10.1016/j.eja.2005.02.002.

Lundin, O., and Raderschall, C. A. (2021). Landscape complexity benefits bumble bee visitation in faba bean (*Vicia faba minor* L.) but crop productivity is not pollinator-dependent. *Agriculture, Ecosystems & Environment*, 314, 107417. https://doi.org/10.1016/j.agee.2021.107417.

Maalouf, F., Hu J, O'Sullivan DM, Zong X, Hamwieh A, Kumar S, Baum M (2019) Breeding and genomics status in faba bean (*Vicia faba*). *Plant Breed* 138:465–473.

Mahapatra, S. S., Swain D, Beura SK, Rout GR (2022) Identification of mungbean germplasm for resistance against *Cercospora canescens* (CLS) and to study the association of biochemical parameters with defence mechanisms. *Agron Jour* 114:1184–1199. https://doi.org/10.1002/agj2.20987.

Mahmoud, A. F., Ab El-Fatah BES (2020) Genetic diversity studies and identification of molecular and biochemical markers associated with fusarium wilt resistance in cultivated faba bean (*Vicia faba*). *Plant Pathol J* 36(1):11–28.

Maurya, A. K., Navathe S, Mohapatra C, Chand R (2018) Antioxidants elevates the resistance to Cercosporacanescens in interspecific cross of *Vigna radiata* (Kopergaon)× *Vigna mungo* (Pant Urd 31). *Indian Phytopathol* 71:519–528.

Merga, B., Egigu MC, Wakgari M (2019) Reconsidering the economic and nutritional importance of faba bean in Ethiopian context. *Cogent Food Agri* 5(1):1683938.

Mínguez, I., Rubiales D. (2020) Faba bean. In: Sadras V. O., Calderini D. F. (eds) *Crop physiology: case histories for major crops*. Academic Press, London, pp 453–482. https://doi.org/10.1016/B978-0-12-819194-1.00015-3.

Mínguez, M. I., Rubiales D. (2021) *Faba bean. Crop physiology case histories for major crops.* Elsevier, pp 452–481.

Mo, Z., Tang Y., Ashraf U, Pan S, Duan M, Tian H, Wang S, Tang X (2019) Regulations in 2-acetyl-1-pyrroline contents in fragrant rice are associated with water-nitrogen dynamics and plant nutrient contents. *J Cereal Sci* 88:96102. https://doi.org/10.1016/j.jcs.2019.05.013.

Mobini, S., Khazaei H, Warkentin TD, Vandenberg A (2020) Shortening the generation cycle in faba bean (*Vicia faba*) by application of cytokinin and cold stress to assist speed breeding. *Plant Breed* 139:1181–1189.

Mokhtar, M. M., Hussein EHA, El-Assal SES, Atia MAM (2020) VfODB: a comprehensive database of ESTs, EST-SSRs, mtSSRs, microRNA-target markers and genetic maps in *Vicia faba*. *AoB Plants* 12(6):plaa064.

Nassib, A.M., Ibrahim, A.A., Khalil, S.A., 1982. Breeding for resistance to Orobanche. In: Hawtin, G., Webb, C. (Eds.), Faba Bean Improvement. The Hague, The Netherlands, pp. 199–206.

Nuessly, G.S., Hentz, M.G., Beiriger, R., Scully, B.T., 2004. Insects associated with faba bean, *Vicia faba* (Fabales: Fabaceae), in Southern Florida. Florida Entomol. 87, 204–211.

O'Sullivan, D. M., Angra D, Harvie T, Tagkouli V, Warsame A (2019) A genetic toolbox for *Vicia faba* improvement. In *International conference on legume genetics and genomics,* May 13–17 Dijon: France.

Papan, P., Chueakhunthod W, Poolsawat O, Arsakit K, Tharapreuksapong A, Tantasawat, PA (2021) Validation of molecular markers linked to cercospora leaf spot Disease resistance in mungbean (*Vigna radiata* [L.] Wilczek). *SABRAO Jour Breed & Genetic* 53:749–758. https://doi.org/10.54910/sabrao2021.53.4.16.

Punia, S. S., Duhon, A. and Sharma, A. R. (2018). Management of Orobanche in mustard crop through glyphosate application. *Indian Farming* 68(08), 14–15.

Rathore, S. S., Shekhawat, K., Premi, O. P., Kandpal, B. K., Chauhan, J. S., 2014. Biology and management of the fast-emerging threat of broomrape in rapeseed– mustard. *Weed Biology and Management* 14(3), 145–158.

Rubiales, D., Khazaei, H. (2022) Advances in disease and pest resistance in faba bean. *TheorAppl Genet* 135, 3735–3756. https://doi.org/10.1007/s00122-021-04022-7.

Satovic, Z., Avila CM, Cruz-Izquierdo S, Díaz-Ruíz R, García-Ruíz GM, Palomino C, Gutiérrez N, Vitale S, Ocaña-Moral S, Gutiérrez MV, Cubero JI, Torres AM (2013) A reference consensus genetic map for molecular markers and economically important traits in faba bean (*Vicia faba* L.). *BMC Genom* 14:932.

Seidenglanz, M., Huňady I (2016) Effects of faba bean (*Vicia faba*) varieties on the development of *Bruchus rufimanus*. Czech *J Genet Plant Breed* 52:22–29.

Sheoran, P., Punia, S.S., Singh, S., Singh, D., 2014. Orobanche weed management in mustard: Opportunities, possibilities and limitations. *Journal of Oilseed Brassica* 5(2), 96–101.

Shtaya, M. J. Y., Emeran AA, Fernández-Aparicio M, Qaoud HA, Abdallah J, Rubiales D (2021) Effects of crop mixtures on rust development on faba bean grown in Mediterranean climates. *Crop Prot* 146:105686.

Sillero, J. C., Rojas-Molina MM, Emeran AA, Kharrat M, Winkler J, Khan HR, Flores F, Rubiales D (2017) Identification and multi-environment validation of resistance to rust (*Uromyces viciae-fabae*) in *Vicia faba*. Crop Pasture *Sci* 68:1013–1023.

Smith, R. F., 1972, The impact of the green revolution on plant protection in tropical and subtropical areas, *Bull. Entomol. Soc. Am.* **18**:7–14.

Smith, R. F., and R. van den Bosch. 1967. Integrated control. Chap. 9. In *Pest Control: Biological Physical and Selected Chemical Methods*, W. W. Kilgore and R. L. Doutt (eds.). Academic Press, N.Y. 477 pp.

Snowdon, R. J., Wittkop, B., Chen, T. W., Stahl, A. (2021). Crop adaptation to climate change as a consequence of long-term breeding. *Theor. Appl. Genet.* 134 (6), 1613–1623. doi: 10.1007/s00122-020-03729-3.

Sudheesh, S., Kimber RBE, Braich S, Forster JW, Paull JG, Kaur S (2019) Construction of an integrated genetic linkage map and detection of quantitative trait loci for ascochyta blight resistance in faba bean (*Vicia faba* L.). *Euphytica* 215:42.

Tantasawat, P. A., Poolsawat O, Arsakit K, Papan P (2020) Identification of ISSR, ISSR-RGA and SSR markers associated with Cercospora leaf spot resistance gene in mungbean. *Int Jour Agricul & Biol* 23:447–453.

Thomas, J., Kenyon, D., 2004. Evaluating resistance to downy mildew (Peronos-pora viciae) in field peas (*Pisum sativum* L.) and field beans (*Vicia faba* L.). In: AEP (Ed.), Proceedings of the 5th European Conference on Grain Legumes and 2nd International Conference on Legume Genomics and Genetics, Dijon, France, pp. 81–82. European Association for Grain Legume Research AEP, Paris, France.

Tiwari, N., Kumar T, Saxena DR, Swain N, Maalouf F, Ahmed S, Sarker A (2021) Evaluation of disease resistant and high yielding faba bean germplasm in India. *J Genet* 100:34. https://doi.org/10.1007/s12041-021-01287-1.

Van Emden, H.F., Ball, S.L., Rao, M.R., 1988. Pest, disease and weed problems in pea, lentil, faba bean and chickpea. In: Summerfield, R.J. (Ed.), World Crops: Cool Season Food Legumes. Kluwer Academic Publishers, pp. 519–534.

Varshney, R. K., Cowling WA, Chitikineni A, Lam H-M, Hickey LT, Croser JC, Bayer PE, Edwards D, Crossa J, Weckwerth W, Millar H, Kumar A, Bevan MW, Siddique KHM (2021) Fast-forward breeding for a food-secure world. *Trends Genet* 37:1124–1136.

Varshney, R. K., Pandey MK, Bohra A, Singh VK, Thudi MT, Saxena RK (2019) Toward the sequence-based breeding in legumes in the post-genome sequencing era. *Theor Appl Genet* 132:797–816.

Villegas-Fernández, A. M., Krajinski F, Schlereth A, Madrid E, Rubiales D (2014) Characterisation by transcription factor expression profiling of the interaction *Medicago truncatula–Botrytis* spp. interactions. *Plant Mol Biol Rep* 32:1030–1040.

Wang, C., Liu R, Liu Y, Hou W, Wang X, Miao Y, He Y, Ma Y, Li W, Wang D, Ji Y, Zhang H, Li M, Yan X, Zong X, Yang T (2021) Development and application of the Faba_bean_130K targeted next-generation sequencing SNP genotyping platform based on transcriptome sequencing. *Theor Appl Genet* 134:3195–3207.

Wang, Y. Y., Ren, J. B., Zhang, Y., Zheng, Y., Tang, L. (2020). Effect of wheat and faba bean intercropping on improving rhizosphere microflora and reducing fusarium wilt of faba bean. *Chin. J. Soil Sci.* 51 (05), 127–1133. doi: 10.19336/j.cnki.trtb.2020.05 .16.

Warsame, A. O. (2021) Towards genetic mapping of nutritional quality traits in faba bean (*Vicia faba* L.). *PhD thesis.* University of Reading, UK.

Watson, A., Ghosh S, Williams MJ, Cuddy WS, Simmonds J, Rey MD, Asyraf Md Hatta M, Hinchliffe A, Steed A, Reynolds D (2018) Speed breeding is a powerful tool to accelerate crop research and breeding. *Nat Plants* 4:23–29.

Webb, A., Cottage A, Wood T, Khamassi K, Hobbs D, Gostkiewicz K, White M, Khazaei H, Ali M, Street D, Duc G, Stoddard FL, Maalouf F, Ogbannaya F, Link W, Thomas J, O'Sullivan DM (2016) A SNP-based consensus genetic map for synteny-based trait targeting in faba bean (*Vicia faba* L.). *Plant Biotechnol J* 14(1):177–185.

Wettberg, E. J. B. and Guerra-Garcia A (2023). Huge broad-bean genome could improve yields of an underused crop. Nature. *Nature* 615, 587-588. doi: https://doi.org/10.1038/d41586-023-00461-y.

Wondale, E. and Berhanu Hi (2020). The distribution, frequency of occurrence, and the status of stored faba bean insect pests in relation to food security in Farta District, North West Ethiopia. *Endshaw & Hiruy, Cogent Food & Agriculture*. 6: 1832400 https://doi.org/10.1080/23311932.2020.1832400].

Yitayih, G., Azmeraw Y (2017) Adaptation of faba bean varieties for yield components and against faba bean gall (*Olpidium viciae* Kusano) disease in South Gondar, Ethiopia. *Crop J* 5:560–566.

You, M. P., Eshete BB, Kemal SA, van Leur J, Barbetti MJ (2021) *Physoderma*, not *Olpidium*, is the true cause of faba bean gall disease of *Vicia faba* in Ethiopia. *Plant Pathol* 70:1180–1194.

Yu, H., Yang F, Hu C, Yang X, Zheng A, Wang Y, Tang Y, He Y, Lv M. (2023). Production status and research advancement on root rot disease of faba bean (*Vicia faba* L.) in China. *Front Plant Sci*. 14:1165658. doi: 10.3389/fpls.2023.1165658.

Zeleke, E. A., Legesse, N. H (2023). Management of chocolate spot of Faba bean (*Botrytis fabae*) through fungicides and host resistance in North Shewa, Central Highland of

Ethiopia. *Indian Phytopathology* 76, 531–540 https://doi.org/10.1007/s42360-023-00636-7.

Zhang, Y., Bai, X. J., Fan, J. X., Guo, Y. P., Liu, Y., Xia, Z. L. (2020). Effects of different intercropping modes on root rot and crop growth of broad bean. *Agric. Sci. Technol. Communication.* 1, 128–131. doi: 10.3969/j.issn.1000-6400.2020.01.045.

Zhang, Z., Yang, W., Li, Y., Zhao, Q., Dong, Y. (2023). Wheat-faba bean intercropping can control fusarium wilt in faba bean under *F.commune* and ferulic acid stress as revealed by histopathological analysis. *Physiol. Mol. Plant Pathology.* 124, 101965. doi: 10.1016/j.pmpp.2023.101965.

Zhou, Y., Yao, M. N., Miao, Y. M., Jin, J. H., Gu, C. Y., Zhao, N (2022). The developmental of fresh broad bean industry in China. *J. Agriculture.* 12 (2), 80–84. doi: 10.11923/j.issn.2095-4050.cjas2021-0137.

Chapter 10

Faba Bean as an Important Food Legume: Abiotic Stresses and Their Mitigation

Muhammad Tanveer Altaf[1,*]
Waqas Liaqat[2]
Amna Jamil[3]
Muhammad Azhar Nadeem[1]
Celaleddin Barutçular[2]
Faheem Shehzad Baloch[4]
and Heba I. Mohamed[5]

[1]Department of Plant Production and Technologies,
Faculty of Agricultural Sciences and Technologies,
Sivas University of Science and Technology, Sivas, Turkey
[2]Department of Field Crops, Faculty of Agriculture,
Institute of Natural and Applied Sciences, Çukurova University, Adana, Turkey
[3]Department of Horticulture, MNS University of Agriculture, Multan, Pakistan
[4]Department of Biotechnology, Faculty of Science, Mersin University, Yenişehir, Mersin, Turkey
[5]Department of Biological and Geological Sciences, Faculty of Education,
Ain Shams University, Cairo, Egypt

Abstract

Faba bean is an important source of protein, dietary fiber, mineral nutrients, vitamins, and various bioactive compounds, making it valuable for both people and animal diets. The crop is subjected to a range of

---

[*] Corresponding Author's Email: tanveermalik759@gmail.com

In: Faba Beans
Editors: Aamir Raina and Samiullah Khan
ISBN: 979-8-89530-447-1

abiotic stressors, such as drought, extreme temperature, salinity, soil acidity, and waterlogging, which pose a detrimental impact on crop yield and hinder its potential for growth in new environments. This study focuses on the significance of enhanced agronomic techniques, including planting schedules, biochar application, and seed priming, as these are essential in optimizing the yield and quality of faba bean produce. The genetic enhancement of abiotic stress resistance in faba beans through conventional and molecular breeding methods is hindered by the extensive and predominantly unexplored genome and the unidentified wild ancestors. The limited availability of molecular and biotechnology tools for faba beans has resulted in a lack of transgenic research advancements from model plant species being applied to this crop. Genomics, genomic selection, and speed breeding techniques have recently become feasible to expedite the genetic advancements in faba beans. Using genomic resources with other breeding tools can expedite the breeding process and augment the genetic gain of this species. Genome-wide association studies can identify effective markers that can be used in breeding programs. The chapter pertains to the reaction of faba bean cultivation to abiotic stress factors, wherein emphasis is laid on the enhancement of agronomic and breeding methodologies in faba beans.

Keywords: drought, high temperature, waterlogging, seed priming, speed breeding

Introduction

Faba bean, a leguminous plant, is classified taxonomically as a member of the Fabaceae family, the Faboideae subfamily, the Vicieae tribe, the Vicia genus, and the *V. faba* species. According to scholarly sources, this food legume was among the first to be domesticated globally, likely during the late Neolithic era (Metayer, 2004), following the domestication of chickpeas and peas (Torres et al., 2006). It is believed to have originated in North Africa and Southwest Asia and has been widely grown in other regions (Zohary and Hopf, 2000). The wild ancestor of the faba bean remains unidentified (Emshwiller et al., 2015). The faba bean possesses several benefits and can be consumed in various forms, including dry seeds, green vegetables, and processed food (Tadele, 2020). The commodities derived from this source significantly contribute to quality protein in human food, with a protein content ranging from 24-35% (Crépon et al., 2010). This makes it one of the most protein-rich pulses and is also utilized as animal feed (Sainte, 2011). It has the potential to

serve as a replacement for meat in situations where there is a desire for protein sources that are not derived from animals (Crépon et al., 2010).

Like other leguminous crops, faba bean and its associated rhizobacteria have the ability to perform biological nitrogen fixation under diverse environmental conditions. This legume exhibits a remarkable ability to fix nitrogen, surpassing many of its counterparts. It can fix nitrogen even in soil environments with elevated nitrogen levels (Herridge et al., 2008). Also, the legume leaves a substantial residue that can effectively reduce the need for inorganic nitrogen fertilizer in following crops (Hauggaard-Nielsen et al., 2009). Therefore, the significance of this lies in its dual role of contributing to residual nitrogen in crop rotation (Watson et al., 2017) and having potential as a green manure (Baddeley et al., 2017).

According to FAOSTAT (2023), the worldwide cultivation of faba beans has experienced a reduction of nearly 50% from 5.4 million hectares in 1961 to 2.7 million hectares in 2021. The decline in production is attributed to various factors, including competition from soybeans, abandoning cultivation in regions where parasitic weeds have become prevalent, and yield instability due to the crop's susceptibility to numerous biotic and abiotic stresses (Adhikari et al., 2021). Regardless of the manifold advantages and the existence of cultivars exhibiting higher yield (MoALR, 2017), its production has persisted at a markedly suboptimal level relative to its maximum potential. The reduced yield is attributed to its sensitivity to environmental and biotic stressors (Mussa et al., 2008) and suboptimal crop production techniques (Gemechu et al., 2016). The faba bean is regarded as susceptible to water deficit within the category of cultivated grain legumes (Muktadir et al., 2020). According to Oweis et al. (2005), the yield of faba bean in irrigated agriculture surpasses that of rainfed systems by a significant margin. However, it is worth noting that most faba bean cultivation occurs outside irrigated regions. Enhancing drought tolerance is crucial for enhancing yield stability and performance. However, the mixed breeding system and the lack of efficient selection protocols present significant obstacles to effective plant breeding programs (Drayner, 1956; Stoddard et al., 2006). While notable advancement has been in addressing biotic stresses, investigating methods to overcome abiotic stresses remains confined (Adhikari et al., 2021). A yield potential of less than 50% has been achieved in faba beans (Duc, 1997; Mulugeta et al., 2019).

The progress of genomic research in faba bean has been hindered by its tremendous genome size of 13.1 Gb and insufficient investment in foundational research compared to other crops (Adhikari et al., 2021). Pulses

have received little attention from academia, particularly compared to other major crops. Faba bean, compared to the common bean, pea, lentil, cowpea, and chickpea, has received less attention and is considered an orphan crop among the already neglected pulse crops. It is suggested that utilizing the findings from genetics and genomics research is imperative for achieving substantial enhancements in this particular crop (Adhikari et al., 2021). Phenotypic selection has served as the primary driver of genetic advancement. However, recent advancements in DNA marker technology have facilitated the use of marker-assisted selection (MAS) to augment phenotypic selection. DNA markers that exhibit strong linkage to traits with significant impact and high heritability have been identified. Nevertheless, attributes such as yield exhibit complexity and are influenced by environmental variables and numerous other traits, thereby diminishing the efficacy of MAS. According to Liu et al. (2019), the effectiveness of genomic selection was significantly superior to phenotypic selection, and its efficiency could be further enhanced by incorporating speed breeding techniques.

The use of gene editing as a novel breeding strategy in numerous crops is currently being explored, thereby facilitating the expeditious attainment of genetic advancements. To date, there has been no documentation of a CRISPR/Cas9 system for faba bean (Adhikari et al., 2021). The lack of an annotated reference genome for this particular crop presents significant obstacles to using CRISPR/Cas gene editing, specifically in developing particular gRNA-targeted genes of interest (Bhowmik et al., 2021).

Significance of Faba Bean

Numerous academic sources have confirmed that faba bean possesses a wide range of applications. Maalouf (2011) reported that this crop serves as a protein source in the diets of humans, as well as fodder for animal husbandry. Additionally, it is utilized for its nitrogen availability in the soil. The faba bean seed is widely acknowledged for its high nutritional value in human and animal diets. According to Duc et al. (2011), the beneficial and easily digestible essential seed components of this crop are starch and proteins. Furthermore, they reported a significant genetic diversity among faba bean genotypes regarding their starch, fibre, and protein contents. Likewise, it is utilized for the purpose of producing green manure or as a leguminous crop in a rotational system with cereals (McVicar et al., 2013). The cultivation of rotation crops that fix atmospheric nitrogen is a valuable means of restoring

soil fertility, resulting in reduced fertilizer usage and cost savings for smallholder farmers (IFPRI, 2010). Hence, this legume crop holds significant value in enhancing the sustainability of cropping systems by ameliorating the health of the soil and reducing the incidence of diseases, pests, and weeds (Jensen et al., 2010).

Faba bean has garnered attention to serve as a pulse crop due to its superior environmental performance compared to other leguminous crops. According to Mesfin et al. (2020), the faba bean is recognized for its superior biological nitrogen fixation potential compared to other legumes such as lentils and field peas. This quality makes it a significant means of augmenting subsequent crop yields and mitigating environmental impacts. Faba bean also possesses environmental sustainability and can be utilized as a supplement to malted barley in the beer production process (Black et al., 2019). The integration of faba beans in cropping systems is anticipated to yield various benefits for the ecosystem (Kopke and Nemecek, 2010). Currently, there is a reassessment of the incorporation of legumes in various agricultural practices, driven by economic metrics encompassing soil quality and ecological considerations arising from the application of synthetic nitrogen fertilizer. The definitive determination of the faba bean's capacity to biologically fix nitrogen remains uncertain. However, faba beans are widely categorized as a highly effective source of nitrogen. As per the findings of Etemadi et al., (2019), the faba bean exhibits promising characteristics to function as an economical cover crop in conjunction with non-legume crops.

In cropping systems, this crop offers several advantages, such as serving as a break crop in a rotational program (Abera et al., 2015), providing a feed source for pollinators and beneficial insects (Etemadi et al., 2015), enhancing soil microbial activity (Wahbi et al., 2016), and controlling soil-borne diseases (Jensen et al., 2010). The cultivation of faba bean for its nitrogen contribution is most advantageous when the succeeding crop has a comparatively extended growth duration, thereby facilitating the optimal utilization of nitrogen mineralization that takes place toward the end of the growing season (Jensen et al., 2010). The residues of faba bean, including root and shoot, exhibit a lower carbon-to-nitrogen ratio, resulting in rapid nitrogen mineralization (Shi, 2013). The practice of intercropping faba bean with cereals, including barley and wheat, has been found to be a productive approach. This is attributed to the complementary utilization of nitrogen resources and other essential nutrients (Tosti and Guiducci, 2010).

Nutritional Profile and Health Benefits

Faba bean holds particular importance in the domains of sustenance, nourishment, and financial security, as well as in the livelihood opportunities among small-scale farming communities in West Asia, North Africa, and East Africa (Maalouf et al., 2021). Faba bean seeds are abundant in easily digestible proteins and contain a significant amount of energy, vitamins, and minerals (Crépon et al., 2010). According to Koivunen et al. (2016), faba bean is abundant in essential amino acids, including arginine, lysine, and leucine. It has been recognized as a potential functional food because it provides macro, micro, and non-nutrient phytochemicals (Maalouf et al., 2021). Baloch et al. (2014) found that there is a broad range of variability for microelements, including iron, manganese, copper, and zinc, in faba bean landraces. Interestingly, this variability does not have an impact on the grain yield of the crop. These findings suggest that there is potential to improve biofortification for higher concentrations of mineral elements, which might contribute to preventing malnutrition. A considerable percentage of individuals consume faba beans with cereals as a means of achieving a well-balanced diet (Maalouf et al., 2021). Similarly, the faba bean has potential applications in the bakery industry. The mixture of faba bean and wheat flour has been found to enhance the nutritious value of bread (Belghith-Fendri et al., 2016; Coda et al., (2017).

The assessment of the nutritional profile of faba bean is a crucial factor to consider when considering its potential use in food products. The nutritional profile, phytonutrient composition, and anti-nutritional factors of the faba bean are contingent upon the cultivar and environmental factors. Faba bean has been reported to contain high levels of protein (25-40%), carbohydrate (47-68%), and fiber (11-30%) (Hall et al., 2017; Macarulla et al., 2001; Valente et al., 2019). It encompasses a variety of phytochemicals, among which certain components such as vicine, convicine, and phytates exhibit anti-nutritive characteristics, whereas others, such as phenolic acid and polyphenols, possess bioactive attributes (Valente et al., 2019). Similarly, faba bean seeds exhibit low levels of fat, cholesterol, and sodium (Adamu et al., 2015). The insufficient intake of essential micronutrients is commonly referred to as "hidden hunger" and has been acknowledged as a global issue (Zandvakili et al., 2019). Legume seeds are commonly abundant in micronutrients such as iron and zinc (Khazaei et al., 2017), and these nutrients are essential for the nourishment and optimal biological functioning of living organisms (Bailey et al., 2015).

The elevated levels of calcium and magnesium present in the leaves may hold significant nutritional value for human consumption in regions where leaves are commonly consumed (Etemadi et al., 2019). It accumulates a significant quantity of L-Dopa (a chemical compound serving as a dopamine precursor) in its different components (Etemadi et al., 2015). It is a key component in treating Parkinson's disease and hormonal imbalances (Inamdar et al., 2013; Hu et al., 2015; Etemadi et al., 2018a). L-Dopa is an amino acid that is naturally obtained from different legumes (Inamdar et al., 2013) and other crops, including bananas (Gautam et al., 2012). Previous studies (Miller et al., 2009; Peaston and Weinkove, 2004) suggested that L-Dopa serves as a significant precursor for the synthesis of various neurotransmitters, including dopamine, noradrenaline, and adrenaline, which are collectively referred to as catecholamines (Gautam et al., 2012). Synthetic L-Dopa is a common treatment for patients with Parkinson's disease. The artificial production of L-Dopa is associated with high costs and a range of adverse effects such as hypotension, vomiting, drowsiness, and agitation (Patil et al., 2013). The consumption of natural sources of L-Dopa has been suggested to prevent potential side effects (Singh et al., 2012b; Patil et al., 2013). Cultivating crops with high levels of organic L-Dopa is a justifiable approach to addressing the costly production issues and side effects associated with synthetic L-Dopa (Patil et al., 2013).

The concentration of L-Dopa is influenced by various factors such as genotypes, environmental conditions, growth stage, and organs (Etemadi et al., 2014a, 2014b; Multari et al., 2015; Geng, 2012). Cao (2010) revealed significant variability in the L-Dopa content across different faba bean genotypes, while Geng (2012) observed varying rates of L-Dopa accumulation in different organs. Prior studies have primarily examined the concentration of L-Dopa in seeds, with comparatively less emphasis on quantifying L-Dopa levels in other plant components. According to Etemadi et al. (2018a), during the initial stages of growth, the amount of L-Dopa in various organs attains its maximum level. The availability of nitrogen and phosphorus may influence the synthesis of L-Dopa. Ramakrishna and Ravishankar (2011) documented that the insufficiencies in nitrogen and phosphorus directly impact the buildup of phenylpropanoids. However, Etemadi et al. (2018b) reported that nitrogen stress did not significantly impact the quantity of L-Dopa.

Abiotic Stresses

Effect of Drought Stress on Faba Bean

Drought is a significant abiotic stressor that poses a global threat to crop production, causing detrimental effects on plant growth and yield. Other abiotic stresses, in addition to global warming, can also lead to drought (Blum, 2011). Drylands constitute more than 40% of the Earth's surface (Mortimore et al., 2009). Drought stress is responsible for losing over 50% of the average yield of major crops (Zlatev and Lidon, 2012). According to Khazaei (2014), global climate change is expected to result in elevated temperatures and altered precipitation regimes across the globe. The faba bean is subject to the effects of climate change and is comparatively vulnerable to water scarcity compared to other legumes (Khan et al., 2010). This crop exhibits a notable degree of sensitivity to water supply (Oweis et al., 2005). However, in most production areas, its cultivation is primarily based on natural precipitation and soil moisture reserves for the crop's growth and development (Tadele, 2020). It experiences various physiological, biochemical, and molecular disturbances as a result of drought-induced stress (Abid et al., 2017; Siddiqui et al., 2015). Furthermore, Kabbadj et al. (2017) documented that drought conditions disrupted multiple enzymatic pathways and resulted in compromised photosynthetic activity.

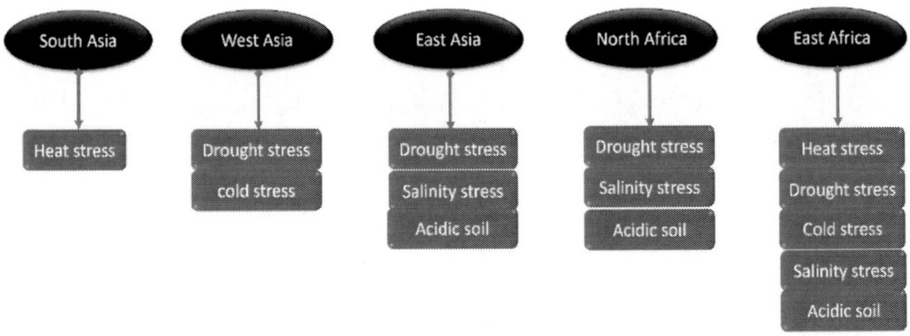

Figure 1. Major abiotic stresses in various regions for faba bean (Maalouf et al., 2021).

Drought stress has been observed to impact the process of photosynthesis negatively. This is mainly due to the alteration of pathway regulation caused by stomatal closure, which decreases the flow of carbon dioxide into

mesophyll tissues. Additionally, the activity of the enzyme RuBisCo has been found to be impaired under such conditions (Bota et al., 2004). Drought stress can disrupt various physiological processes such as respiration, nutrient assimilation, translocation, ion uptake, and carbohydrate metabolism (Farooq, 2008). The ability of plants to endure abiotic stresses is a complex characteristic that encompasses various molecular and physiological mechanisms (Razmjoo et al., 2008). The reaction of plants against stressors is contingent upon the genotypes, duration, intensity of water scarcity, and growth phase (Barnabás et al., 2008). Cultivating drought-resistant faba bean genotypes is a crucial objective in enhancing the productivity and quality of this crop. Nevertheless, the process of field phenotyping to evaluate drought response incurs significant costs. Additionally, this process is time-intensive and frequently provides unreliable results due to seasonal fluctuations.

According to Monneveux et al. (2014), an optimal phenotyping technique should possess the following attributes: rapidity, affordability, non-invasiveness, precision, and the ability to operate on multiple samples. Hence, to enhance the yield and stability of faba bean under drought stress, a comprehensive approach to breeding for drought stress is required, which involves the incorporation of various traits related to drought stress, such as early maturation and other relevant attributes. Studies have shown that the faba bean exhibits susceptibility to water deficit throughout its growth stages, ranging from seedling to maturity (Khan et al., 2007; Khan et al., 2010). In regions characterized by Mediterranean-type climates, the faba bean is typically sown during autumn and harvested in spring, coinciding with the rise in temperature and decrease in rainfall. However, it is common for the faba bean to encounter "terminal" droughts during this period (Amede and Schubert, 2003). In addition to terminal drought, the faba bean is frequently subjected to transient droughts in subtropical and temperate regions. Gnanasambandam et al. (2012) reported that faba beans experience a decrease in plant height and a reduction in the quantity of reproductive nodes during transient drought. Unlike escaping, managing unpredictable and intermittent transient droughts lacks a comparable and straightforward strategy (Muktadir et al., 2020). Effective regulation and function of stomata, root morphology, plasticity, and function, as well as epicuticular wax and osmotic adjustment, are crucial factors in managing transient droughts by maintaining higher relative water contents under water deficits (Khazaei et al., 2013; Belachew et al., 2018; Khan et al., 2010).

Whilst the faba bean has its origins in a semi-arid region, it is not typically known for its drought adaptation (Muktadir et al., 2020). The root system of

this plant is relatively shallow and is incapable of extracting water from deep soil (Manschadi et al., 1998). As a result, moisture stress occurs during the reproductive phase when the topsoil becomes dry. The presence of genetic resources that exhibit adaptability to fluctuations in soil moisture levels has been documented in various studies (Maalouf et al., 2015; Khan et al., 2007; Elshafei et al., 2019). This suggests there is a possibility of developing breeding strategies tailored to moisture-limited environments. The presence of extensive genotypic variations provides an opportunity to create faba bean genotypes that are tolerant to drought using conventional breeding methods (Muktadir et al., 2020). Water deficit is a factor that can also impact the quality of seeds. The regulation of seed protein contents is primarily influenced by genotypic background and external factors such as drought and heat (Rharrabti et al., 2001). According to Duc et al. (1999), there was a slightly higher protein concentration in winter faba beans as compared to spring beans. This could suggest water deficits tend to elevate protein levels.

The protein and sulfur contents of faba bean crops are enhanced when grown under water-deficit conditions as opposed to those grown with sufficient water (Alghamdi, 2009; Schumacher et al., 2011). Under optimal circumstances, the faba bean typically exhibits a substantial fixation of nitrogen. However, drought conditions can significantly hinder the fixation of nitrogen (Kabbadj et al., 2017; Neugschwandtner et al., 2015), leading to a reduction in nitrogenase activity. A decrease in soil moisture levels significantly impacted the quantity of infection filaments in faba beans. The exposure of faba bean plants to drought resulted in the deformation of outer cortical cells. The major abiotic stresses for faba beans in various regions of the world are shown in Figure 1.

Effect of High Temperature Stress on Faba Bean

The physiological response of faba bean is impacted by an episode of heat stress through various mechanisms. Gounaris et al. (1984) demonstrated that the rearrangement of chloroplast thylakoid membranes in faba beans in response to heat stress is reliable evidence of its ability to tolerate high temperatures. McDonald and Paulsen (1997) demonstrated that the photosynthetic activity of the entire chain in thylakoids of faba bean crops was susceptible to heat stress. This observation indicated that photosystem II was the most fragile. According to Hamada (2001) findings, exposure to a temperature of 42°C reduced the growth rate, membrane stability, and levels

of photosynthetic pigments (namely, chlorophyll a, chlorophyll b, and carotenoids) in faba bean. The net photosynthetic assimilation rate of three Mediterranean faba bean genotypes was considerably reduced when the leaf temperature exceeded 32°C (Avola et al., 2008). The reproductive phase of faba bean plants is particularly vulnerable to the negative effects of heat stress (Bishop et al., 2016). The literature indicates that legumes and various other plant species exhibit considerable pollen sensitivity to higher temperatures (Prasad et al., 2006; Wang et al., 2006).

According to Patrick and Stoddard (2010), the optimal temperature range for the flowering of faba bean is between 22 and 23°C. The delay of faba bean flowering can be attributed to sub-optimal temperatures and photoperiods during the initial stages of plant growth. The artificial imposition of terminal heat stress can result in a notable decrease in the yield components (Abdelmula and Abuanja, 2007). The negative impact observed can be primarily ascribed to elevated temperatures experienced during the vegetative phase. This resulted in growth inhibition and the production of a small crop with limited branching and pod development (Maalouf et al., 2021). The impact of high temperatures exceeding 32°C on the faba bean crop is significant, resulting in reduced seed germination, a rise in abnormal seedling occurrence, early flowering, degeneration of nodules that affect nitrogen fixation efficiency, and a decrease in plant biomass (Maalouf et al., 2021). The study carried out at ICARDA has determined that a diverse spectrum of genetic variability has been detected in faba bean with regards to its ability to withstand high temperatures. The pollen viability of tested faba bean lines exhibited a range of 0 to 50% when subjected to a temperature of 32°C and a range of 0 to 33% when exposed to a temperature of 35°C. Seven accessions were chosen for their heat stress tolerance based on the findings obtained. Seven accessions that exhibited increased rates of pollen viability under 35°C and increased numbers of pod and seed per plant were identified. These accessions were IG11743, IG11843, VF351, VF420, VF522, VF626, and INRA1631 (Maalouf et al., 2018).

Effect of Salinity Stress on Faba Bean

Faba bean cultivation is possible across a diverse range of agroecosystems; however, the crop's productivity is significantly impeded by various environmental stressors, such as salinity (Latef et al., 2021). The phenomenon of soil salinity is recognized as a prominent abiotic stressor, which has a

considerable economic impact due to the loss of a substantial amount of agricultural land and a reduction in crop yield (Mbarki et al., 2018; Fariduddin et al., 2019). The issue of salinity poses a significant challenge in agriculture, particularly for crops grown in areas with inadequate rainfall to facilitate the removal of mineral salts from the vicinity of the roots (Safdar et al., 2019). The global extent of salt-contaminated lands exceeds 800 million hectares (Munns and Tester, 2008; Osman et al., 2020). According to Shrivastava and Kumar (2015), approximately 45 million hectares of irrigated land are affected by salt stress-related issues globally. Furthermore, it is projected that over 50% of arable land may be impacted by salinity within the next three decades. The phenomenon of soil salinization on a global scale is being exacerbated by a combination of factors, including the rise in sea levels resulting from global warming, the use of saline-enriched water for irrigation, and extensive soil erosion (Shrivastava and Kumar, 2015; Nachshon, 2018).

Srivastava et al. (2019) reported that the security of people's livelihoods in over 100 countries is jeopardised by soil salinity. The cultivation of *Vicia faba* is significantly limited by the presence of saline soil, particularly in semi-arid areas (Manchanda and Garg, 2008). Salinity stress has been observed to impede the molecular, biochemical, and physiological mechanisms of crop plants. This is primarily due to the negative impact on photosynthetic activity, alteration of osmotic adjustment and ionic homeostasis, generation of toxic oxidants and radicals, modulation of phytohormonal functions, disruption of crucial metabolic pathways, and manipulation of gene expression patterns (Yang and Guo, 2018; Morton et al. (2019). The presence of excessive quantities of Na^+ and Cl^- can lead to ionic imbalance, osmotic stress, water scarcity, and nutrient deficiencies (including K, Ca, Mg, Fe, Zn, N, and P) in crop and vegetable plants, resulting in physiological dysfunction (Rehman et al., 2019). The disruption of redox homeostasis results in the excessive generation of reactive oxygen species (ROS), which in turn causes oxidative stress to cellular biomolecules (Abdel Latef et al., 2019; Attia et al., 2021).

In addition, it has been observed that salt stress has a negative impact on various physiological processes such as osmotic equilibrium, nutrient balance, stomatal activity, hydraulic conductance, and photosynthesis (Füzy et al., 2019). The physicochemical properties of soil can be modified by salinity, leading to a reduction in the diversity of microbial species. Faba beans are exposed to Na^+ toxicity and subsequent K^+ toxicity under conditions of salt stress. This can result in damage to cellular organs and a reduction in the efficiency of photosynthesis in faba beans (Ahmad et al., 2010; Pozo et al., 2015). Earlier findings indicate that an elevated NaCl concentration has an

adverse impact on plant height (Ziche et al., 2023). However, Qados (2011) and Janda et al. (2016) observed that the application of NaCl at low and medium concentrations resulted in an increase in certain morphological parameters, such as plant height. The authors attributed this phenomenon to the osmotic adjustment mechanism, which can enhance its growth. The yield of faba beans was significantly impacted by the concentration of salt in the soil (Bimurzayev et al., 2021). Nevertheless, the growth of faba bean seed during germination is more vulnerable to salinity. Faba bean exhibits sensitivity to salt (Ahmad et al., 2018). According to Tavakkoli et al. (2010), the application of salt treatments led to a decrease in the yield of biomass and water intake in various genotypes of faba beans. The growth and biomass yield of faba bean were impeded by the presence of NaCl, which can be attributed to the inhibition of cell division and elongation (Ahmad et al., 2016). The reduction of mineral intake, generation of reactive oxygen species, inhibition of enzyme activity, and induction of hormonal imbalances are among the mechanisms by which NaCl reduces growth and biomass yield (Ashraf et al., 2010). Desouky et al. (2023) reported that elevated salinity levels resulted in a substantial decline in dry matter accumulation and reductions in shoot height, root length, leaf count, and branch count. Furthermore, the salinity increment was found to cause a significant reduction in the levels of the photosynthetic pigments, namely chlorophyll a and b.

Effect of Soil Acidity Stress on Faba Bean

The host plant cultivar, the Rhizobium strain, and the relationship of these symbionts with the pedoclimatic components and the ambient conditions are necessary for symbiotic nitrogen fixation. Extremes in soil acidity impact nodulation by reducing rhizobial colonization of the soil. Low levels of phosphorus, calcium, and molybdenum are frequently found in highly acidic soils (pH less than 4.0), while excessive levels of aluminium and manganese found in acidic soils are frequently hazardous to both partners. The detrimental effects of soil acidity on crop growth and soil biodiversity are linked to nutrient deficiencies and metal toxicity (Chimdi, 2015). Rhizobia, a symbiotic microbe found in root nodules, is used by most species of the legume family to fix atmospheric nitrogen.

The symbiotic fixation of nitrogen is a crucial component of soil fertility. Nodulation and the amount of nitrogen fixed by legumes in acidic soils are decreased due to direct impacts on the growth of the host flora and the survival

and proliferation of rhizobia (Giller, 2001). The issue of soil acidity is a prevalent global concern that is linked to the increased presence of Al^{3+} ions, in addition to the fundamental aspect of lower levels of pH (Atemkeng et al., 2011). The natural phenomenon of soil acidity poses significant challenges to the growth of plants. According to Alemu et al. (2016), soil with a pH value ranging from 6.6 to 7.3 is classified as neutral, while soil with a pH value ranging from 5.6 to 6.5 is categorized as moderately acidic. Soil with a pH value below 5.5 is considered to be strongly acidic. When the pH level of soil decreases below 4.5, the production of food crops becomes challenging due to the reduction in the supply of most plant nutrients and the increased solubility and toxicity of aluminium and a few micronutrients to plants (Harter, 2007). According to Jensen et al. (2010), most chemical and biological processes in crops are significantly affected by soil acidity. The sensitivity of faba beans to soil acidity is widely recognized, and this characteristic restricts the utilization of faba beans in certain agricultural systems. Faba bean is commonly recognized to be susceptible to soil acidity and has been found to thrive in mildly acidic soil conditions (Burns et al., 2017; Chris and Stephen, 2009).

Phosphorus is a crucial element in the process of symbiotic nitrogen fixation and the development of seeds and fruits in legumes. The inadequacy of this nutrient leads to suboptimal crop productivity and inferior seed and fruit quality (Fageria, 2009). Faba bean necessitates a substantial amount of phosphorus to facilitate energy expenditure for the purpose of nodule formation (Kopke and Nemecek, 2010). Nevertheless, the process of phosphorus fixation occurs in acidic soils, resulting in a significant decrease in its accessibility. The primary limitation to the cultivation of faba beans in the highlands of Ethiopia is soil acidity (Endalkachew et al., 2018). Leguminous crops generally exhibit a preference for a pH level greater than 5.0, particularly at a depth of 20 cm or more, and layers below 5 cm can have a negative impact on root growth, nodulation, plant vigor, and the potential for nitrogen fixation in acid-sensitive pulses (Burns et al., 2017). The optimal pH range for the growth of faba bean is between 6.5 and 9.0 (Jensen et al., 2010), while French and White (2005) have reported that the crop exhibits poor performance when grown in soils with pH values of 5 or lower. Singh et al. (2012a) have reported that certain accessions exhibit productivity even in low pH soils, i.e., 4.5. The yield of faba bean is negatively impacted by low soil pH, which is commonly linked to poor soil fertility (Endalkachew et al., 2018). The presence of aluminum and manganese toxicities creates an unfavorable condition for the development of roots (Ouertatani et al., 2011).

Mesfin (2019) reported that in the central highlands of Ethiopia, faba bean genotypes experienced a significant reduction in mean yield of 32.34% at soil pH levels ranging from 4.49 to 4.96, which was attributed to soil acidity. The susceptibility of faba bean to chocolate spot infection is heightened in acidic soil, which can also lead to decreased plant vigour and ultimately results in reduced yield (Elliot and Whittington, 2009; Getachew et al., 2005). The early-stage response to soil acidity exhibited a range of outcomes in faba bean genotypes. Certain genotypes demonstrated a root length reduction of less than 3%, while others experienced a notable hindrance in root growth of 30-40%. The cultivars Dosha, NC 58, and Kassa exhibited tolerance to acidity, as evidenced by a reduction of < 3% in the length of the taproot (Kiflemariam and Stoddard, 2017). Consequently, it is imperative to mitigate soil acidity in order to establish a more favorable cultivation ecosystem for a better crop yield from faba bean.

Effect of Waterlogging Stress on Faba Bean

Abiotic stresses such as drought, high temperatures, frost, and water logging have been identified as significant factors that impact the productivity of faba beans in various production regions globally (Figure 2). Drought and heat have a worldwide impact on crop production, whereas waterlogging and frost are contingent upon the ecological and geographical characteristics of specific localities (Adhikari et al., 2021). The issue of waterlogging is confined to specific geographic regions, such as those with high levels of rainfall and dominated by vertisols in Ethiopia, as well as the irrigated Nile River basin in Egypt (Adhikari et al., 2021). In East Africa, the constraints that significantly limit agricultural productivity are waterlogging and acidic soils, particularly in the elevated regions of Ethiopia (Keneni et al., 2010). According to Shahin (2019), crops can suffer significant harm due to two distinct forms of waterlogging: transient and continuous. The phenomenon of waterlogging has the potential to cause soil nutrient depletion and structural disturbance. Waterlogging can lead to the depletion of nitrogen in the soil, resulting in insufficient nitrogen availability for plant growth. The yield of faba beans can be restricted by waterlogging, particularly during the flowering stage (Pampana et al., 2016).

The condition of waterlogging during the maturity stage is a significant abiotic element that influences faba bean growth and developmental processes (Redden et al., 2014). The enduring consequences of waterlogging on growth

and physiological characteristics, such as chlorophyll a and b, persist even following the termination of soil flooding (Pociecha et al., 2008). It has been observed that faba bean exhibits tolerance towards short episodes of waterlogging (Tekalign et al., 2016). Although faba beans have been found to have a higher tolerance for waterlogging compared to other legumes (Solaiman et al., 2007), waterlogging during the blooming and pod setting phases can lead to a considerable reduction in pods and flower numbers. This reduction can lead to a substantial decrease in growth and ultimately result in yield losses (Pampana et al., 2016; Tadele, 2020). Furthermore, an extended period of waterlogging has a detrimental impact on the growth of plants and exacerbates foliar diseases at any point in the plant's life cycle. According to Keneni et al. (2001), effective management of excess water can be achieved through the use of broad beds, well-aerated soil and furrow cultivation for drainage. However, the most optimal solution to address waterlogging issues may lie in the development of productive genotypes through breeding efforts. The issue of waterlogging has been tackled in Ethiopia through the development and subsequent release of faba bean varieties (Wayu, Selale, Didi'a, Hachalu, Ashebeka, Wolki) that have demonstrated the capability to retain high yields even in waterlogged conditions (Tadele, 2020). Sallam and Ul-Allah (2019) propose that the impact of waterlogging is anticipated to intensify as a result of alterations in climatic conditions.

Agronomic Strategies to Mitigate Abiotic Stresses

Agricultural practices such as tillage, selection of appropriate planting time and crop varieties, intercropping, mulching, seed priming, and utilizing plant growth-promoting rhizobacteria and arbuscular mycorrhizal fungal inoculation, among others, play a significant role in ameliorating the effect of abiotic stresses (Figure 3).

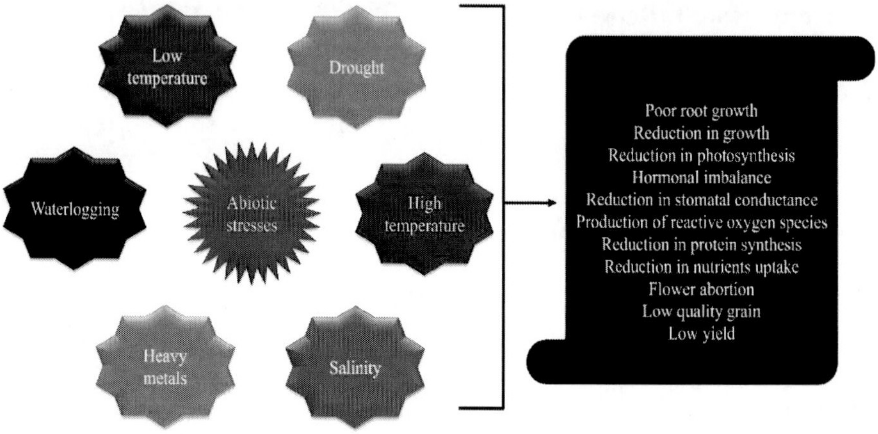

Figure 2. Major abiotic stresses and their effects on plants.

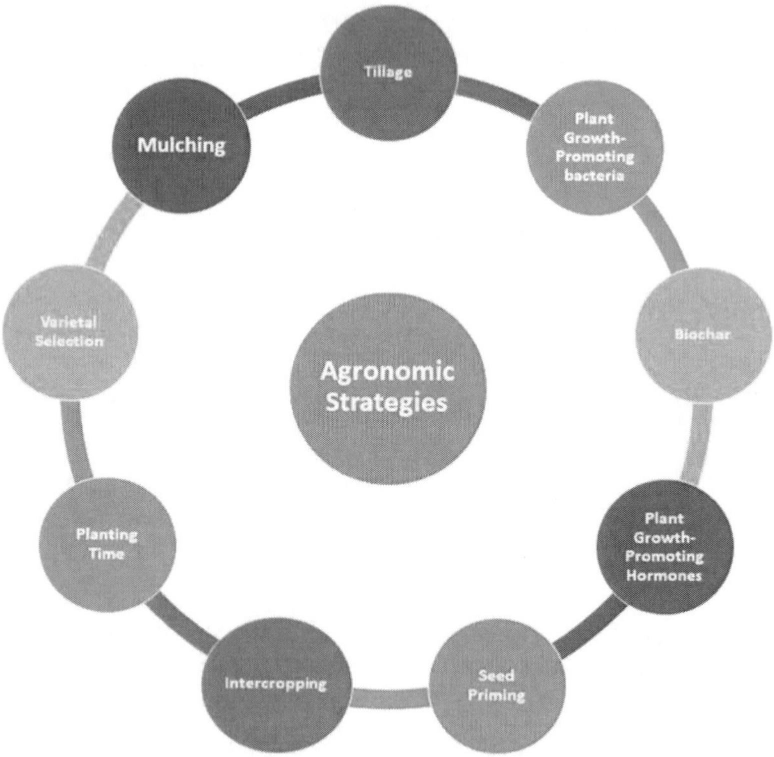

Figure 3. Agronomic strategies to mitigate abiotic stress.

Conservation Tillage

Conservation tillage and planting systems involve covering at least one-third of the upper surface of the soil with crop residues after sowing, leading to improved water availability for crops by preserving more soil moisture. In regions with a Mediterranean climate, sowing faba beans after the first autumn rains with a conservation tillage approach helped to reduce the effect of drought with a seed yield of 4.2 t ha^{-1}, which depends on the amount of seasonal precipitation (Loss et al., 1997). One of the positive impacts of conservation tillage on plant water status can be attributed to its effects on soil biota. Tillage practices can significantly influence the abundance of mycorrhizal fungi (Säle et al., 2015). Compared to intensive tillage practices, conservation tillage promotes the abundance of fungi in the soil by minimizing the mechanical disruption of fungal networks (Zhang et al., 2014). However, the effects of tillage on the mycorrhizal community appeared to be specific to the soil type and crop species involved (Hannula et al., 2021). According to Madejón et al. (2023), conservation tillage, including no-tillage and reduced tillage, has a positive impact on plant ecophysiology and promotes microbial symbiosis, resulting in increased crop resistance to water stress. This symbiosis includes enhanced colonization of arbuscular mycorrhizal fungi and nodulation by rhizobia bacteria.

Planting Time and Varietal Selection

Sowing time plays a crucial role in insect-pest phenological development, crop growth, and weed dynamics, as well as final crop productivity. It is considered one of the most significant nonmonetary inputs with substantial effects in the context of abiotic stresses. The sowing date is a crucial factor in determining crop yields. Terminal heat and drought inhibit the growth of vegetative parts of plants, leading to pre-maturity and reduced yields (Reddy, 2009). For cool-season pulses, i.e., faba beans, field peas, chickpeas, and lentils, high temperatures above 25°C can cause flower drop and pod abortion, resulting in yield reductions ranging from 20% to 70% (Kumar et al., 2016). Research conducted by Abdel Aziz (2007) in the Syrian coastal region revealed that an early sowing date of 11/15 for the Cypriot cultivar resulted in a higher number of branches per plant and a shorter century length compared to a later sowing date of 12/15. The early sowing date promoted increased internode numbers

and stimulated lateral bud growth due to the earlier arrival of photosynthesis products. On the contrary, productivity decreased with late planting due to a reduced number of plants remaining at harvest. In Syria, Al-Othman and Al-Assaf (2009) observed that delaying the planting date from 10/15 to 5/11 led to a decrease in plant height but an increase in the number of pods and overall production per unit area in Karbala, Iraq. Al-Anbari et al. (2009) found that early planting of broad beans on 9/10 resulted in increased leaf area, percentage of leaf, number of pods, number of seeds per plant, and weight of 100 seeds.

The choice of crop varieties also plays a key role in reducing the effects of environmental stresses. Certain varieties are better suited to tolerate the effect of stress and produce good yield. Selecting crops and varieties that are well-adapted to the local environment and capable of thriving with limited water supply is crucial for effectively dealing with drought stress (Singh et al., 2014). In areas prone to drought, it is advisable to encourage the cultivation of short-duration crops and early maturing cultivars that require less water. Increasing profitability and water use efficiency can be achieved by substituting low-demand water crops with high-demand water crops (Bobojonov et al., 2013).

Intercropping

Under rainfed conditions, it is recommended to employ intercropping systems to attain greater and more steady yields (Singh et al., 2014). According to Gautam and Bana (2014), intercropping in specific ratios has the potential to mitigate the risk of complete yield loss and enhance water utilization in rainfed regions. The practice of intercropping has been found to enhance the production of crops concerning the amount of water supplied, as reported by Sharma et al. (2017). The practice of intercropping has been found to have several benefits, including improved soil water conservation, reduced drainage, and increased water supply. Additionally, intercropping has been shown to enhance water use efficiency (Raza et al., 2021). Intercropping has the ability to reduce evaporation from inter-row spaces, regulate excessive transpiration, facilitate water storage in the root zone, and create a conducive micro-climate for plant development (Zhang et al., 2012). Caihong and colleagues (2015) documented that faba bean can be successfully intercropped with wheat.

Seed Priming and Foliar Application

The process of seed priming is a refined agricultural technique that entails the controlled hydration of seeds prior to planting, intending to enhance the long-term productivity of crops (Seleiman et al., 2021; Taylor et al., 1990). By subjecting seeds to restricted hydration, priming cultivates resilience to stress during the seedling stage (Kubala et al., 2015) and throughout the growing season (Iqbal et al., 2006). The process of priming stimulates faster germination and more consistent emergence (Aliloo et al., 2014; Kumar et al., 2017), enabling farmers to overcome time lost due to drought conditions (Harris et al., 2001).

Zinc priming alleviated the negative impact of drought stress on faba bean plants at the vegetative phase. This priming technique results in several beneficial effects, including an increase in total biomass by 99.8%, area of leaf surface by 23%, soluble sugar content by 54.7%, activity of α-amylase by 85%, total chlorophyll content by 48.7%, and zinc concentration in the leaves by 79.8%. Additionally, it leads to a decrease in malondialdehyde (MDA) content by 42.7% and antioxidant activities by 35.2% under water-deficit conditions (Farooq et al., 2021). The study conducted by Abid et al. (2020) revealed that the application of 1(mM) β-aminobutyric acid as a priming agent on faba bean seeds can mitigate the adverse impacts of drought stress. The utilization of this priming technique has been observed to increase the rate of leaf photosynthesis, stomatal conductance, leaf relative water content, and transpiration rate. Furthermore, it reduced the production of H_2O_2, levels of MDA, electrolyte leakage, and damage to the cell.

The application of $CaCl_2$ as a priming agent resulted in a significant enhancement of both nodule number and dry weight in faba bean, with increases of up to 20 and 34%, respectively, under 50 μM cadmium (Cd) stress conditions when compared to non-stress conditions. Furthermore, it effectively alleviates the detrimental Cd effects on the physiological and biochemical attributes of the plants (Nouairi et al., 2021). Similarly, seed priming with Monopotassium phosphate (KH_2PO_4) reduces the detrimental effects of phosphorus deficiency in faba beans. The utilization of this priming technique results in an augmentation of overall biomass, phosphorus uptake, and the activation of antioxidant mechanisms (M'Sehli et al., 2020). The findings of M'Sehli et al. (2020) indicate that seed priming is a viable approach in situations where there is a lack of nutrients. The application of salicylic acid (SA) and abscisic acid (ABA), which are important plant hormones, can protect plants against abiotic and biotic stresses while

promoting plant growth and development (Hayat et al., 2010). The mitigation effect of SA against abiotic stresses has been studied through foliar spray (Khodary, 2004), seed soaking (Al-Hakimi, 2006), or application to the rooting medium (Arfan et al., 2007). The application of ABA at a concentration of 10 μg l−1 has enhanced drought resistance in stressed faba bean plants (Abdel and El-Hamadany, 2010). Furthermore, Abdelaal (2015) stated that the exogenous application of SA and ABA minimizes the harmful effects of drought stress and improves the morpho-physiological characteristics, stem and leaflet anatomy, as well as yield components of faba bean plants. The study conducted by Abdel Latef et al. (2021) revealed that the application of indole 3 acetic acid (IAA) resulted in a decrease in proline levels and an increase in the contents of soluble protein, total free amino acid and soluble sugar in the stems, seeds and roots of the plants under salt stress. Likewise, IAA resulted in an augmentation in the number of nodules. Furthermore, IAA exhibited a positive impact on mineral homeostasis through the augmentation of calcium, potassium, and magnesium levels while impeding the disproportionate buildup of sodium in the roots of faba bean.

Bouallègue et al. (2017) reported that priming with SA is an efficacious physiological technique for enhancing faba bean germination in the presence of salt stress. The observed effects included an elevation in the antioxidant activities of superoxide dismutase and peroxide, as well as an increase in total lipid content and, notably, the C18:3 fatty acid. It has also been found to be efficacious in reducing MDA levels. The utilization of ascorbic acid in seed priming was found to be efficacious in alleviating the detrimental impact of salt stress on the growth of wheat (Azzedine et al., 2011). This was attributed to the augmentation of the leaf area, enhancement of chlorophyll and carotenoid contents, and increased accumulation of proline. The study conducted by Azooz et al. (2013) demonstrated that the application of ascorbic acid and nicotinamide, either separately or in combination, through seed priming, resulted in an improvement in the salt tolerance of faba bean plants. The application of this particular treatment resulted in a notable augmentation of photosynthetic pigments, a reduction in ion leakage, and a decrease in the levels of Na^+ and Cl^- contents. Additionally, there was a significant accumulation of osmotically soluble carbohydrates, proteins, total free amino acids, and proline within the plant specimens. Acetylsalicylic acid (ASA), also known as aspirin, acts as a hormone-like substance in plants and plays a vital role in regulating various physiological processes while protecting plants against biotic and abiotic stresses (Azooz and Al-Fredan 2009). Ehtaiwesh (2022) reported that the application of ASA alleviated the negative effects of

NaCl toxicity in faba bean plants. Melatonin exhibits solubility in both water and lipids, enabling it to function as a versatile antioxidant with hydrophilic and hydrophobic properties (Janas and Posmyk, 2013). In a study conducted by Dawood and El-awadi (2015), the application of melatonin treatments (100 mM and 500 mM) through seed priming resulted in improved growth parameters. Additionally, it reduced the levels of compatible solutes, Na^+, and Cl^- in the leaf tissues of faba bean plants under salt stress. Another study by Abbas et al. (2021) highlighted the benefits of incorporating plant compost as an organic amendment and using biostimulants (VIUSID® agro) through foliar application. The aforementioned method demonstrated noteworthy outcomes in enhancing the productivity and grain yield of faba beans grown in circumstances where the irrigation water was affected by salinity.

Role of Plant Growth-Promoting Rhizobacteria and Arbuscular Mycorrhizal Fungal Inoculation in Mitigating Stress

The utilization of plant growth-promoting rhizobacteria (PGPR) is a crucial measure to mitigate the adverse drought stress effects on leguminous plants (Khan et al., 2019a). The coping strategy employed by the organism to mitigate the effects of drought stress encompasses a range of mechanisms, including but not limited to phosphorus solubilization, organic acid production, nitrogen fixation, siderophore production, and the synthesis of enzymes such as ACC deaminase. This has been documented in studies conducted by Glick et al. (2007) and Niu et al. (2018). According to Dimkpa et al. (2009), PGPR has the ability to synchronize phytohormones such as ABA, ethylene, gibberellins, cytokinins and auxins. Chanway et al. (1989) reported that the utilization of *Pseudomonas putida* in lentils resulted in improved nodulation and plant growth, leading to enhanced drought resistance. In conditions of drought-induced stress, the arbuscular mycorrhizal fungi (AMF) facilitate the absorption of water and nutrients (Auge et al., 2002). Habibzadeh et al. (2014) observed that the augmentation of osmoprotectants and antioxidant potential leads to a reduction in lipid peroxidation. The formation of soil aggregates, facilitated by AMF, is a crucial factor in enhancing soil structure and its capacity to retain water (Smith et al., 2010). Arbuscular mycorrhizal (AM) fungi function as growth regulators and alleviate the deleterious impacts of salt stress on plants. AM fungi provide

plants with essential mineral nutrients, particularly phosphorus. These nutrients are precipitated by ions such as calcium, magnesium, and zinc (Alqarawi et al., 2014). AM fungi stabilize membranes, increase photosynthetic pigments, stimulate plant defense chemicals, and maintain cell osmotic and ionic balance. According to Abeer et al. (2014), the presence of AM fungi has been found to alleviate the negative impact of NaCl stress on faba beans, resulting in enhanced growth and yield.

The utilization of PGPR, specifically the co-inoculation of *P. putida* and *R. leguminosarum*, has been observed to mitigate the adverse impacts of drought stress on faba bean crops. This approach has also been found to enhance seed yield and crop water productivity by increasing the expression of photosynthetic pigments, root parameters, plant-water relations, and enzymatic and non-enzymatic defense system components (Mansoor et al., 2021). According to Mohammed (2018), the biofilm formation and exopolysaccharides production of bacteria that promote plant growth have been found to improve the resilience of faba bean plants to varying levels of salt concentration. The study conducted by Hafez et al. (2021) demonstrated that the utilization of PGPR and potassium silicate in combination could be regarded as a lucrative approach for mitigating the adverse effects of salt stress while simultaneously enhancing the growth and productivity of faba beans. When *Rhizobium leguminosarum* bv. viciae was used to inoculate faba beans under alkaline soil or salt stress conditions, which led to an increase in growth and production in faba beans (Abd-Alla et al., 2014; Nour el din et al., 2011). Another PGPR, *Bacillus circulans*, has numerous advantages for inoculated plants, such as its capacity to improve potassium ion availability in salt-affected faba bean plants (Khalil et al., 2019). The utilization of PGPR has been observed to have a positive effect on the yield-related characteristics and overall productivity of faba beans (Hafez et al., 2021). The positive effects of PGPR on plant yield can be attributed to their ability to produce phytohormones and polysaccharides, solubilize phosphate and engage in symbiotic nitrogen fixation. These processes enhance physiological functions and improve biochemical constituents, ultimately leading to an increase in yield-related traits and productivity in faba bean plants that are irrigated with saline water (Paul et al., 2014; Alaa, 2018).

Biochar

The utilization of biochar has demonstrated considerable benefits in enhancing soil properties and agricultural productivity, under both normal and unfavorable conditions (Figure 4). Biochar has exhibited effectiveness in augmenting soil quality and promoting crop development through the regulation of enzyme activity and mitigation of oxidative damage in plants (Abbas et al., 2017; Ali et al., 2019; El-Naggar et al., 2018). The scholarly literature suggests that biochar has the potential to improve soil water retention and increase water availability for plants, as demonstrated by the studies conducted by Ahmed et al. (2018) and Liu et al. (2021a). The observed phenomenon can be ascribed to the substantial surface area, copious microchannels, and presence of hydrophilic functional groups such as -OH and -COOH (Li and Tan, 2021). The application of biochar has been observed to enhance the water status of plants, increase stomatal conductance, and regulate hormonal balance. This is attributed to the biochar's ability to augment soil water-holding capacity and the availability of water content for plants. The aforementioned phenomenon results in an increase in plant growth and subsequent yields (Akhtar et al., 2015; Liu et al., 2021a, 2021b). According to Zhang et al. (2020), biochar can alleviate the unfavourable consequences of drought stress on water use efficiency in leguminous crops, such as soybeans. Furthermore, scholarly research has shown that the application of biochar can yield advantageous outcomes for salt-affected soils by enhancing their water retention capacity and reducing their bulk density (Burrell et al., 2016; Huang et al., 2019; Mansoor et al., 2021). El Nahhas et al. (2021) reported that biochar and jasmonic acid treatments effectively mitigated the negative effects of salinity on faba bean growth. The application of these interventions yielded enhancements in plant attributes, including augmented plant size, expanded leaf surface, amplified relative water content, and increased concentrations of chlorophyll a and b. Furthermore, the utilization of biochar and jasmonic acid on plants subjected to saline environments resulted in a noteworthy improvement in plant productivity, as evidenced by a rise in flower quantity, seed yield per plant, and the weight of 100 seeds over the course of two successive cultivation periods. Egamberdieva et al. (2020) revealed that the utilization of maize-derived biochar has a positive impact on the growth and nutrient absorption of faba beans. This is achieved through the enhancement of nodule formation and soil nutritional composition in sandy loam soils that are subjected to drought stress. Rajhi et al. (2022) have reported that the application of biochar can mitigate the impact

of drought stress on faba beans. The findings of this study validate the beneficial impact of adding biochar into alkaline soils, which enhances the moisture retention capacity of the modified soils. This, in turn, leads to enhancements in the biochemical, photosynthetic and physiological characteristics of faba bean plants. The study conducted by Rezaie and colleagues in 2019 revealed that the inclusion of 2.5% w/w biochar can effectively alleviate the negative effects of salinity stress. This is attributed to the biochar's ability to effectively absorb salt, as well as its capacity to increase the K^+/Na^+ ratio in the soil. As a result, the growth, physiological parameters, and yield of faba bean were significantly improved. The provision of acidified biochar proved to be efficacious in improving the detrimental effects of water scarcity by augmenting the quantum efficiency of a photosystem, water status, soil microbial activity, plant growth characteristics, and enhancing soil properties (Abd El-Mageed et al., 2021). The mixing of biochar into soil resulted in the enhancement of faba bean growth by augmenting the availability of vital nutrients and elevating the potassium: sodium ratio. This was achieved through the amplification of potassium content in the soil (Kim et al., 2016). Additionally, Abbas et al. (2017) found that biochar addition increased water availability and stomatal density, thereby contributing to the growth of faba beans. Furthermore, the accumulation of ROS was reduced because of biochar addition.

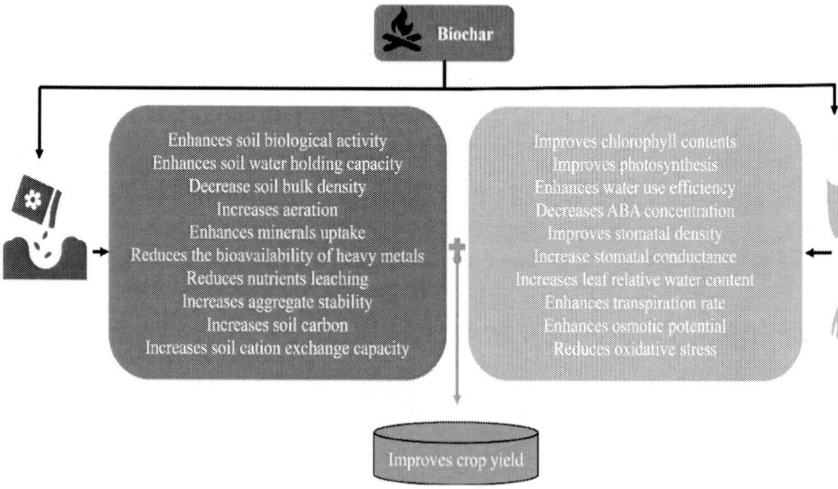

Figure 4. Effects of biochar on soil properties and plants.

Breeding Strategies for Abiotic Stresses

The current era of crop production is faced with noteworthy obstacles, encompassing biotic and abiotic stressors that curtail crop productivity (Cushman et al., 2022). To enhance crop productivity, it is imperative to comprehend the various ways in which plants react to diverse abiotic stresses. The integration of traditional breeding techniques with biotechnological methods has been proposed as a viable means of achieving this objective (Singh et al., 2012a). The practice of plant breeding is of utmost importance in augmenting crop productivity and acclimatizing to environmental stressors (Moose and Mumm, 2008). The process entails the generation, curation, and maintenance of advantageous plant traits to cultivate enhanced varieties. In terms of cost, especially for farmers with limited resources who are unable to afford production inputs, genetic modification is preferable to continuous manipulation of the growing conditions in order to increase crop productivity. Therefore, breeding efforts should focus on selecting varieties suitable for the target production environments (Gemechu, 2007). For example, drought-resistant genotypes should be bred in drought-prone environments, and genotypes that exhibit tolerance to acid soil and waterlogging should be developed and used in environments that are susceptible to acidification and waterlogging, respectively. The breeding strategy aims to exploit the genetic potential of cultivars in optimal environments while developing varieties that consistently perform well in various conditions. It is important to the development of crop cultivars that are well-suited to resource-limited conditions, such as regions characterized by drought, nutrient deficiencies, and acidic soils. The goal of particular adaptation is to create varieties that fit well into the present environment, have low resource requirements, and are resilient to stressed situations, in addition to maximizing the genetic capability of crops (Gemechu et al., 2016). The principal objectives of faba bean breeding are to enhance yields, seed quality (such as reducing vicine convicine and tannin levels), and acclimatization to abiotic and biotic stressors (Torres et al., 2012). Apart from improving yield and stability, other challenges in faba bean cultivation include poor mechanization, vulnerability to diseases and pests, and the presence of non-nutritional factors (Torres et al., 2012). Faba bean is affected by various stresses, both from living organisms (biotic) and the environment (abiotic), which lead to decreased yield and instability. The main abiotic stressor that influences the faba bean yield is heat and drought. Previous studies have identified sources of tolerance to frost, heat, acidic soils

and drought through screening for these abiotic stresses (Torres et al., 2012; Maalouf et al., 2019a).

Traditional Breeding Methods

Faba bean breeders are interested in identifying the most effective breeding strategy to attain greater genetic gains, which may ultimately result in improved outcomes. Faba bean is a type of plant that relies on both cross-pollination and insect pollinators. It has the potential to contribute significantly to sustainable agriculture and the preservation of pollinators (wild) in the natural environment (Maalouf et al., 2019a). The outcrossing rate, which can range from 4 to 84% (Suso et al., 1999), measures the level of heterosis observed in the crop. Higher heterosis leads to increased stability, resistance to environmental stresses and yield (Gasim and Link, 2007). However, there has been a lack of research comparing different breeding strategies for faba beans in recent years. Although synthetic cultivars and pedigree selection are widely employed breeding techniques for this partially cross-pollinated crop, the recurrent selection (RS) method is rarely utilized in faba bean populations (Maalouf et al., 2002; Ibrahim, 2015).

The significant degree of interbreeding in faba beans holds crucial importance in the development of composite cultivars and in enhancing the open breeding population (Link et al., 1994a). One solution is to create synthetic cultivars employing self-fertilizing lines, which assures some yield even in the absence of insect pollinators. The utilization of pollinators to enhance yield and stability via hybrid vigour is also a feature of this approach (Cubero and Moreno, 1984). The utilization of hybrid vigour in faba bean cultivars can be achieved through the implementation of both recurrent and synthetic breeding methods, resulting in enhanced yield and stability (Arbaoui and Link, 2008; Gharzeddin et al., 2019) and the ability to tolerate abiotic stressors (Gasim and Link, 2007).

Recurrent selection (RS) has been implemented as a means of enhancing the quality of cross-pollinated crops, such as faba beans, that are either fully or partially cross-pollinated. Rowland (1987) reported the initial utilization of the recurrent selection method in faba bean. The author expounded that an RS program involves cultivating better lines in open-pollinated random mating nurseries. The study conducted by Rowland (1987) involved the selection of robust plants and the subsequent assessment of their progeny's yield. The lines that were chosen exhibited a favourable genetic advancement of 1.8%

annually. It is plausible that a majority of individual plants chosen under open pollination circumstances were of hybrid origin, given that hybrid faba bean plants exhibit greater levels of self-fertility than inbred plants (Drayner, 1959). Consequently, they are expected to yield higher pods per plant. The current primary selection criterion employed in the ICARDA breeding program is akin to this approach (Rowland et al., (1986). Gharzeddin et al. (2019) reported that recurrent selection resulted in higher yields compared to the pedigree method, indicating a stronger response in recurrent line selection. Synthetic varieties of crops, such as faba beans, can achieve higher yields compared to other open-pollinated varieties (Gharzeddin et al., 2019). This is because recurrent varieties, which undergo traditional breeding methods, have not been specifically bred for high-yield performance. Developing synthetic varieties takes less time compared to classical breeding studies that focus on selecting individual clones (Flajoulot et al., 2005). Synthetic varieties yield was greater than those lines which were developed through RS methods (Gharzeddin et al., 2019), as they can better exploit heterosis (hybrid vigour) and heterogeneity. Various studies have shown that faba bean synthetic varieties have the potential to maximize yield compared to recurrent cultivars (Poulsen, 1981; Link et al., 1994a,1994b; Stelling et al., 1994). The pedigree method is adopted to create pure lines with high auto-fertility in faba bean. Some researchers have suggested transforming the mating system of faba beans towards autogamy to develop inbred lines for organic farms (Bozzini and Chiaretti, 1999; Ghaouti et al., 2008). However, achieving higher yield through autogamy is challenging due to strong inbreeding depression (Lawes et al., 1983). Various studies have compared the pedigree method, mass selection and single seed descent for faba bean breeding, and the pedigree method was found to be the most suitable. Selecting a high self-fertility degree can simplify breeding and seed production (Ahmed et al., 2008).

Advancing Faba Bean Improvement Through Mutation Breeding

de Vries (1901) initially introduced the concept of mutation breeding. A mutation is a phenomenon characterized by a permanent alteration in the number or sequence of nucleotides, which occurs relatively infrequently. Mutation breeding is the process of genetically enhancing crop plants for diverse economic traits utilizing induced mutations. This practice is frequently

employed in species that undergo self-pollination or asexual propagation. The theory of evolution through mutation was proposed by de Vries in the year 1901. As per his assertion, mutation denotes an abrupt heritable change and holds an essential role in the process of evolution. Muller (1927) was the first to demonstrate the mutagenic effects of X-rays in Drosophila. The implementation of crop enhancement schemes via induced mutations dates back approximately eighty years, following the identification of the mutagenic properties of X-rays on Drosophila melanogaster by Muller in 1927 and on maize, barley, and wheat by Stadler in 1928 and 1930. The application of induced mutations and mutation breeding techniques to improve food legumes in agricultural programs was only achieved with success during the mid-1950s. The distinction of being the inaugural legume mutant variety worldwide is attributed to N.C. 4-X, a groundnut variety that was generated through X-ray induction by Gregory (1955) and subsequently introduced in the United States in 1959. Over the past 80 years, substantial progress has been achieved in comprehending the characteristics of mutations, refining techniques for inducing mutations and applying induced mutations in agricultural enhancement initiatives. Physical mutagens such as X-rays, gamma rays, fast and thermal neutrons, as well as heavy ion beams, and chemical agents such as EMS, DES, and SA have been employed to induce mutations.

The introduction of novel mutations on specific genes of the faba bean is expected to enhance its genetic diversity, thereby facilitating its breeding. Chemical mutagenesis or irradiation can induce such mutations. The efficacy of faba bean mutation induction has been demonstrated through the application of physical mutagens such as gamma-rays and chemical mutagens such as ethyl methane sulphonate-EMS (Khursheed et al., 2019; Nurmansyah et al., 2020). According to Khazaei et al. (2018), ion beam irradiation has been identified as a distinctive and efficient method for the induction of mutations in faba beans. Sjödin (1971) reported the initial mutagenesis on faba bean, which entailed the identification of determinate mutants, ti1 and ti2. These mutations resulted in a considerable reduction in the number of flowering nodes, ultimately leading to the emergence of a plant type characterized as "topless." Van Norel and Hoogendoorn (1989) subsequently documented the discovery of a mutant, dw1, which resulted in a nearly 50% reduction in internode length.

Ramsay et al. (1991) identified a gene, vcr, that is responsible for decreased vicine-convicine content. Duc (1995) discovered five nodulation mutants through EMS mutagenesis, one of which was a supermodulating

mutant (f32) that produced 3-5 times more nodules than the normal type. Bhatia and colleagues (2001) have provided an extensive account of 265 cultivars produced via induced mutation on grain legumes, encompassing 13 genotypes of faba bean. Mao et al. (2019) detected several mutations that confer resistance to imazapyr via EMS in South Australia. These mutations led to the creation of the initial imidazolinone herbicide-resistant cultivar, PBA Bendoc, in Australia. The exploration of morphological diversity within mutant populations of faba bean has been investigated (Nurmansyah et al., 2020). Huyghe (1998) conducted a review on the mutation of grain legumes, including faba beans. Several mutagenesis projects conducted globally have demonstrated the effectiveness of mutagenesis in grain legumes for expanding genetic diversity (Tadege et al., 2009; Horn et al., 2016; Raina and Khan, 2020).

The identification of mutants that enhance resistance against the root-parasitic weeds broomrape and stem parasitic dodders is crucial for promoting faba bean production in the Mediterranean basin, where these weeds pose a growing challenge. Of interest is the development of faba bean varieties resistant to herbicides used to control these parasitic weeds. This trait is highly sought after, as evidenced by recent studies (Rubiales and Fernández-Aparicio, 2012; Fernández-Aparicio et al., 2016; Rubiales et al., 2016; Mejri et al., 2018). Mutations that result in a reduction of pod wall thickness have been observed to have a positive impact on the overall grain yield (Bhatia et al., 2001). The International Atomic Energy Agency (IAEA) has compiled a database of mutant varieties of faba bean from different countries, which currently amounts to 20. These varieties have potential for use in breeding programs (MVD, 2021). The integration of mutagenesis and biotechnology techniques can expedite the introduction of innovative faba bean germplasm and enhanced cultivars.

Speed Breeding

In numerous research systems, the prevailing breeding techniques for faba bean programs entail a single generation per season, resulting in a protracted 8-9-year duration for the development of improved lines. The extended duration of the breeding cycle has a detrimental effect on the enhancement of genetic traits and may not be commensurate with the anticipated increase in population and consumption levels by the year 2030 (Maalouf et al., 2019b). To address this matter, it is crucial to revolutionize faba bean breeding efforts

to accelerate genetic progress by reducing the breeding cycle (Kole, 2022). Norman Borlaug proposed a technique for decreasing the breeding cycle in the 1950s at CIMMYT. This method entailed cultivating two generations annually in Mexico (Ortiz-Monasterio et al., 2007). The ICARDA faba bean program has implemented a comparable methodology, whereby two generations are produced annually through the utilization of summer and winter seasons at the Terbol station, commencing in 2012. The implementation of a speed breeding approach can lead to an additional reduction of the breeding cycle. Speed breeding involves increasing the number of generations per year and can be done in various ways, such as extending plants' exposure to light and harvesting seeds early (Ghosh et al., 2018). Cold treatment after flowering onset has also been found to promote pod formation and faster pod set. It's important to note that extra light is not a single method for fast generation advancement in plants (Mobini et al., 2020). This approach shows promise for legume crops, including faba bean, and has the potential to attain 6-8 generations in a year.

Genetic and Genomics Research

Inadequate work on faba bean genetics was conducted after Erith's initial report in 1930. However, in 1970 and 1971, synaptic trisomic mutants were discovered and studied (O'Sullivan and Angra, 2016), potentially allowing the identification of genetic markers on physical chromosomes (Patto et al., 1999). Previous research has identified specific genes associated with rhizobia symbiosis and colour composition in faba beans (Duc and Picard,1986). The genetic transmission of a seed dormancy gene was determined through the utilization of recombinant inbred lines (Ramsay, 1997). Genomics research is crucial for improving faba bean breeding programs, enabling the selection of vast populations at initial stages. Tragically, the considerable genome size of faba beans, which is approximately 13 Gb, poses a significant challenge in acquiring comprehensive genomic data. The significance of the genomic study in faba beans has been underscored in recent literature due to the dearth of genomic resources available for this crop. This has been highlighted by various authors, including O'Sullivan and Angra (2016), Webb et al. (2016), and Maalouf et al. (2019a).

The utilization of recombinant inbred lines (RILs) has made it easier the detection of associated molecular markers, resulting in the creation of linkage maps with more precision and the identification of quantitative trait loci

(QTLs) (Torres et al., 2010; O'Sullivan and Angra, 2016). Webb et al. (2016) used linkage-based sequencing to create maps that show synteny with other leguminous plants. The linkage of 687 *Vicia faba* specimens was established through the utilization of SNP markers, which were able to assign a linkage group that corresponds to one of the six chromosomes present in the faba bean genome. Torres et al. (2010) conducted the initial assessment of the significant advancements in marker-assisted selection (MAS). Current research has concentrated on creating molecular markers to aid in selecting crops resilient to heat stress. This has resulted in the discovery of two types of markers, namely, single nucleotide polymorphism (SNP) and simple sequence repeat markers (Hu et al., 2018; Babiker et al., 2018), which are linked to yield components in hot environments. On the contrary, Ali and colleagues (2016) discovered a gene linked to freezing tolerance, specifically the SNP locus Vf_Mt3 g086600, which corresponds to one of the potential QTLs. The utilization of genome sequences from various legume species such as *M. truncatula, Lotus japonicas, Glycine max, Lens culinaris, and Cicer arietinum* (Satovic et al., 2013; Varshney et al., 2014; Webb et al., 2016) has become a crucial aspect in the field of faba bean research.

The initial investigation into the transcriptome of faba bean was conducted by Ray and Georges (2010), who generated around 5,000 expressed sequence tags (EST) from the developing embryos of the 'Windsor' variety. The augmentation of transcriptomic and genomic sequence datasets has facilitated the creation of numerous markers, resulting in a substantial enhancement in marker density and map coverage. The drought tolerance has been assessed in faba beans through the utilization of transcriptome analysis. Khan et al. (2019b) conducted a study wherein they utilized RNA sequencing to profile the transcriptome of the faba bean drought-tolerant cultivar (Hassawi-2) under drought stress (DS) conditions and a total of 606.35 million high-quality pair-end clean reads were obtained from leaf tissues, resulting in the assembly of 164,679 unigenes. The aforementioned outcome has the potential to facilitate the development of superior faba bean cultivars that exhibit heightened resistance to drought conditions. Further, it can generate germplasm that is resilient to drought in various crops.

The field of proteomics is utilized to investigate alterations in the proteome of diverse crop species, including rice, maize, wheat, and faba bean, in response to varying environmental factors such as vernalization, chilling treatment, and drought stress. Numerous research endeavours have been undertaken to examine the proteomic reactions of said crops. Cao et al. (2017) and Li et al. (2018) conducted investigations on the proteomic reactions of

faba beans towards vernalization and drought stress, respectively. The authors utilized sophisticated methodologies such as Isobaric Tag for Relative and Absolute Quantification (iTRAQ) LC-MS/MS to detect and measure proteins linked to vernalization in faba bean. The identification and analysis of protein abundance in faba beans under DS conditions were initially conducted by Li et al. (2019). The researchers discovered that the proteins which participate in cellular defense pathways are subject to regulation through overlap signalling mechanisms. This finding is of great significance for comprehending and enhancing the drought tolerance of faba beans. The investigation also unveiled distinct proteins that were either downregulated or upregulated. The upregulated proteins were found to be associated with protein folding, aggregation, and photosynthesis, whereas the downregulated proteins were implicated in stress defense, energy metabolism, cytoskeleton, and oxidation. The study revealed that the heat shock protein 81-2 plays a crucial role in facilitating the synthesis of operative proteins and aiding in the breakdown of misfolded proteins as a countermeasure to drought-induced stress.

ICARDA (International Center for Agricultural Research in the Dry Areas) and Washington State University conducted studies on the genetic makeup of plants using a technique called genome-wide association studies. They used a program called Universal Network Enabled Analysis Kit to analyze the genetic markers in the plants' DNA sequences. Through this analysis, they identified 10,950 different loci in the genome (Hu et al., 2018) and SNPs that are effective in hot regions (Maalouf et al., 2018). They found effective single nucleotide polymorphisms (SNPs) under heat-prone conditions, which can help in developing heat-tolerant varieties. Genomic selection, a method that involves selecting plants based on their genetic information, was found to be effective in RS within synthetic populations (Müller et al., 2017). This approach provided perfect and precise predictions for selecting individual plants rather than predicting yield potential. Recent advancements include the discovery of gene-based SNP markers specifically related to heat tolerance (Maalouf et al., 2018) and the creation of a consensus map with high density (Carrillo-Perdomo et al., 2020). Additionally, a new assembly of the faba bean exome (the protein-coding portion of the genome) was created using transcriptome data from 4 different accessions named Hiverna, Silian, Nova Gradiska, and Quasar. This resulted in the detection of numerous informative SNPs found within genes (Maalouf et al., 2022).

Breeding Success Stories

Breeders prefer working in favorable conditions because it leads to higher heritability and expected genetic gains from selection (Singh, 1990). However, breeding productive genotypes that can withstand waterlogging can be a stable solution for managing excessive moisture (Gemechu and Mussa, 2002). To address this issue in vertisol areas, where waterlogging is a problem, a direct selection method was used to screen genotypes under waterlogging conditions. Different cultivars capable of enduring this stressful condition have been introduced in various years. The national faba bean research in Ethiopia has successfully produced and disseminated six distinct varieties of faba bean, namely Wayu, Ashebeka, Wolki, Selale, Hachalu and Didi'a at the national level (Tadele, 2020). These varieties have high yield and resistance to rot, foot rot and black rot, which are factors contributing to low productivity in waterlogged environments. Breeding success was achieved through the utilization of various mechanisms to create genetic variabilities, such as collecting local varieties from vertisol regions and screening under sick plots for hybridization and *Fusarium solani*.

The identification of tolerance sources for the faba bean in Ethiopia, specifically with regard to abiotic production constraints such as acidity of soil or poor soil fertility, low pH, and frost, remains unknown (Mussa et al., 2008). However, it is expected that there is genetic variability for most of these stresses. Breeding success has been reported for different biotic and abiotic stresses in faba beans, such as chocolate spot disease and high moisture stress (waterlogging) (Tamene et al., 2015; Asnakech et al., 2016; Mussa et al., 2008). This has been achieved by identifying materials for tolerance from the ICARDA and collecting them from areas with water logging issues. In cases where soil pH has dropped below 6.0, plant breeding or biotechnology is necessary to improve resistance to acidity (Conyers et al., 2005). Genetic manipulation of crops is considered a better strategy than constantly manipulating the growing environment, as it is more cost-effective for resource-poor farmers who cannot afford to purchase inputs. Selecting appropriate varieties for the target production environment is crucial (Gemechu, 2007). In order to cultivate faba bean genotypes that are tolerant to acid soil, breeding activities should be carried out in environments that are prone to acidity. Additionally, it is crucial to identify parents that possess genes for effective utilization under conditions of low soil fertility and resilience to stress caused by soil acidity, which can then be used for hybridization purposes. Previous studies have shown significant variation

among faba bean genotypes in terms of their tolerance to soil acidity, with a hindrance in taproot growth ranging from 3% to 40% at pH 4.5 (Kiflemariam et al., 2018). Therefore, it is important to evaluate faba bean genotypes for soil acidity stress tolerance and explore the potential of existing genetic resources to find tolerant genotypes.

Conclusion and Future Remarks

The faba bean holds significant importance in human and animal nutrition and the ecosystem. The growth and yield of faba beans are significantly limited by abiotic stresses, with reproductive processes particularly vulnerable. Furthermore, the anticipated effects of climate change on crop production underscore the necessity of further investigation into the nutritional, biotic, and abiotic limitations that impede the development of nutritious and climate-resilient cultivars. It is imperative to develop 'omics' techniques in conjunction with tissue culture methodologies to achieve frequent regeneration and genetic modification of faba beans. It is necessary to systematically preserve and employ the extensive germplasm assortment of faba bean as a foundational source of genes that confer tolerance to abiotic stresses. The utilization of next-generation genotyping and sequencing technologies can facilitate the production of molecular markers for the purpose of evaluating genetic diversity, constructing genetic linkage maps, mapping QTLs, and accelerating genetic enhancement against abiotic stresses via genomics-assisted breeding in the faba bean. Given the current developing state of faba bean genome analysis, it is imperative that collaborative global initiatives be undertaken to expedite the molecular research pertaining to this crop.

References

Abbas, M. S., Badawy, R. A., Abdel-Lattif, H. M., & El-Shabrawi, H. M. (2021). Synergistic effect of organic amendments and biostimulants on faba bean grown under sandy soil conditions. *Scientia Agricola,* 79.

Abbas, T., Rizwan, M., Ali, S., Zia-ur-Rehman, M., Farooq Qayyum, M., Abbas, F., Hannan, F., Rinklebe, J., Sik Ok, Y., 2017. Effect of biochar on cadmium bioavailability and uptake in wheat (Triticum aestivum L.) grown in soil with aged contamination. *Ecotoxicol. Environ. Saf.* 140, 37–47.

Abd El-Mageed, T. A., Belal, E. E., Rady, M. O., Abd El-Mageed, S. A., Mansour, E., Awad, M. F., & Semida, W. M. (2021). Acidified biochar as a soil amendment to drought-stressed (*Vicia faba* L.) plants: Influences on growth and productivity, nutrient status, and water use efficiency. *Agronomy*, 11(7), 1290.

Abd-Alla, M. H., El-Enany, A. W. E., Nafady, N. A., Khalaf, D. M., Morsy, F. M. Synergistic interaction of Rhizobium leguminosarum bv. viciae and arbuscular mycorrhizal fungi as plant growth promoting biofertilizers for faba bean (*Vicia faba* L.) in alkaline soil. *Microbiol. Res.* 2014, 169, 49–58.

Abdel Aziz, M. 2007. Effect of planting date and plant density on growth, maturity, earliness and componentsYield in fava beans under the conditions of the Syrian coast, *Tishreen University Journal for Studies and Scientific Research -Biological Sciences Series*, Volume (29), Issue (4) (30-46).

Abdel Latef, A. A. H., Tahjib-Ul-Arif, M., & Rhaman, M. S. (2021). Exogenous auxin-mediated salt stress alleviation in faba bean (*Vicia faba* L.). *Agronomy,* 11(3), 547.

Abdel Latef, A. A., Kordrostami M, Zakir A, Zaki H, Saleh OM (2019). Eustress with H2O2 facilitates plant growth by improving tolerance to salt stress in two wheat cultivars. *Plants* 8:303.

Abdel, C. G. and S. Y. H. Al-Hamadany (2010). Response of five water stressed Fababean (*Vicia faba* L.) cultivars to exogenous absiscic acid application. *Dohuk Univ. J. Agric. Vet.,* 13(1), 12.

Abdelaal, K. A. (2015). Effect of salicylic acid and abscisic acid on morpho-physiological and anatomical characters of faba bean plants (*Vicia faba* L.) under drought stress. *Journal of Plant Production*, 6(11), 1771-1788.

Abdelmula, A. A., Abuanja IK (2007). Genotypic responses, yield stability, and association between characters among some of Sudanese Faba bean (*Vicia faba* L.) genotypes under heat stress. *Proceedings of the Tropical German Conference* (Utilization of diversity in land use systems: Sustainable and organic approaches to meet human needs), Witzenhausen, 9–11 Oct 2007.

Abeer, H., Abd_Allah, E. F., Alqarawi, A. A., El-Didamony, G., Alwhibi, M., Egamberdieva, D., & Ahmad, P. (2014). Alleviation of adverse impact of salinity on faba bean (*Vicia faba* L.) by arbuscular mycorrhizal fungi. *Pak. J. Bot,* 46(6), 2003-2013.

Abera, T., Semu, E., Debele, T., Wegary, D., Kim, H., 2015. Effects of faba bean break crop and N rates on subsequent grain yield and nitrogen use efficiency of highland maize varieties in Toke Kutaye, western Ethiopia. *Am J Res Commun*, 3: 32–72.

Abid, G., Hessini K, Aouida M, Aroua I, Baudoin JP, Muhovski Y, Mergeai G, Sassi K, Machraoui M, Souissi F, Jebara M (2017). Agro-physiological and biochemical responses of faba bean (*Vicia faba* L. var. 'minor') genotypes to water deficit stress. *Biotechnol Agron Soc Environ* 21:146–159.

Abid, G., Ouertani, R. N., Jebara, S. H., Boubakri, H., Muhovski, Y., Ghouili, E., ... & Jebara, M. (2020). Alleviation of drought stress in faba bean (*Vicia faba* L.) by exogenous application of β-aminobutyric acid (BABA). *Physiology and Molecular Biology of Plants*, 26, 1173-1186.

Adamu, G. O. L., Ezeokoli, O. T., Dawodu, A. O., Adebayo-Oyetoro, A. O., Ofodile, L. N., 2015. Macronutrients and micronutrients profile of some underutilized beans in southwestern Nigeria. *Int J Biochem Res Rev*, 7: 80–89.

Adhikari, K. N., Khazaei, H., Ghaouti, L., Maalouf, F., Vandenberg, A., Link, W. and O'Sullivan, D. M., (2021). Conventional and molecular breeding tools for accelerating genetic gain in faba bean (*Vicia faba* L.). *Frontiers in plant science*, p.2174.

Ahmad, P., Abdel Latef, A. A., Hashem, A., Abd_Allah, E. F., Gucel, S., Tran, L. S. P. Nitric oxide mitigates salt stress by regulating levels of osmolytes and antioxidant enzymes in chickpea. *Front. Plant Sci.* 2016, 7, 347.

Ahmad, P., Alyemeni, M., Ahanger, M., Egamberdieva, D., Wijaya, L., Alam, P. Salicylic acid (SA) induced alterations in growth, biochemical attributes and antioxidant enzyme activity in faba bean (*Vicia faba* L.) seedlings under NaCl toxicity. Russ. *J. Plant Physiol*. 2018, 65, 104–114.

Ahmad, P., Jaleel, C. A., Salem, M. A., Nabi, G., Sharma, S. Roles of Enzymatic and Nonenzymatic Antioxidants in Plants during Abiotic Stress. *Crit. Rev. Biotechnol.* 2010, 30, 161–175.

Ahmed, F., Arthur, E., Plauborg, F., Razzaghi, F., Kørup, K., & Andersen, M. N. (2018). Biochar amendment of fluvio-glacial temperate sandy subsoil: Effects on maize water uptake, growth and physiology. *Journal of Agronomy and Crop Science,* 204(2), 123-136.

Ahmed, M. S. H., Abd-El-Haleem, SHM, Bakheit MA et al., (2008) Comparison of three selection methods for yield and components of three faba bean (*Vicia faba* L.) crosses. *World J Agric Sci* 4(5):635–639 264.

Akhtar, S. S., Andersen, M. N., & Liu, F. (2015). Residual effects of biochar on improving growth, physiology and yield of wheat under salt stress. *Agricultural Water Management,* 158, 61–68.

Alaa, F. M. (2018). Effectiveness of exopolysaccharides and biofilm-forming plant growth promoting rhizobacteria on salinity tolerance of faba bean (*Vicia faba* L.). *African Journal of Microbiology Research,* 12(17), 399–404.

Al-Anbari, M. A. A., Hakeed Abed Khashan, and Ali Saleh Mahdi. 2009. Response of the bean crop to planting date and foliar fertilization with boron. *Karbala University Journal*, vol7:No3: 99-103.

Alemu, L., Tekalign M., Wassie H. and Hailu S. 2016. Assessment and Mapping of Status and Spatial Distribution of Soil Macronutrients in Kambata Tembaro Zone, Southern Ethiopia. *Advances in Plants and Agriculture Research*, 4(4): 1-14.

Alghamdi, S. S. Chemical composition of faba bean (*Vicia faba* L.) genotypes under various water regimes. *Pak. J. Nutr*. 2009, 8, 477–482.

Al-Hakimi, A. M. A. (2006). Counteraction of drought stress on soybean plants by seed soaking in salicylic acid. *International Journal of Botany.*

Ali, S., Rizwan, M., Noureen, S., Anwar, S., Ali, B., Naveed, M., ... & Ahmad, P. (2019). Combined use of biochar and zinc oxide nanoparticle foliar spray improved the plant growth and decreased the cadmium accumulation in rice (Oryza sativa L.) plants. *Environmental Science and Pollution Research,* 26, 11288–11299.

Ali, M. B., Welna, G. C., Sallam, A., Martsch, R., Balko, C., Gebser, B. & Link, W. (2016). Association analyses to genetically improve drought and freezing tolerance of faba bean (*Vicia faba* L.). *Crop Science*, *56*(3), 1036-1048.

Aliloo, A. A., Alahyari, S., Mousavi, S. B. Micronutrient priming improves germination and seedling establishment in lentils. *Adv. Appl. Agric. Sci.* 2014, 11, 37–44.

Al-Othman, M. K. and Ibrahim Al-Assaf. 2009. The effect of planting date and plant density on the productivity of faba bean. L faba Vicia. *Damascus University Journal of Agricultural Sciences* (25): 66-92.

Alqarawi, A. A., Abd Allah, E. F., & Hashem, A. (2014). Alleviation of salt-induced adverse impact via mycorrhizal fungi in Ephedra aphylla Forssk. *Journal of Plant Interactions*, 9(1), 802-810.

Amede, T., Schubert, S. Mechanisms of drought resistance in grain legumes I: Osmotic adjustment. *Ethiop. J. Sci.* 2003, 26, 37–46.

Arbaoui, M., Link W (2008) Effect of hardening on frost tolerance and fatty acid composition of leaves and stems of a set of faba bean (*Vicia faba* L.) genotypes. *Euphytica* 162(2):211–219.

Arfan, M., R. Athar and M. Ashraf, 2007. Does exogenous application of salicylic acid through the rooting medium modulate growth and photosynthetic capacity in two differently adapted spring wheat cultivars under salt stress? *J. Plant Physiol.,* 164: 685–694.

Ashraf, M. A., Ashraf, M., Ali, Q. Response of two genetically diverse wheat cultivars to salt stress at different growth stages: Leaf lipid peroxidation and phenolic contents. *Pak. J. Bot.* 2010, 42, 559–565.

Asnakech, T., Derera, J., Sibiya, J. and Asnake F. 2016. Participatory assessment of production threats, farmers' desired traits and selection criteria of faba bean (*Vicia faba* L.) varieties: opportunities for faba bean breeding in Ethiopia. *Indian Journal of Agriculture Research*, 50 (4): 295-302.

Atemkeng, M. F., Muki, T. J., Park, J. and Jifon, J. 2011. Integrating molecular tools with conventional breeding strategies for improving phosphorus acquisition by legume crops in acid soils of sub-Saharan Africa. *Biotechnology and Molecular Biology Reviews,* 6:142–154.

Attia, M. S., Osman MS, Mohamed AS, Mahgoub HA, Garada MO, Abdelmouty ES, Abdel Latef AA (2021). Impact of foliar application of chitosan dissolved in different organic acids on isozymes, protein patterns and physiobiochemical characteristics of tomato grown under salinity stress. *Plants* 10:388.

Auge, R. M. Water relations, drought and vesicular-arbuscular mycorrhizal symbiosis. *Mycorrhiza* 2001, 11, 3–42.

Avola, G., Cavallaro V, Patane` C, Riggi E (2008) Gas exchange and photosynthetic water use efficiency in response to light, CO2 concentration and temperature in *Vicia faba*. *J Plant Physiol* 165:796–804.

Azooz, M. M., Al-Fredan MA (2009) The inductive role of vitamin C and its mode of application on growth, water status, antioxidant enzyme activities and protein patterns of *Vicia faba* L cv. Hassawi grown under seawater irrigation. *Amer J Plant Physiol* 4: 38-51.

Azooz, M. M., Alzahrani, A. M., & Youssef, M. M. (2013). The potential role of seed priming with ascorbic acid and nicotinamide and their interactions to enhance salt tolerance in broad bean ('*Vicia faba*'L.). *Australian Journal of Crop Science,* 7(13), 2091-2100.

Azzedine, F., Gherroucha, H., & Baka, M. (2011). Improvement of salt tolerance in durum wheat by ascorbic acid application. *J. Stress Physiol. Biochem*, 7(1), 27-37.

Babiker, Z., Maalouf F, Hamwieh A, Baum M, Omer O, Alhashmi A (2018) New SSR markers related to heat tolerance in faba bean under diverse environments. In: *Seventh international food legumes research conference,* Marrakesh, Morocco, May 06–08, 2018.

Baddeley, J., Pappa, V., Pristeri, A., Bergkvist, G., Monti, M., Reckling, M., Schläfke, N., Watson, C. Legume-based green manure crops. In *Legumes in Cropping Systems*; Murphy-Bokern, D., Stoddard, F. L., Watson, C. A., Eds., CAB International: Oxford, UK, 2017; pp. 125–138.

Bailey, R. L., West Jr., K. P., Black, R. E., 2015. The epidemiology of global micronutrient deficiencies. *Ann Nutr Metab,* 66: 22–33.

Baloch, F. S., Karakoy, T., Demirba¸s, A., Toklu, F., Ozkan, H., Hatipoglu, R., 2014. Variation of some seed mineral contents in open pollinated faba bean (*Vicia faba* L.) landraces from Turkey. *Turk. J. Agric. For.* 38, 591-602.

Barnabás, B., K Jäger and A Fehér, 2008. The effect of drought and heat stress on reproductive processes in cereals. *Plant Cell Environ.* 31, 11–38.

Belachew, K. Y., Nagel, K. A., Fiorani, F., Stoddard, F. L. Diversity in root growth responses to moisture deficit in young faba bean (*Vicia faba* L.) plants. *PeerJ* 2018, 6, e4401.

Belghith-Fendri, L., Chaari, F., Kallel, F., Zouari-Ellouzi, S., Ghorbel, R., Besbes, S., et al., 2016. Pea and broad bean pods as a natural source of dietary fiber: the impact on texture and sensory properties of cake. *J. Food Sci.* 81, C2360C2366.

Bhatia, C. R., Maluszynski, M., Nichterlein, K., and van Zanten, L. (2001). Grain Legume Cultivars Derived from Induced Mutations, and Mutations Affecting Nodulation. Vienna: *IAEA.* Available online at: https://www.osti.gov/etdeweb/biblio/20169549.

Bhowmik, P., Konkin, D., Polowick, P., Hodgins, C. L., Subedi, M., Xiang, D., et al., (2021). CRISPR/Cas9 gene editing in legume crops: opportunities and challenges. *Legum. Sci.* 1–16.

Bimurzayev, N., Sari, H., Kurunc, A., Doganay, K. H., Asmamaw, M. Effects of different salt sources and salinity levels on emergence and seedling growth of faba bean genotypes. *Sci. Rep.* 2021, 11, 18198.

Bishop, J., Potts, S. G., Jones, H. E., 2016. Susceptibility of faba bean (*Vicia faba* L.) to heat stress during floral development and anthesis. *J. Agron. Crop. Sci.* 202 (6), 508-517.

Black, K., Barnett, A., Tziboula-Clarke, A., White, P. J., Iannetta, P. P. M., & Walker, G. (2019). Faba bean as a novel brewing adjunct: Consumer evaluation. *Journal of the Institute of Brewing,* 125, 310–314.

Blum, A., 2011. Drought resistance. Is it really a complex trait? *Funct. Plant Biol.,* 38, 753–757.

Bobojonov, I., Lamers, J. P. A., Bekchanov, M., Djanibekov, N., Franz-Vasdeki, J., Ruzimov, J., Martius, C. 2013. Options and constraints for crop diversification: a case study in sustainable agriculture in Uzbekistan. *Agroecology and Sustainable Food Systems,* 37(74): 788-811.

Bota, J., H Medrano and J Flexas, 2004. Is photosynthesis limited by decreased Rubisco acivity and RuBP content under progressive water stress? *New Phytol.* 162, 671–681.

Bouallègue, A., Souissi, F., Nouairi, I., Souibgui, M., Abbes, Z., & Mhadhbi, H. (2017). Salicylic acid and hydrogen peroxide pretreatments alleviate salt stress in faba bean (*Vicia faba*) seeds during germination. *Seed Science and Technology*, 45(3), 675-690.

Bozzini, A., Chiaretti D (1999) The genetic improvement of the Mediterranean faba bean (*Vicia faba* L.). III. Development of obligate self fertile lines. *J Genet Breed* 53:207–214.

Burns, H., Norton, M. and Tyndall, P. 2017. Improving yield potential of legumes on acidic soils. *Australian government grain research and development program corporation.* https://grdc.com.au/resources-andpublications/grdc-update-papers.

Burrell, L. D., Zehetner, F., Rampazzo, N., Wimmer, B., & Soja, G. (2016). Long-term effects of biochar on soil physical properties. Geoderma, 282, 96-102.

Caihong, Y., Qiang, C., Guang, L., Fuxue, F., Li, W. 2015. Water use efficiency of controlled alternate irrigation on wheat/faba bean intercropping. *African Journal of Agricultural Research,* 10(48): 4348-4355.

Cao, X., Fan G, Dong Y, Zhao Z, Deng M, Wang Z, Liu W (2017) Proteome profiling of Paulownia seedlings infected with phytoplasma. *Front Plant Sci* 8:342.

Cao, Y., 2010. *Study on L-DOPA content of Vicia faba L. Fujian Agriculture and Forestry University*, Fuzhou.

Carrillo-Perdomo, E., Vidal A, Kreplak J, Duborjal H, Leveugle M, Duarte J, Tayeh N (2020) Development of new genetic resources for faba bean (*Vicia faba* L.) breeding through the discovery of gene-based SNP markers and the construction of a high-density consensus map. *Sci Rep* 10(1):1–14.

Chanway, C. P., Hynes, R. K., Nelson, L. M. Plant growth-promoting rhizobacteria: Effects on growth and nitrogen fixation of lentil (Lens esculenta Moench) and pea (Pisumsativum L.). *Soil Biol. Biochem.* 1989, 21, 511–517.

Chimdi, A., 2015. Assessment of the severity of acid saturations on soils collected from cultivated lands of East Wollega Zone, Ethiopia. *Science, Technology and Arts Research Journal, 3*(4), pp.42-48.

Chris, G. and Stephen, D. 2009. Soil acidity a guide for West Australia farmers and consultants. *Bulletin* 4784 ISSN: 1833-7236.

Coda, R., Varis, J., Verni, M., Rizzello, C. G., Katina, K., 2017. Improvement of the protein quality of wheat bread through faba bean sourdough addition. *LWT Food Sci. Technol.* 82, 296-302.

Conyers, M., Helyar, K. and Moroni, J. S. 2005. The Carbon Cost of Protecting the Root Apex from Soil Acidity: A Theoretical Framework. *Plant and Soil,* 278: 195-204.

Crépon, K., Marget, P., Peyronnet, C., Carrouee, B., Arese, P., Duc, G., 2010. Nutritional value of faba bean (*Vicia faba* L.) seeds for feed and food. *Field Crop. Res.* 115 (3), 329-339.

Cubero, J. I., Moreno MT (1984) Breeding for self-fertility. In: Hebblethwaite PD, Dawkins TCK, Heath MC, Lockwood G (eds) *Vicia faba: agronomy, physiology and breeding*. Springer, Dordrecht, pp 209–217.

Cushman, J. C., Denby, K., & Mittler, R. (2022). Plant responses and adaptations to a changing climate. *The Plant Journal,* 109(2).

Dawood, M. G., & El-Awadi, M. E. (2015). Alleviation of salinity stress on *Vicia faba* L. plants via seed priming with melatonin. *Acta Biológica Colombiana,* 20(2), 223-235.

de Vries, H. (1901*): Die Mutations theorie* I. Von viet and co. Leipiz, 1-648.63-69.

Desouky, A. F., Ahmed, A. H. H., Reda, A. S. A., Stützel, H. and Hanafy, M. S., 2023. Physiological and biochemical responses of two faba bean (*Vicia faba* L.) varieties grown *in vitro* to salt stress. *Journal of Crop Science and Biotechnology,* 26(2), pp.151-160.

Dimkpa, C., Weinand, T., Asch, F. Plant–rhizobacteria interactions alleviate abiotic stress conditions. *Plant Cell Environ.* 2009, 32, 1682–1694.

Drayner, J. M. (1959) Self-and cross-fertility in field beans (*Vicia faba* Linn.). *J Agric Sci* 53(3):387–403.

Drayner, J. M. Regulation of outbreeding in field beans (*Vicia faba*). *Nature* 1956, 177, 489–490.

Duc, G. (1995). Mutagenesis of faba bean (*Vicia faba* L.) and the identification of five different genes controlling no nodulation, ineffective nodulation or supernodulation. *Euphytica* 83, 147–152.

Duc, G. (1997). Faba bean (*Vicia faba* L.). *Field Crops Res.* 53, 99–109.

Duc, G., Marget, P., Esnault, R., Le Guen, J., Bastianelli, D. Genetic variability for feeding value of faba bean seeds (*Vicia faba*): Comparative chemical composition of isogenics involving zero-tannin and zero-vicine genes. *J. Agric. Sci.* 1999, 133, 185–196.

Duc, G., Pascal, M. and Paolo, A. 2011. Breeding priorities for improved nutritional value of *Vicia faba* seeds. *Grain legumes,* (56): 17-18.

Duc, G., Picard J (1986) Fertility components in a faba bean line near isogenic for male sterility. *Can J Plant Sci* 66:235–239.

Egamberdieva, D., Zoghi, Z., Nazarov, K., Wirth, S., & Bellingrath-Kimura, S. D. (2020). Plant growth response of broad bean (*Vicia faba* L.) to biochar amendment of loamy sand soil under irrigated and drought conditions. *Environmental Sustainability*, 3, 319-324.

Ehtaiwesh, A. (2022). Effect of acetyl salicylic acid (aspirin) on growth and yield of faba bean (viciafaba l.) under salinity stress. *Azzaytuna University Journal*, 42-415.

El Nahhas, N., AlKahtani, M. D., Abdelaal, K. A., Al Husnain, L., AlGwaiz, H. I., Hafez, Y. M., ... & Elkelish, A. (2021). Biochar and jasmonic acid application attenuates antioxidative systems and improves growth, physiology, nutrient uptake and productivity of faba bean (*Vicia faba* L.) irrigated with saline water. *Plant Physiology and Biochemistry,* 166, 807-817.

Elliott, J. E. and Whittington, W. J. 2009. The effect of soil pH on the severity of Chocolate Spot infection on field bean varieties. *Journal of Agricultural Science*, 91:563- 567.

El-Naggar, A., Awad, Y. M., Tang, X. Y., Liu, C., Niazi, N. K., Jien, S. H. & Lee, S. S. (2018). Biochar influences soil carbon pools and facilitates interactions with soil: A field investigation. *Land degradation & development*, 29(7), 2162-2171.

Elshafei, A. A. M., Amer, M. A. E., Elenany, M. A. M., Helal, A. G. A. E. Evaluation of the genetic variability of faba bean (*Vicia faba* L.) genotypes using agronomic traits and molecular markers. *Bull. Natl. Res. Cent.* 2019, 43, 106.

Emshwiller, E., Calberto-Sánchez, G., Girma, G., Jansky, S., Sardos, J., Staver, C., Stoddard, F. L., Roux, N. Unavailability of wild relatives. In *Crop Wild Relatives and Climate Change;* Redden, R., Yadav, S. S., Maxted, N., Dulloo, M. E., Guarino, L., Smith, P., Eds., John Wiley & Sons: Oxford, UK, 2015; pp. 222–258.

Endalkachew, F., Kibebew K., Asmare M. and Bobe B. 2018. Yield of faba bean (*Vicia faba* L.) as affected by lime, mineral P, farmyard manure, compost and rhizobium in acid soil of lay gayint district, northwestern highlands of Ethiopia. *Agriculture and Food Security*, 7(16):1-11.

Erith, A. G. (1930) The inheritance of colour, size and form of seeds, and flower colour in *Vicia faba* L. *Genetica* 12(6):562–562.

Etemadi, F., Hashemi, M., Autio, W., Mangan, F., Zandvakili, O., 2018a. Accumulation and distribution trend of L-Dopa in different parts of eight varieties of faba bean plant through its growth period. *J Crop Sci,* 6: 426–434.

Etemadi, F., Hashemi, M., Barker, A. V., Zandvakili, O. R. and Liu, X., 2019. Agronomy, nutritional value, and medicinal application of faba bean (*Vicia faba* L.). *Horticultural Plant Journal*, *5*(4), pp.170-182.

Etemadi, F., Hashemi, M., Mangan F., Weis, S., 2015. *Faba beans; Growers guide in New England.* http://ag.umass.edu/sites/ag.umass.edu/files/research-reports/faba_bean_guide_2.pdf.

Etemadi, F., Hashemi, M., Mangan, F., Xing, B., 2014a. Distribution of L-Dopa in different parts of faba Beans, in: Northeast Section of the American Society of Plant Biologists (NEASPB) Conference. *The University of Rhode Island*, RI 2.

Etemadi, F., Hashemi, M., Randhir, R., Zandvakili, O., Ebadi, A., 2018b. Accumulation of L-Dopa in various organs of faba bean and influence of drought, nitrogen stress, and processing methods on L-Dopa yield. *Crop J,* 6: 426–434.

Etemadi, F., Hashemi, M., Xing, B., Mashayekhi, H., 2014b. Accumulation trend of L-Dopa in different parts of faba beans varieties. In: *ASA, CSSA, & SSSA International Annual Meetings.* California, p. 281.

Fageria, N. K. 2009. *The Use of Nutrients in Crop Plants*. Taylor and Francis Group, Boka Raton, London: CRC Press.

FAOSTAT (2023). *Food and Agriculture Organization of the United Nations.* Available online at: http://faostat.fao.org (accessed on June 3, 2023).

Fariduddin, Q., Zaid A, Mohammad F (2019). Plant growth regulators and salt stress: Mechanism of tolerance trade-off. In: Akhter M (Ed). *Salt Stress, Microbes and Plant Interactions: Causes and Solution.* https://doi.org/10.1007/978-981-13-8801-9_4.

Farooq, M., Almamari, S. A. D., Rehman, A., Al-Busaidi, W. M., Wahid, A., & Al-Ghamdi, S. S. (2021). Morphological, physiological and biochemical aspects of zinc seed priming-induced drought tolerance in faba bean. *Scientia Horticulturae*, 281, 109894.

Farooq, M., SMA Basra, A Wahid, ZA Cheema, MA Cheema, A Khaliq, 2008. Physiological role of exogenously applied glycinebetaine in improving drought tolerance of fine grain aromatic rice (*Oryza sativa* L.). *J. Agron. Crop Sci.* 194, 325–333.

Fernández-Aparicio, M., Flores, F., and Rubiales, D. (2016). The Effect of Orobanche crenata Infection Severity in Faba Bean, Field Pea, and Grass Pea Productivity. *Front Plant Sci.* 7:1409.

Flajoulot, S., Ronfort J, Baudouin P, Barre P, Huguet T, Huyghe C, Julier B (2005) Genetic diversity among alfalfa (Medicago sativa) cultivars coming from a breeding program, using SSR markers. *Theor Appl Genet* 111(7):1420–1429.

French, B. and White, P. 2005. Soil and environmental factors affecting pulse adaptation in Western Australia. *Australia Journal of Agricultural Research*, 50:375–387.

Füzy, A., Kovács, R., Cseresnyés, I., Parádi, I., Szili-Kovács, T., Kelemen, B., Rajkai, K., Takács, T. Selection of Plant Physiological Parameters to Detect Stress Effects in Pot Experiments Using Principal Component Analysis. *Acta Physiol. Plant.* 2019, 41, 56.

Gasim, S., Link W (2007) Agronomic performance and the effect of self-fertilization on German winter faba beans. *J Central Eur Agri* 8(1):121–128.

Gautam, M., Chandel, M., Azmi, W., 2012. Therapeutic role of l-DOPA produced as a secondary metabolite from different legumes and plant sources. *Ann Phytomed,* 1: 1–8.

Gautam, R. C., Bana, R. S. 2014. Drought in India: its impact and mitigation strategies–a review. *Indian Journal of Agronomy*, 59(2): 179-190.

Gemechu, K. and J Mussa, 2002. Comparision of three secondary traits as determinants of grain yield in faba bean on waterlogged Vertisols. *Journal of Genetics and Breeding,* 56: 317-326.

Gemechu, K., 2007. Concerns on Mismatches between Environments of Selection and Production of Crop Varieties in Ethiopia. *East African Journal of Sciences*. Volume 1 (2) 93-103.

Gemechu, K., F Asnake and E Million, 2016. Reflections on Highland Pulses Improvement Research in Ethiopia. *Ethiopian Journal of Agricultural Sciences.* EIAR 50th Year Jubilee Anniversary Special Issue: 17-50.

Geng, G., 2012. *Analysis of l-DOPA contents in six organs of a faba bean cultivar.* North Hortic, 2: 26–27.

Getachew, A., Taye B. and Agajie T. 2005. P fertilizer and FYM effect on the growth and yield of faba bean and some chemical properties in acidic nitosols of centeral high land of Ethiopia. *Ethiopian Journal of Natural Resources*, 7: 23-39.

Ghaouti, L., Vogt-Kaute W, Link W (2008) Development of locally-adapted faba bean cultivars for organic conditions in Germany through a participatory breeding approach. *Euphytica* 162(2):257–268.

Gharzeddin, K., Maalouf F, Khoury B, Abou-Khater L, Christmann S, El Dine NAJ (2019) Efficiency of different breeding strategies in improving the faba bean productivity for sustainable agriculture. *Euphytica* 215(12):203.

Ghosh, S., Watson A, Gonzalez-Navarro OE, Ramirez-Gonzalez RH, Yanes L, Mendoza-Suárez M et al., (2018) Speed breeding in growth chambers and glasshouses for crop breeding and model plant research. *Nat Protoc* 13(12):2944–2963.

Giller, K. E., 2001. *Nitrogen fixation in tropical cropping systems.* Cabi.

Glick, B. R., Cheng, Z., Czarny, J., Duan, J. Promotion of plant growth by ACC deaminase producing soil bacteria. In *New Perspectives and Approaches in Plant Growth-*

Promoting Rhizo-Bacteria Research; Springer: Dordrecht, The Netherlands, 2007; pp. 329–339.

Gnanasambandam, A., Paull, J., Torres, A., Kaur, S., Leonforte, T., Li, H., Zong, X., Yang, T., Materne, M. Impact of molecular technologies on faba bean (*Vicia faba* L.) breeding strategies. *Agronomy* 2012, 2, 132–166.

Gounaris, K., Brain ARR, Quinn PJ, Williams WP (1984) Structural reorganization of chloroplast thylakoid membranes in response to heat stress. *Biochim Biophys Acta* 766:198–208.

Gregory, W. C. (1955): X-ray breeding of peanut (Arachis hypogaea L.). *Agron. J.,* 47: 396-399.

Habibzadeh, Y., Evazi, A. R., Abedi, M. Alleviation drought stress of mungbean (Vigna radiata L.) plants by using arbuscular mycorrhizal fungi. *Int. J. Agric. Sci. Nat. Res.* 2014, 1, 1–6.

Hafez, E. M., Osman, H. S., El-Razek, U. A. A., Elbagory, M., Omara, A. E. D., Eid, M. A., & Gowayed, S. M. (2021). Foliar-applied potassium silicate coupled with plant growth-promoting rhizobacteria improves growth, physiology, nutrient uptake and productivity of faba bean (*Vicia faba* L.) irrigated with saline water in salt-affected soil. *Plants,* 10(5), 894.

Hall, C., Hillen, C., & Robinson, J. G. (2017). Composition, nutritional value, and health benefits of pulses. *Cereal Chemistry,* 94, 11–31.

Hamada, A. M. (2001) Alteration in growth and some relevant metabolic processes of broad bean plants during extreme temperatures exposure. *Acta Physiol Plant* 23:193–200.

Hannula, S. E., Di Lonardo, D. P., Christensen, B. T., Crotty, F. V., Elsen, A., van Erp, P. J., ... & Termorshuizen, A. J. (2021). Inconsistent effects of agricultural practices on soil fungal communities across 12 European long-term experiments. *European Journal of Soil Science*, 72(4), 1902-1923.

Harris, D. B. S. R., Raghuwanshi, B. S., Gangwar, J. S., Singh, S. C., Joshi, K. D., Rashid, A., Hollington, P. A. Participatory evaluation by farmers of on-farm seed priming in wheat in India, Nepal and Pakistan. *Exp. Agric.* 2001, 37, 403–415.

Harter, R. D. 2007. Acid Soils of the Tropics. *ECHO technical note, ECHO*, 11.

Hauggaard-Nielsen, H., Mundus, S., Jensen, E. S. Nitrogen dynamics following grain legumes and subsequent catch crops and the effects on succeeding cereal crops. *Nutr. Cycl. Agroecosyst.* 2009, 84, 281–291.

Hayat, Q., Hayat S, Irfan M, Ahmad A (2010) Effect of exogenous salicylic acid under changing environment: a review. *Environ Exp Bot* 68(1):14–25. https://doi.org/10.1016/j.envexpbot.2009.08.005.

Herridge, D. F., Peoples, M. B., and Boddey, R. M. (2008). Global inputs of biological nitrogen fixation in agricultural systems. *Plant Soil* 311, 1–18.

Horn, L. N., Ghebrehiwot, H. M., and Shimelis, H. A. (2016). Selection of novel cowpea genotypes derived through gamma irradiation. *Front. Plant Sci*. 7:262.

Hu J, Maalouf F, Dong H, Hawkins C, Zhang Z, Ma Y, Coyne C, Jin D, Yu LX, Babiker Z, Hamwieh A, Abou-Khater L, Baum M (2018) Enhancing faba bean (*Vicia faba* L.) germplasm for resilience to temperature Extremes. In: *Seventh international food legumes research conference*, Marrakesh, Morocco.

Hu, J., Kwon, S. J., Park, J. J., Landry, E., Mattinson, D. S., Gang, D. R., 2015. LC-MS determination of L-DOPA concentration in the leaf and flower tissues of six faba bean (*Vicia faba* L.) lines with common and rare flower colors. *Funct Foods Health Dis*, 5: 243–250.

Huang, Q., Song, S., Chen, Z., Hu, B., Chen, J., & Wang, X. (2019). Biochar-based materials and their applications in removal of organic contaminants from wastewater: state-of-the-art review. *Biochar,* 1, 45-73.

Huyghe, C. (1998). Genetics and genetic modifications of plant architecture in grain legumes: a review. *Agronomie 18*, 383–411.

Ibrahim, H. M. (2015) Effectiveness of breeding methods for production of superior genotypes and maintenance of genetic variance in faba bean (*Vicia faba* L.). *Amer J Life Sci* 3(1):11–16.

IFPRI (International Food Policy Research Institute). 2010. Fertilizer and Soil Fertility Potential in Ethiopia. *Working Paper.* Washington: IFPRI. Institute. UK. p.26-47.

Inamdar, S. A., Surwase, S. N., Jadhav, S. B., Bapat, V. A., Jadhav, J. P., 2013. *Statistically optimized biotransformation protocol for continuous production of l-DOPA using Mucuna monosperma callus culture*. Springer Plus, 2: 570.

International Atomic Energy Agency (2021). Available online at: https://bit.ly/3yVEReM.

Iqbal, M., Ashraf, M. Wheat seed priming in relation to salt tolerance: Growth, yield and levels of free salicylic acid and polyamines. *Ann. Bot. Fenn.* 2006, 43, 250–259.

Janas, K. M., Posmyk MM. Melatonin, an under estimated natural substance with great potential for agricultural application. *Acta Physiol Plant.* 2013;35(12):3285-3292. doi: http://dx.doi.org/10.1007/s11738-013-1372-0.

Janda, T., Darko É, Shehata S, Kovács V, Pál M, Szalai G (2016) Salt acclimation processes in wheat. *Plant Physiol Biochem* 101:68–75.

Jensen, E. S., Peoples, M and Hauggaard-Nielsen, H. 2010. Faba bean in cropping systems. *Field Crops Research*, 115 (3): 203–216.

Kabbadj, A., Makoudi, B., Mouradi, M., Pauly, N., Frendo, P., Ghoulam, C. Physiological and biochemical responses involved in water deficit tolerance of nitrogen-fixing *Vicia faba. PLoS ONE* 2017, 12, e0190284.

Keneni, A., Assefa F, Prabu PC (2010) Characterization of acid and salt tolerant rhizobial strains isolated from faba bean fields of Wollo, Northern Ethiopia. *J Agr Sci Technol* 12:365–376.

Keneni, G., Asmamaw B, Jarso M (2001) Efficiency of drained selection environments for improving grain yield in faba bean under undrained target environments on vertisol. *Euphytica* 122(2): 279–285.

Khalil, M. M. N. Effect of mineral fertilizers and biofertilization on some soil properties and faba bean productivity under saline soil conditions. *J. Soil Sci. Agric. Eng.* 2019, 10, 889–897.

Khan, H. R., Paull, J. G., Siddique, K. H. M., Stoddard, F. L. Faba bean breeding for drought-affected environments: A physiological and agronomic perspective. *Field Crops Res*. 2010, 115, 279–286.

Khan, H. R., W Link, TJH Hocking and FL Stoddard, 2007. Evaluation of physiological traits for improving drought tolerance in faba bean (*Vicia faba* L.). *Plant Soil* 292:205–217.

Khan, M. A., Alghamdi SS, Ammar MH, Sun Q, Teng F, Migdadi HM, Al-Faifi SA (2019b) Transcriptome profiling of faba bean (*Vicia faba* L.) drought-tolerant variety hassawi-2 under drought stress using RNA sequencing. *Elect J Biotechnol* 39:15–29

Khan, N., Bano, A., Babar, M. A. Metabolic and physiological changes induced by plant growth regulators and plant growth promoting rhizobacteria and their impact on drought tolerance in Cicer arietinum L. *PLoS ONE* 2019a(, 14, e0213040.

Khazaei, H, 2014. Leaf traits associated with drought adaptation in faba bean (*Vicia faba* L.). *Doctoral thesis.* University of Helsinki Department of Agricultural Sciences Faculty of Agriculture and Forestry. Helsinki.

Khazaei, H., Mäkelä, P. S., and Stoddard, F. L. (2018). Ion beam irradiation mutagenesis in rye (Secale cereale L.), linseed (Linum usitatissimum L.) and faba bean (*Vicia faba* L.). *Agric. Food Sci.* 27, 146–151.

Khazaei, H., Podder, R., Caron, C. T., Kundu, S. S., Diapari, M., Vandenberg, A., Bett, K. E., 2017. Marker–trait association analysis of iron and zinc concentration in lentil (Lens culinaris Medik.) seeds. *Plant Genome,* 10: 1–8.

Khazaei, H., Street, K., Santanen, A., Bari, A., Stoddard, F. L. Do faba bean (*Vicia faba* L.) accessions from environments with contrasting seasonal moisture availabilities differ in stomatal characteristics and related traits? *Genet. Resour. Crop Evol.* 2013, 60, 2343–2357.

Khodary, S. E. A., 2004. Effect of salicylic acid on the growth, photosynthesis and carbohydrate metabolism in salt stressed maize plants. *Int. J. Agric. Biol.,* 6: 5-8.

Khursheed, S., Raina, A., Parveen, K., and Khan, S. (2019). Induced phenotypic diversity in the mutagenized populations of faba bean using physical and chemical mutagenesis. *J. Saudi Soc. Agric. Sci.* 18, 113–119.

Kiflemariam, Y. B. and Stoddard, F. L. 2017. Screening of faba bean (*Vicia faba* L.) accessions to acidity and aluminium stresses. *Peer Journal*, 2963:1-19.

Kiflemariam, Y. B., Nagel, K. A., Fiorani, F., & Stoddard, F. L. (2018). Diversity in root growth responses to moisture deficit in young faba bean (*Vicia faba* L.) plants. PeerJ, 6, e4401.

Kim, H. S., Kim, K. R., Yang, J. E., Ok, Y. S., Owens, G., Nehls, T., Wessolek, G., Kim, K. H., 2016. Effect of biochar on reclaimed tidal land soil properties and maize (Zea mays L.) response. *Chemosphere* 142, 153–159.

Koivunen, E., Partanen, K., Perttila¨, S., Palander, S., Tuunainen, P., Valaja, J., 2016. Digestibility and energy value of pea (*Pisum sativum* L.), faba bean (*Vicia faba* L.) and blue lupin (narrow-leaf)(Lupinus angustifolius) seeds in broilers. *Anim. Feed. Sci. Technol.* 218, 120-127.

Kole, C. (Ed.). (2022). *Genomic Designing for Abiotic Stress Resistant Pulse Crops*. Springer International Publishing.

Kopke, U. and Nemecek, T. 2010. Ecological services of faba bean. *Field Crops Research,* 115:217–33.

Kubala, S., Wojtyla, L., Quinet, M., Lechowska, K., Lutts, S., Garnczarska, M. Enhanced expression of the proline synthesis gene P5CSA in relation to seed osmo-priming improvement of Brassica napus germination under salinity stress. *J. Plant Physiol.* 2015, 183, 1–12.

Kumar, N., Hazra, K. K., Singh, S., Nadarajan, N. Constraints and Prospects of growing pulses in rice fallows of India. *Indian Farming* 2016, 66, 13–16.

Kumar, P. M., Chaurasia, A. K., Michael Bara, B. M. Effect of osmo-priming on seed germination behaviour and vigour of chickpea (Cicer arietinum L.). *Int. J. Sci. Nat.* 2017, 8, 330–335.

Latef, A. A. A., Hasanuzzaman, M. and Tahjib-Ul-Arif, M., 2021. Mitigation of salinity stress by exogenous application of cytokinin in faba bean (*Vicia faba* L.). *Notulae Botanicae Horti Agrobotanici Cluj-Napoca*, 49(1), pp.12192-12192.

Lawes, D. A., Bond DA, Poulsen MH (1983) Classification, origin, breeding methods and objectives. In: Habblewaite PD (ed) *The Faba Bean* (*Vicia faba* L.). Butterworth, London, pp 23–76.

Li, H., & Tan, Z. (2021). Preparation of high water-retaining biochar and its mechanism of alleviating drought stress in the soil and plant system. *Biochar,* 3(4), 579-590.

Li, L., Yuan TZ, Setia R, Raja RB, Zhang B, Ai Y (2019) Characteristics of pea, lentil and faba bean starches isolated from air-classified flours in comparison with commercial starches. *Food Chem* 276:599–607.

Li, P., Zhang Y, Wu X, Liu Y (2018) Drought stress impact on leaf proteome variations of faba bean (*Vicia faba* L.) in the Qinghai–Tibet Plateau of China. 3 *Biotechnology* 8(2):110.

Link, W., Ederer W, Metz P, Buiel H, Melchinger AE (1994a) Genotypic and environmental variation for degree of cross-fertilization in faba bean. *Crop Sci* 34(4):960–964.

Link, W., Stelling D, Ebmeyer E (1994b) Factors determining the performance of synthetics in *Vicia faba* L. Heterogeneity, heterozygosity, and degree of cross-fertilization. *Euphytica* 75(1–2):77–84.

Liu, H., Tessema, B. B., Jensen, J., Cericola, F., Andersen, J. R., and Sørensen, A. C. (2019). ADAM-Plant: a software for stochastic simulations of plant breeding from molecular to phenotypic level and from simple selection to complex speed breeding programs. *Front. Plant Sci.* 9:1926.

Liu, X., Wei, Z., Ma, Y., Liu, J., & Liu, F. (2021a). Effects of biochar amendment and reduced irrigation on growth, physiology, water-use efficiency and nutrients uptake of tobacco (Nicotiana tabacum L.) on two different soil types. *Science of the Total Environment,* 770, 144769.

Liu, X., Wei, Z., Manevski, K., Liu, J., Ma, Y., Andersen, M. N., & Liu, F. (2021b). Partial root-zone drying irrigation increases water-use efficiency of tobacco plants amended with biochar. *Industrial Crops and Products,* 166, 113487.

Loss, S. P., & Siddique, K. H. M. (1997). Adaptation of faba bean (*Vicia faba* L.) to dryland Mediterranean-type environments I. Seed yield and yield components. *Field Crops Research*, 52(1-2), 17-28.

M'Sehli, W., Kallala, N., Jaleli, K., Bouallegue, A., & Mhadhbi, H. (2020). Monopotassium phosphate (KH2PO4) and salicylic acid (SA) as seed priming in *Vicia faba* L. and Vicia sativa L. *Biosci. j.(Online),* 2078-2091.

Maalouf, F. S., Suso MJ, Moreno MT (2002) Comparative performance of faba bean synthetics developed from different parental number (*Vicia faba* L). *J Genet Breed* 56(3):521–258

Maalouf, F., Ahmed, S. and Bishaw, Z., 2021. Faba bean. *In The beans and the peas* (pp. 105-131). Woodhead Publishing.

Maalouf, F., Ahmed, S., Somanagouda, P., 2018. Developing improved varieties of faba bean. In: Sivasankar, S., et al., (Eds.), *Achieving sustainable cultivation of grain legumes* Volume 2: improving cultivation particular grain legumes. Burleigh Dodds Science Publishing, Cambridge, UK, ISBN:978.1786761408.

Maalouf, F., Hu, J., O'Sullivan, D. M., Zong, X., Hamwieh, A., Kumar, S., & Baum, M. (2019a). Breeding and genomics status in faba bean (*Vicia faba*). *Plant Breeding,* 138(4), 465-473.

Maalouf, F., Khalil, S., Ahmed, S., Kharrat, M., Hajjar, S., and El Shama'a, K. (2011). Yield stability of faba bean lines under diverse broomrape prone production environments. *Field Crops Res.* 124, 288–294.

Maalouf, F., Khater, L. A., Babiker, Z., & Mohamed, A. (2022). Genetic and Genomic Research for Abiotic Stresses in Faba Bean. In *Genomic Designing for Abiotic Stress Resistant Pulse Crops* (pp. 249-270). Cham: Springer International Publishing.

Maalouf, F., Nachit, M., Ghanem, M. E., Singh, M. Evaluation of faba bean breeding lines for spectral indices, yield traits and yield stability under diverse environments. *Crop Pasture Sci.* 2015, 66, 1012–1023.

Maalouf, F., Tadesse DW, Wery J, Boutfiras M, Aw-Hassan A (2019b) Legume-based rotations have clear economic advantages over cereal monocropping in dry areas. *Agron Sustain Dev* 39(6):58 270.

Macarulla, M. T., Medina, C., De Diego, M. A., Ch′ avarri, M., Zulet, M. A., Martínez, J. A., et al., (2001). Effects of the whole seed and a protein isolate of faba bean (*Vicia faba*) on the cholesterol metabolism of hypercholesterolaemic rats. *British Journal of Nutrition*, 85(5), 607–614.

Madejón, P., Fernández-Boy, E., Morales-Salmerón, L., Navarro-Fernández, C. M., Madejón, E., & Domínguez, M. T. (2023). Could conservation tillage increase the resistance to drought in Mediterranean faba bean crops?. *Agriculture, Ecosystems & Environment,* 349, 108449.

Manchanda, G., Garg, N. Salinity and its effects on the functional biology of legumes. *Acta Physiol. Plant*. 2008, 30, 595–618.

Manschadi, A. M., Sauerborn, J., Stützel, H., Göbel, W., Saxena, M. C. Simulation of faba bean (*Vicia faba* L.) root system development under Mediterranean conditions. *Eur. J. Agron*. 1998, 9, 259–272.

Mansoor, S., Kour, N., Manhas, S., Zahid, S., Wani, O. A., Sharma, V., ... & Ahmad, P. (2021). Biochar as a tool for effective management of drought and heavy metal toxicity. *Chemosphere,* 271, 129458.

Mao, D., Michelmore, S., Paull, J., Preston, C., Sutton, T., Oldach, K., et al., (2019). Phenotypic and molecular characterisation of novel *Vicia faba* germplasm with tolerance to acetohydroxyacid synthase-inhibiting herbicides (AHAS) developed through mutagenesis techniques. *Pest Manag. Sci.* 75, 2698–2705.

Mbarki, S., Sytar O, Cerda A, Zivcak M, Rastogi A, He X, Zoghlami A, Abdelly C, Brestic M (2018). Strategies to mitigate the salt stress effects on photosynthetic apparatus and productivity of crop plants. *Salinity Responses and Tolerance in Plants* 1:85-136.

McDonald, G. K., Paulsen GM (1997) High temperature effects on photosynthesis and water relations of grain legumes. *Plant Soil* 196:47–58.

McVicar, R., Panchuk, D., Brenzil, C., Hartley, S., Pearse, P. and Vandenberg, A. 2013. Faba bean Gov. Saskatchewan, Agriculture, *Crops.*

Mejri, S., Mabrouk, Y., Belhadj, O., and Saidi, M. (2018). Orobanche foetida resistance in two new faba bean genotypes produced by radiation mutagenesis. *Int. J. Radiat. Biol.* 94, 671–677.

Mesfin, S., Gebresamuel, G., Haile, M., Zenebe, A., & Desta, G. (2020). Mineral fertilizer demand for optimum biological nitrogen fixation and yield potentials of legumes in Northern Ethiopia. *Sustainability,* 12, 6449.

Mesfin, T. 2019. Genotype x Location x Management and Genetic Variability of Faba Bean (*Vicia faba* L.) Genotypes Based on Agro-Morphology Traits. *MSc Thesis,* Haramaya University, Haramaya, Ethiopia.

Metayer, N., 2004. *Vicia faba* breeding for sustainable agriculture in Europe: Identification of regional priorities and definition of target genotypes. *GIE Févérole, Paris, 5.*

Miller, M. L., Soufi, B., Jers, C., Blom, N., Macek, B., Mijakovic, I., 2009. NetPhosBac—a predictor for Ser/Thr phosphorylation sites in bacterial proteins. *Proteomics, 9*: 116–125.

MoALR (Ministry of Agriculture and Livestock Resource), 2017. Plant variety release, protection and seed quality control directorate. *Crop variety register,* Issue No. 20. Addis Ababa, Ethiopia.

Mobini, S., Khazaei H, Warkentin TD, Vandenberg A (2020) Shortening the generation cycle in faba bean (*Vicia faba*) by application of cytokinin and cold stress to assist speed breeding. *Plant Breed* 139(6):1181–1189.

Mohammed, A. F. (2018). Effectiveness of exopolysaccharides and biofilm forming plant growth promoting rhizobacteria on salinity tolerance of faba bean (*Vicia faba* L.). *African Journal of Microbiology Research, 12*(17), 399-404.

Monneveux, P., JM Ribaut and A Okono, 2014. Drought phenotyping in crops: from theory to practice. *Frontier Ebooks.*

Moose, S. P. and RH Mumm, 2008. Molecular plant breeding as the foundation for 21st century crop improvement. *Plant Physiol* 147:969–977.

Mortimore, M., S Anderson, L Cotula, J Davies, K Faccer, C Hesse, J Morton, W Nyangena, J Skinner and C Wolfangel, 2009. *Dryland Opportunities: A new paradigm for people, ecosystems and development,* IUCN, Gland, Switzerland; IIED, London, UK and UNDP/DDC, Nairobi, Kenya. 86 p.

Morton, M. J., Awlia M, Al-Tamimi N, Saade S, Pailles Y, Negrão S, Tester M (2019). Salt stress under the scalpel–dissecting the genetics of salt tolerance. *The Plant Journal* 97(1):148-163.

Muktadir, M. A., Adhikari, K. N., Merchant, A., Belachew, K. Y., Vandenberg, A., Stoddard, F. L. and Khazaei, H., 2020. Physiological and biochemical basis of faba bean breeding for drought adaptation—A review. *Agronomy,* 10(9), p.1345.

Müller, B. S., Neves LG, de Almeida Filho JE, Resende MF, Muñoz PR, dos Santos PE et al., (2017) Genomic prediction in contrast to a genome-wide association study in explaining heritable variation of complex growth traits in breeding populations of Eucalyptus. *BMC Genomics* 18(1):524.

Muller, H. J. (1927): Artificial transmutation of the gene. *Sci.* 66:84-87.

Multari, S., Stewart, D., Russell, W. R., 2015. Potential of fava bean as future protein supply to partially replace meat intake in the human diet. *Compr Rev Food Sci Food Saf,* 14: 511–522.

Mulugeta, W., Tesfaye, K., Getnet, M., Ahmed, S., Nebiyu, A., and Mekuanint, F. (2019). Quantifying yield potential and yield gaps of faba bean in Ethiopia. *Ethiop. J. Agric. Sci.* 29, 105–120.

Munns, R., Tester M (2008). Mechanisms of salinity tolerance. *Annual Review of Plant Biology Annual Reviews* 59:651- 681.

Mussa, J., G Dereje and K Gemechu, 2008. Procedures of Faba Bean Improvement through Hybridization. *Technical Manual* No. 21, Ethiopian Institute of Agricultural Research, 48pp.

MVD (2021). www.iaea.org/resources/databases/mutant-varieties-database

Nachshon, U. (2018). Cropland soil salinization and associated hydrology: Trends, processes and examples. *Water* 10(8):1030.

Neugschwandtner, R., Ziegler, K., Kriegner, S., Wagentristl, H., Kaul, H.-P. Nitrogen yield and nitrogen fixation of winter faba beans. *Acta Agric. Scand. B* 2015, 65, 658–666.

Niu, X., Song, L., Xiao, Y., Ge, W. Drought-tolerant plant growth-promoting rhizobacteria associated with foxtail millet in a semi-arid agro-ecosystem and their potential in alleviating drought stress. *Front. Microbiol.* 2018, 8, 2580.

Nouairi, I., Jalali, K., Benmoussa, S., Kammoun, M., Zribi, K., & Mhadhbi, H. (2021). CaCl2seed priming stimulate nodulation and oleosome lipids formation in the root nodules of cadmium-treated faba bean plants. *Rhizosphere,* 18, 100326.

Nour El-Din, M., El-Shanshoury, A. R., El-Akhdar, I. A. Isolation and evaluation of salt-tolerant microorganisms and their impact in adaptation of faba bean to salinity stress. *J. Soil Sci. Agric. Eng.* 2011, 2, 977–999.

Nurmansyah, A., S. S., and Migdadi, H. H. (2020). Morphological diversity of faba bean (*Vicia faba* L.) M2 mutant populations induced by gamma radiation and diethyl sulfate. *J. King Saud Univ. Sci.* 32, 1647–1658.

O'Sullivan, D. M., Zong X, Hamwieh A, Kumar S, Baum M (2019) Breeding and genomics status in faba bean (*Vicia faba*). *Plant Breed* 138(4):465–473.

Ortiz-Monasterio, J. I., Palacios-Rojas N, Meng E, Pixley K, Trethowan R, Pena RJ (2007) Enhancing the mineral and vitamin content of wheat and maize through plant breeding. *J Cereal Sci* 46(3):293–307.

Osman, M. S., Badawy AA, Osman AI, Latef AAHA (2020). Ameliorative impact of an extract of the halophyte Arthrocnemum macrostachyum on growth and biochemical parameters of soybean under salinity stress. *Journal of Plant Growth Regulation* 1-12.

Ouertatani, S., Regaya, K., Ryan, J. and Gharbi, A. 2011. Soil liming and mineral fertilization for root nodulation and growth of faba beans in an acid soil in Tunisia. *Journal of Plant Nutrition,* 34:850–860.

Oweis, T., A Hachum and M Pala, 2005. Faba bean productivity under rainfed and supplemental irrigation in northern Syria. *Agr Water Manag* 73:57–72.

Pampana, S., Masoni A, Arduini I (2016) Response of cool-season grain legumes to waterlogging at flowering. *Can J Plant Sci* 96(4):597–603.

Patil, S. A., Apine, O. A., Surwase, S. N., Jadhav, J. P., 2013. Biological sources of L-DOPA: an alternative approach. *Adv Parkinson's Dis*, 2: 81–87.

Patrick, J. W., Stoddard FL (2010) Physiology of flowering and grain filling in faba bean. *Field Crops Res* 115:234–242.

Patto, M. V., Torres AM, Koblizkova A, Macas J, Cubero JI (1999) Development of a genetic composite map of *Vicia faba* using F2 populations derived from trisomic plants. *Theor Appl Genet* 98(5):736–743.

Paul, D., Lade, H. Plant-growth-promoting rhizobacteria to improve crop growth in saline soils: A review. *Agron. Sustain. Dev.* 2014, 34, 737–752.

Peaston, R. T., Weinkove, C., 2004. Measurement of catecholamines and their metabolites. *Ann Clin Biochem,* 41: 17–38.

Pociecha, E., Ko´scielniak, J., and Filek, W. (2008). Effects of root flooding and stage of development on the growth and photosynthesis of field bean (*Vicia faba* L. minor). *Acta Physiol. Plant*. 30, 529–535.

Poulsen, M. H. (1981) Survey of the breeding work on *Vicia faba* at VEG Saatzucht Gotha/Friedrichswerth. In: Thompson R. (ed) *Vicia faba: physiology and breeding.* Springer, Dordrecht, pp 259–265.

Pozo, M. J., López-Ráez, J. A., Azcón-Aguilar, C., García-Garrido, J. M. Phytohormones as Integrators of Environmental Signals in the Regulation of Mycorrhizal Symbioses. *New Phytol.* 2015, 205, 1431–1436.

Prasad, P. V. V., Boote KJ, Allen LH Jr (2006) Adverse high temperature effects on pollen viability, seed-set, seed yield and harvest index of grain-sorghum [*Sorghum bicolor* (L.) Moench] are more severe at elevated carbon dioxide due to higher tissue temperatures. *Agric For Meteorol* 139:237–251.

Qados, A. M. A. (2011) Effect of salt stress on plant growth and metabolism of bean plant *Vicia faba* (L.). *J Saudi Soc Agric Sci* 10(1):7–15.

Raina, A., and Khan, S. (2020). *Mutagenic effectiveness and efficiency of gamma rays and sodium azide in M2 generation of Cowpea* [Vigna unguiculata (L.) Walp.]. bioRxiv 2020:983486.

Rajhi, I., Baccouri, B., Frchichi, N., Damegi, B., Zribi, F., Hammami, J., & Mhadhbi, H. (2022). *Impact of biochar soil amendment on alleviation of single and combined effect of drought and salt stresses on Vicia faba seedlings planted in alkaline soil.*

Ramakrishna, A., Ravishankar, G. A., 2011. Influence of abiotic stress signals on secondary metabolites in plants. *Plant Signal Behav*, 1: 1720–1731.

Ramsay G, Grifiths DW, Dow ND 1991 Spontaneous and induced variation in levels of vicine and convicine in faba beans. *Asp Appl Biol* **27** 43–47.

Ramsay, G. (1997) Inheritance and linkage of a gene for testa-imposed seed dormancy in faba bean (*Vicia faba* L.). *Plant Breed* 116(3):287–289.

Ray, H., Georges F (2010) A genomic approach to nutritional, pharmacological and genetic issues of faba bean (*Vicia faba*): prospects for genetic modifications. *GM Crops* 1(2):99–106.

Raza, M. A., Gul, H., Wang, J., Yasin, H. S., Qin, R., Khalid, M. H. B., Yang, W. 2021. Land productivity and water use efficiency of maize-soybean strip intercropping systems in semi-arid areas: A case study in Punjab Province, Pakistan. *Journal of Cleaner Production*, 308: 127282.

Razmjoo, K., P Heydarizadeh and MR Sabzalian, 2008. Effect of salinity and drought stresses on growth parameters and essential oil content of Matricaria. chamomile. *Int. J. Agric. Biol.,* 10, 451–454.

Redden, R., Paull J, Zong X, Sass O, Yang T, Ling L (2014) Faba Bean. In: Singh M, Singh I, Dutta M (eds) *Broadening the genetic base of grain legumes*. Springer, New Delhi, pp 75–90.

Reddy, A. A. Pulses production technology: Status and way forward. Econ. *Polit. Wkly.* 2009, 44, 73–80.

Rehman, S., Abbas G, Shahid M, Saqib M, Farooq ABU, Hussain M, et al., (2019). Effect of salinity on cadmium tolerance, ionic homeostasis and oxidative stress responses in conocarpus exposed to cadmium stress: Implications for phytoremediation. *Ecotoxicology and Environmental Safety* 171:146-153.

Rezaie, N., Razzaghi, F., & Sepaskhah, A. R. (2019). Different levels of irrigation water salinity and biochar influence on faba bean yield, water productivity, and ions uptake. *Communications in Soil Science and Plant Analysis*, 50(5), 611-626.

Rharrabti, Y., Villegas, D., García del Moral, L., Aparicio, N., Elhani, S., Royo, C. Environmental and genetic determination of protein content and grain yield in durum wheat under Mediterranean conditions. *Plant Breed.* 2001, 120, 381–388.

Rowland, G. G. (1987) A recurrent selection scheme for faba bean. *Can J plant sci* 67(1):79–85.

Rowland, G., Duc G, Picard J (1986) Fertility components in a faba bean line near isogenic for male sterility. *Can J Plant Sci* 66:235–239.

Rubiales, D., and Fernández-Aparicio, M. (2012). Innovations in parasitic weeds management in legume crops. A review. *Agron. Sustain. Dev.* 32, 433–449.

Rubiales, D., Rojas-Molina, M. M., and Sillero, J. C. (2016). Characterization of Resistance Mechanisms in Faba Bean (*Vicia faba*) against Broomrape Species (Orobanche and Phelipanche spp.). *Front. Plant Sci.* 7:1747.

Safdar, H., Amin, A., Shafiq, Y., Ali, A., Yasin, R., Shoukat, A., Hussan, M. U., Sarwar, M. I. A Review: Impact of Salinity on Plant Growth. *Nat. Sci.* 2019, 17, 34–40.

Sainte, M., 2011. *The magazine of the European Association for Grain Legume Research.* Issue No. 56 Model Legume Congress, France, 15-19 May 2011.

Säle, V., Aguilera, P., Laczko, E., Mäder, P., Berner, A., Zihlmann, U., ... & Oehl, F. (2015). Impact of conservation tillage and organic farming on the diversity of arbuscular mycorrhizal fungi. *Soil Biology and Biochemistry*, 84, 38-52.

Sallam, A. and Ul-Allah, S., 2019. Genomics-aided breeding for climate-smart traits in faba bean. *Genomic designing of climate-smart pulse crops,* pp.359-395.

Satovic, Z., Avila CM, Cruz-Izquierdo S, Díaz-Ruíz R, García-Ruíz GM, Palomino C et al., (2013) A reference consensus genetic map for molecular markers and economically important traits in faba bean (*Vicia faba* L.). *BMC Genomics* 14(1):932.

Schumacher, H., Paulsen, H., Gau, A., Link, W., Jürgens, H., Sass, O., Dieterich, R. Seed, protein amino acid composition of important local grain legumes Lupinus angustifolius L., Lupinus luteus L., Pisum sativum L. and *Vicia faba* L. *Plant Breed.* 2011, 130, 156–164.

Seleiman, M. F., Al-Suhaibani, N., Ali, N., Akmal, M., Alotaibi, M., Refay, Y., Dindaroglu, T., Abdul-Wajid, H. H., Battaglia, M. L. Drought stress impacts on plants and different approaches to alleviate its adverse effects. *Plants* 2021, 10, 259.

Shahin, U. Z. (2019) Waterlogging tolerance at germination in pea (*Pisum sativum* L.) *Thesis university of western Australia.*

Sharma, N. K., Singh, RJ., Mandal, D., Kumar, A., Alam, N. M., Keesstra, S. 2017. Increasing farmer's income and reducing soil erosion using intercropping in rainfed maize-wheat rotation of Himalaya, India. *Agriculture, ecosystems and Environment,* 247: 43-53. Shui, X. R., Chen, Z. W., Li, J. X. 2013.

Shi, J., 2013. *Decomposition and nutrient release of different cover crops in organic farm systems.* University of Nebraska – Lincoln, NE.

Shrivastava, P., Kumar R (2015). Soil salinity: a serious environmental issue and plant growth promoting bacteria as one of the tools for its alleviation. *Saudi Journal of Biological Sciences* 22(2):123-131.

Siddiqui, M. H., Al-Khaishany MY, Al-Qutami MA, Al-Whaibi MH, Grover A, Ali HM, Al-Wahibi MS, Bukhari NA (2015) Response of different genotypes of faba bean plant to drought stress. *Int J Mol Sci* 16:10214–10227.

Singh, A. K., Bhatt, B. P., Upadhyaya, A., Kumar, S., Sundaram, P. K., Singh, B. K., Chandra, N., Bharati, R. C. 2012a. Improvement of faba bean (*Vicia faba* L.) yield and quality through biotechnological approach: a review. *African Journal of Biotechnology,* 11:15264–15271.

Singh, A. K., Verma, N., Singh, C. S., Kumar, S., Gupta, A. K., 2012b. Faba Bean (*Vicia faba* L.): a potential leguminous crop of India, in: Singh, A. K., Bhatt, B. P. (Eds.), Ethnobotany of Faba Bean (*Vicia faba* L.). *ICAR, RC for ER, Patna,* pp. 431–450.

Singh, B. D., 1990. *Plant breeding: Principles and methods.* Kalyani Publishers, New Delhi Ludhiana, India.

Singh, L., Beg, M. K. A., Akhter, S., Qayoom, S., Lone, B. A., Singh, P., Singh, P. 2014. Efficient techniques to increase water use efficiency under rainfed ecosystems. *Journal of Agriculture Search*, 1(4):193-200.

Sjödin, J. (1971). Induced morphological variation in *Vicia faba* L. *Hereditas* 67, 155–180.

Smith, S. E., Facelli, E., Pope, S., Smith, F. A. Plant performance in stressful environments: Interpreting new and established knowledge of the roles of arbuscular mycorrhizas. *Plant Soil* 2010, 326, 3–20.

Solaiman, Z., Colmer, T. D., Loss, S. P., Thomson, B. D., and Siddique, K. H. M. (2007). Growth responses of cool-season grain legumes to transient waterlogging. *Aust. J. Agric. Res.* 58, 406–41.

Srivastava, P., Wu QS, Giri B (2019) Salinity: an overview. In Microorganisms in saline environments: strategies and functions. Springer, *Cham*, pp 3–18.

Stadler, L. G. (1928): Genetic effects of X-rays in maize. *Prod Natl. Acad. sci.,* USA, 14:69-75.

Stadler, L. G. (1930): Some genetic effects of X-rays in plants. *J. Heredity*, 21:3-19.

Stelling, D., Link W, Ebmeyer E (1994) Factors determining the performance of synthetics in *Vicia faba* L. 2. Syn-generation. *Euphytica* 75(1–2):85–93.

Stoddard, F. L., Balko, C., Erskine, W., Khan, H. R., Link, W., Sarker, A. Screening techniques and sources of resistance to abiotic stresses in cool-season food legumes. *Euphytica* 2006, 147, 167–186.

Suso, M. J., Moreno MT, Melchinger AE (1999) Variation in outcrossing rate and genetic structure on six cultivars of *Vicia faba* L. as affected by geographic location and year. *Plant Breed 118*(4):347–350.

Tadege, M., Wang, T. L., Wen, J., Ratet, P., and Mysore, K. S. (2009). Mutagenesis and beyond! Tools for understanding legume biology. *Plant Physiol.* 151, 978–984.

Tadele, M. (2020) Impact assessment of abiotic production constrains of Faba Bean (*Vicia faba* L.) and breeding mechanisms for acid soil, drought and waterlogging affected environments. *J Biol Agric Healthc* 10(11):49–56.

Tamene, T., Gemechu K. and Hussein M. 2015. *Genetic progresses from over three decades of faba bean.*

Tavakkoli, E., Rengasamy, P., McDonald, G. K. High concentrations of Na+ and Cl– ions in soil solution have simultaneous detrimental effects on growth of faba bean under salinity stress. *J. Exp. Bot.* 2010, 61, 4449–4459.

Taylor, A. G., Harman, G. E. Concepts and technologies of selected seed treatments. *Annu. Rev. Phytopathol.* 1990, 28, 321–339.

Tekalign, A., Derera, J., Sibiya, J., Fikre, A., 2016. Participatory assessment of production threats, farmers' desired traits and selection criteria of faba bean (*Vicia faba* L.) varieties: opportunities for faba bean breeding in Ethiopia. *Ind J Agric Res,* 50: 295–302.

Torres, A. M., Avila CM, Gutierrez N, Palomino C, Moreno MT, Cubero JI (2010). Marker-assisted selection in faba bean (*Vicia faba* L.). *Field Crops Res* 115(3):243–252.

Torres, A. M., B Roman, CM Avila, Z Satovic, D Rubiales, JC Sillero, JI Cubero and MT Moreno, 2006. Faba bean breeding for resistance against biotic stresses: towards application of marker technology. *Euphytica* 147, 67- 80.

Torres, A. M., CM Avila, FL Stoddard and JI Cubero, (2012). Faba bean. In: Pérez de la Vega M, Torres AM, Cubero JI, Kole C (eds) *Genetics, genomics and breeding in crop plants: cool season food legumes*. Science Pubs Inc, New Hampshire. pp 50–97.

Tosti, G., Guiducci, M., 2010. Durum wheat–faba bean temporary intercropping: effects on nitrogen supply and wheat quality. *Eur J Agron*, 33: 157–165.

Valente, I. M., Cabrita, A. R. J., Malushi, N., Oliveira, H. M., Papa, L., Rodrigues, J. A., et al., (2019). Unravelling the phytonutrients and antioxidant properties of European *Vicia faba* L. seeds. *Food Research International*, 116, 888–896.

Van Norel, A., and Hoogendoorn, J. (1989). Faba bean dwarf selections outyield Dutch top varieties, *FABIS Newsl.* 25, 10–13.

Varshney, R. K., Kudapa H, Pazhamala L, Chitikineni A, Thudi M, Bohra A, Gaur PM, Janila P, Fikre A, Kimurto P, Ellis N (2014). Translational genomics in agriculture: some examples in grain legumes. *Crit Rev Plant Sci* 34(1–3):169–194.

Wahbi, S., Prin, Y., Thioulouse, J., Sanguin, H., Baudoin, E., Maghraoui, T., Oufdou, K., Roux, C. L., Galiana, A., Hafidi, M., Duponnois, R., 2016. Impact of wheat/faba bean mixed cropping or rotation systems on soil microbial functionalities. *Front Plant Sci,* 7: 1364.

Wang, J., Gan YT, Clarke F, McDonald CL (2006). Response of chickpea yield to high temperature stress during reproductive development. *Crop Sci* 46:2171–2178.

Watson, C. A., Reckling, M., Preissel, S., Bachinger, J., Bergkvist, G., Kuhlman, T., Lindström, K., Nemecek, T., Topp, C. F. E., Vanhatalo, A., et al., Grain legume production and use in European agricultural systems. *Adv. Agron.* 2017, 144, 235–303.

Webb, A., Cottage A, Wood T, Khamassi K, Hobbs D, Gostkiewicz K et al., (2016) A SNP-based consensus genetic map for synteny-based trait targeting in faba bean (*Vicia faba* L.). *Plant Biotechnol J* 14(1):177–185.

Yang, Y., Guo Y (2018). Elucidating the molecular mechanisms mediating plant salt-stress responses. *New Phytologist* 217(2):523-539.

Zandvakili, O. R., Barker, A. V., Hashemi, M., Etemadi, F., 2019. Biomass and nutrient concentration of lettuce grown with organic fertilizers. *J Plant Nutr,* 42: 444–457.

Zhang, B., Li, Y., Ren, T., Tian, Z., Wang, G., He, X., & Tian, C. (2014). Short-term effect of tillage and crop rotation on microbial community structure and enzyme activities of a clay loam soil. *Biology and fertility of soils,* 50, 1077-1085.

Zhang, F. Y., Wu, P. T., Zhao, X. N., Cheng, X. F. (2012). Water-saving mechanisms of intercropping system in improving cropland water use efficiency. *Journal of Applied Ecology* 23(5): 1400-1406.

Zhang, Y., Ding, J., Wang, H., Su, L., & Zhao, C. (2020). Biochar addition alleviate the negative effects of drought and salinity stress on soybean productivity and water use efficiency. *BMC Plant Biology,* 20, 1-11.

Ziche, Z. I., Belouchrani, A. S. and Drouiche, N., 2023. Study of the Effect of Salt Stress on a Legume Faba Bean (*Vicia faba* L.). *Gesunde Pflanzen,* pp.1-7.

Zlatev, Z., and FC Lidon, 2012. An overview on drought induced changes in plant growth, water relations and photosynthesis. *Emir. J. Food Agric.* 24, 57–72.

Zohary, D. and M Hopf, (2000). *Domestication of Plants in the Old World,* 3rd eds. Oxford: Clarendon Press.

Chapter 11

Valorization of Faba Bean (*Vicia faba* L.) Hulls and Seeds: Alternative Dietary Resources in Ruminant Nutrition Towards More Sustainable Production Systems

Margarida R. G. Maia[1,*]
Nertila Malushi[1,2]
Ana R. J. Cabrita[1]
Inês M. Valente[1,3]
Hugo M. Oliveira[4]
Lumturi Papa[2]
José A. Rodrigues[3]
and António J. M. Fonseca[1]
[1]Requimte, Laqv, Icbas, School of Medicine and Biomedical Sciences, University of Porto, Porto, Portugal
[2]Department of Animal Production, Agricultural University of Tirana, Koder Kamez, Albania
[3]REQUIMTE, LAQV, Department of Chemistry and Biochemistry, Faculty of Sciences, University of Porto, Porto, Portugal
[4]INL, International Iberian Nanotechnology Laboratory, Braga, Portugal

[*] Corresponding author's email address: mmaia@isa.ulisboa.pt

In: Faba Beans
Editors: Aamir Raina and Samiullah Khan
ISBN: 979-8-89530-447-1

Abstract

Legume seeds, including faba bean (*Vicia faba* L.), are important protein sources in human diets worldwide. The industrial processing of beans generates a major by-product, faba hulls, which are commonly disposed of as waste. Nevertheless, this fibrous agro-food residue may be valued in animal nutrition, particularly of ruminants, where faba seeds, an underexploited source of protein, may contribute to reducing imports and enhancing the sustainability of European farming systems. Therefore, seven European cultivars of broad and horse beans' hulls and seeds were evaluated as alternative fiber and protein sources for ruminants. The chemical composition of hulls and seeds was analysed, and the effects on rumen fermentation were evaluated after 24-h incubations in in *vitro* batch systems. Faba hulls were shown to be a good source of fiber and effective in methane mitigation, although with an impact on feed fermentation. The high protein content of seeds, higher in horse beans than broad beans, makes them potential substitutes or complements to imported protein sources in ruminant diets. Cultivars of European faba beans differently affected *in vitro* methane production without impairing the rumen fermentation. This study suggests that *V. faba* hulls and seeds can simultaneously add value to ruminants' nutrition and contribute to more sustainable farming systems.

Keywords: alternative feeds, rumen fermentation, *Vicia faba* hull, *Vicia faba* seed

Introduction

The sustainability of farming systems requires a reduction of imported feeds, which are subjected to world market price fluctuations and have a high environmental footprint. A key illustrative example is soybean, the most important protein source used in animal feeding, which is almost absent from European crops and represented imports of approximately 20.5 million tons in 2021, the last year with published data (FAOSTAT, 2021). In pursuit of reduced reliance on soybeans imported from non-European countries, the production of protein supplements targeting self-adequacy is needed (European Commission, 2018). This approach will also contribute to preventing the use of genetically modified organisms in organic farming, as required by European Union regulation, and in intensive animal production

(Taheri et al., 2017). This is an aspect of growing importance as consumers' concerns about diet and livestock feeding practices are increasing.

Faba bean (*Vicia faba* L.) edible seeds are widely consumed in human diets worldwide, being good sources not only of protein and starch but also of fibre, minerals, and vitamins. Like most legumes, faba hulls are removed, and seeds are consumed fresh, dry, frozen, or tinned. Large quantities of hulls result from this process, representing nearly 75% of the total faba bean production yield (Renna et al., 2020), which are usually disposed of as waste. This inexpensive agro-food by-product has been underexploited, and available information on fava hulls is scarce. Broad bean hulls have been suggested as functional foods due to their content of macro-, micro-, and non-nutrient phytochemicals (Mateos-Aparicio et al., 2012; Mateos-Aparicio et al., 2010b; Renna et al., 2020). The antioxidant activity and phenolic compounds of broad bean and horse bean hulls have also been characterized (Valente et al., 2018). The nutritive value of this fibrous agricultural by-product may be valued in animal nutrition, particularly in that of ruminant animals.

When faba seeds underperform the standards for human consumption, they can be used in animal nutrition. When compared to the main protein feed used in animal feeding – soybean - the production of faba bean has lower nutrient demands and shows greater soil adaptation (Köpke & Nemecek, 2010). Faba bean is an annual legume usually sown in the spring in cold climates and in the winter in warm climates (Crépon et al., 2010), fixing more nitrogen than the major cool season grain legumes under the same soil conditions (Köpke & Nemecek, 2010). China, Ethiopia, Australia, and Europe are the largest producers of faba beans (Rawal & Navarro, 2019).

Faba seeds present lower protein, methionine and cysteine contents than soybean but higher lysine and starch contents (Palander et al., 2006), as well as higher ruminal protein degradability (Goelema et al., 1998). Heat treatment of faba beans could thus be advisable to decrease both protein degradability at the rumen level (Aguilera et al., 1992) and tannins reactivity, known to be high in faba seeds, particularly in the hulls of colored-flowering varieties (Jansman, 1993). Faba seeds contain low fat levels, but with a relatively high proportion of unsaturated fatty acids (Grela & Günter, 1995). Anti-nutritional compounds found in faba beans, such as trypsin inhibitors, gallic acid, phytic acid, convicine, vicine, and lectins (Dvořák et al., 2006), present no metabolic and health problems in animals with fully developed rumen (Melicharová et al., 2009). The nutritive value of faba seeds has been shown to differ mostly due to cultivars rather than to cultivation methods (Bjerg et al., 1988). Positive

results of faba seeds inclusion in dairy cows (Mordenti et al., 2007) and sheep diets (Liponi et al., 2009) have been reported.

In the face of the need to promote European farming systems' sustainability, the present study aimed to valorize faba bean hulls and seeds as alternative and complementary resources in ruminant nutrition. Therefore, the chemical composition of seven cultivars of faba hulls and seeds was assessed and evaluated the effects on *in vitro* rumen fermentation after 24-h batch incubations, with rumen liquor from cows fed with different diets used as inocula. This study is, to the best of our knowledge, the first to assess the potential of faba hulls as alternative fibrous feed for ruminant animals by evaluating their impact on rumen fermentation.

Case Study Design and Methodology

Faba Beans

Seven commercial faba bean cultivars of broad beans (*V. faba* L. var. *major*) and horse beans (*V. faba* L. var. *minor*) were evaluated in this study. Broad beans included Cekin, Belshi, Skrapari, Aguadulce, and Jögeva cultivars, and horse beans comprised Bauska and Lielplatones cultivars. Broad and horse beans were sowed in the spring at the Agricultural University of Tirana (Lushnja, Albania) experimental farm was harvested, and seeds and hulls were manually separated and collected.

Chemical Analysis of the Beans

The chemical analysis of faba hulls and seeds was analyzed following the methodologies previously described by Maia et al., (2019). Briefly, all faba samples were grounded (1-mm), and the dry matter (DM), ash, ether extract (EE), and nitrogen (N) content were determined. The crude protein (CP) content was calculated as N x 6.25. Fiber content was fractioned and analyzed as neutral detergent fiber (NDF), acid detergent fiber (ADF), and acid detergent lignin (ADL) in all samples. Neutral detergent fiber and ADL values are presented without residual ash and ADF with residual ash. The gross energy of faba hulls and seeds was analyzed using calorimetry. The starch content was only determined by spectrophotometry on faba seeds. Condensed

tannins of faba hulls and seeds were extracted using the butanol-HCl method (Porter et al., 1985) and quantified as leucocyanidin equivalent (Makkar, 2000). All methodologies were run in duplicate replicas for each sample.

Animals, Diets and Rumen Contents Collection

Two adult and healthy dry Holstein cows with permanent cannula in the rumen (Bar Diamond Inc., Parma, ID, USA) provided the rumen content samples. The dairy cows were housed at the agricultural campus (Vairão, Portugal) of the School of Medicine and Biomedical Sciences, University of Porto (ICBAS-UP), and were handled following good animal and welfare practices (2010/63/EU). All experiments, procedures and animal practices were previously approved by the ICBAS-UP Animal Ethics Committee, licensed by the Portuguese General Directorate for Food and Veterinary (DGAV; 0421/000/000/2015). All activities were conducted by laboratory animal science trained researchers (FELASA category C).

To ensure biological replicates, the donor cows were fed one of two complete diets over two experimental periods (n = 4 inocula): i) a haylage-based diet (100 g/kg CP, 602 g/kg NDF, 48.5 g/kg starch, DM basis), and ii) a maize silage-based diet (98.2 g/kg CP, 424 g/kg NDF, 293 g/kg starch, DM basis). Complete diets were supplemented daily with a mineral-vitamin mixture (Premix, Viana do Castelo, Portugal), sodium chloride, dicalcium phosphate, and calcium carbonate. Each diet was offered in two daily meals (09:30 and 17:30 h), and fresh drinking water was available at all times. After an adaptation period of two weeks to the diet, representative samples of rumen digesta were collected from each cow before the morning meal. The ruminal digest of each cow was transported in a pre-heated thermal jug (39°C) to the laboratory and used within less than 1 h to guarantee the viability of the rumen inocula.

Following the rumen samples collection, the complete diets were swapped between cows. A new adaptation period to diets started, and after two weeks, the rumen contents were collected as previously described.

Rumen *In Vitro* Incubation

The effects of faba hulls and seeds on rumen fermentation and total gas and enteric methane production were assessed in 24-h batch incubations.

Immediately before the incubation, the rumen contents of each cow were separately homogenized and filtered through gauze under O_2-free CO_2. One part of the strained rumen fluid was combined with two parts of buffer solution (Mould et al., 2005) under O_2-free CO_2, homogenized, and the buffered rumen inoculum saturated with O_2-free CO_2 at 39°C (Maia et al., 2016). Fifty mL of buffered ruminal inoculum were added anaerobically into serum bottles (125 mL; Sigma-Aldrich Inc., St. Louis, MO, USA) with 250 mg (DM basis) of each faba bean cultivar and fraction. The bottles were flushed with O_2-free CO_2, sealed with butyl stoppers and crimped with aluminum caps (Sigma-Aldrich Inc.), being then incubated in a water-bath at 39°C for 24 h. By the end of the incubation period, the fermentation was halted by placing the bottles in a water bath at 39°C for at least 15 min.

Two replicas of each faba bean cultivar and fraction were incubated per rumen inoculum and incubation. Two incubation runs were carried out, one at the end of each adaptation period of the donor cows to the diet.

Incubation End-Products Analysis

To assess the faba bean cultivars and fractions fermentability, the headspace gas produced during the 24-h incubation was measured using a pressure transducer (Theodorou et al., 1994) in the fermentation bottles after being gradually warmed up to room temperature (25°C). The headspace gas was collected with a gas-tight syringe (SGE International Pty. Ltd., Ringwood Victoria, Australia) and injected into a gas chromatograph with a thermal conductivity detector (GC-4000A; East & West Analytical Instruments Inc., Beijing, China) and a micro packed column (Shincarbon ST 100/120; Restek Corporation, Bellefonte, PA, USA) to evaluate the gas composition. The chromatographic conditions used were described in detail by Maia et al. (2016). The identification of the fermentation gas composition was made by comparison of retention times with a commercial standard (Air Liquide, Lda., Algés, Portugal). Methane production was calculated using CO_2 as a reference gas (Lopez & Newbold, 2007).

After gas sampling, the fermentation bottles were opened, and pH was immediately read (GLP22+ pH meter; Crison Instruments SA, Barcelona, Spain). Fermentation media samples were collected for ammonia-N and volatile fatty acids analysis. For ammonia-N analysis, fermentation media were added to 0.1 N hydrochloric acid solution (1:1, v/v), steam distilled, and the N content was determined by titration. For volatile fatty acids analysis, the

fermentation media was mixed with 25% *o*-phosphoric acid solution containing 3-methyl valeric acid (16 mM, internal standard; Sigma-Aldrich Inc.), at 4:1 (v/v) ratio, vortexed and centrifuged at 19,800 × *g* at 4°C for 15 min. The supernatant was collected in GC vials and analyzed in a gas chromatograph with a flame ionization detector (Shimadzu GC-2010 Plus, Shimadzu Corporation, Kyoto, Japan) and a capillary column (HP-FFAP, 30 m x 0.25 mm x 0.25 µm; Agilent Technologies, Santa Clara, CA, USA), using the chromatographic conditions described by Maia et al., (2016). Individual volatile fatty acids were identified by comparison of retention times with a commercial standard (Volatile Free Acid Mix, Sigma-Aldrich Inc.) and quantified based on the internal standard calculation method.

Calculations

Stoichiometry calculations based on end-fermentation parameters after 24-h in vitro incubations were used to estimate the reducing equivalents generated and consumed, the recovery percentage, and the fermentation efficiency, as proposed by Demeyer (1991) and Chalupa (1977).

Statistical Analysis

The MIXED procedure of the SAS software (2002; version 9.2, SAS Institute Inc., Carry, NC, USA) was used to analyze the data. The statistical model had as fixed effects the faba cultivar (Cekin, Belshi, Skrapari, Aguadulce, Jögeva, Bauska and Lielplatones), the faba fraction (hull and seed), and the interaction between faba cultivar and faba fraction, and as random effects the inoculum nested within the period, and the residual error. Significant effects were considered when $P \leq 0.05$ and a tendency when $0.05 < P \leq 0.10$.

Case Study Outcomes

Chemical Composition

Table 1 presents the proximate composition, condensed tannins, and gross energy content of broad and horse faba beans' cultivars and fractions.

Ash content was higher in hulls than in seeds, achieving 84.5 g/kg DM in Aguadulce. Conversely, CP content was higher in seeds, with Lielplatones presenting the highest value (367 g/kg DM). All broad and horse faba beans presented low EE content, ranging from 3.65 g/kg DM in Lielplatones hulls to 12.7 g/kg DM in Belshi seeds. Neutral detergent fibre content of hulls varied between 559 g/kg DM in Aguadulce and 727 g/kg DM in Cekin, the NDF content of faba seeds being much lower (163-217 g/kg DM in Belshi and Bauska). Similarly, ADF and ADL contents were higher in hulls than in seeds, regardless of the cultivar. Starch content of faba seeds varied from 367 g/kg DM in Skrapari to 431 g/kg DM in Cekin. Gross energy was slightly higher in faba seeds than in faba hulls. Faba seeds presented higher condensed tannins content than faba hulls, the highest value being found in Lielplatones cultivar (1.36 g/kg DM).

Table 1. Proximate composition and condensed tannins content (g/kg dry matter) and gross energy (MJ/kg dry matter) of the broad and horse faba bean cultivars studied

		Broad bean					Horse bean	
	Fraction	Cekin	Belshi	Aguadulce	Jögeva	Skrapari	Bauska	Lielplatones
Ash	Hull	71.9	69.5	84.5	81.3	80.2	84.2	82.9
	Seed	31.2	38.7	34.3	35.7	34.9	32.0	36.2
Crude protein	Hull	56.7	74.9	95.3	74.3	100	66.9	65.7
	Seed	294	300	283	344	302	345	365
Ether extract	Hull	4.48	5.09	5.33	6.26	5.62	3.90	3.65
	Seed	7.81	12.7	6.32	11.0	7.32	9.68	8.51
Neutral detergent fiber	Hull	727	648	558	638	560	637	651
	Seed	209	163	172	164	173	216	172
Acid detergent fiber	Hull	606	548	416	492	496	515	542
	Seed	126	128	117	119	123	114	111
Acid detergent lignin	Hull	113	116	76.4	104	98.2	102	121
	Seed	19.6	17.7	18.0	16.6	10.5	37.3	35.0
Starch	Hull	ND	ND	ND	ND	ND	ND	ND
	Seed	431	395	426	410	367	399	385
Condensed tannins	Hull	0.016	0.038	0.105	0.006	0.060	0.006	0.021
	Seed	0.275	0.319	0.601	0.792	0.715	0.672	1.36
Gross energy	Hull	16.9	17.2	16.8	17.4	17.3	17.4	17.5
	Seed	18.3	18.3	18.6	18.6	18.4	18.4	18.6

ND, not determined.

Rumen Fermentation Parameters

On average, total gas production was *ca.* 31% higher for the incubation of faba seeds when compared to faba hulls ($P < 0.001$, Table 2). Faba cultivars also affected gas production (Table 3). The lowest gas production was found when Cekin and Jögeva were incubated (194 and 197 mL/g DM), and the highest was with Skrapari (215 mL/g DM; $P = 0.042$, Table 3).

Table 2. Effects of faba bean fraction (hull and seed) on *in vitro* fermentation parameters from 24-h batch incubations

	Faba			
	Hull	Seed	SEM	P
Gas (mL/g DM)	176	231	30.1	<0.001
Methane (mL/g DM)	17.8	28.6	6.14	<0.001
pH	6.21	6.02	0.065	<0.001
Ammonia-N (mg/L)	60.3	79.7	3.29	<0.001
Total volatile fatty acids (mmol/g DM)	9.26	11.5	0.603	<0.001
Acetic acid (%)	65.1	60.0	1.52	<0.001
Propionic acid (%)	18.2	18.3	0.67	0.331
iso-Butyric acid (%)	1.03	1.29	0.098	<0.001
Butyric acid (%)	11.7	15.0	0.93	<0.001
iso-Valeric acid (%)	2.00	2.76	0.302	<0.001
Valeric acid (%)	1.16	1.98	0.136	<0.001
iso-Caproic acid (%)	0.0131	0.0112	0.00067	0.025
Caproic acid (%)	0.436	0.619	0.0715	<0.001
Enanthic acid (%)	0.0274	0.0406	0.01039	<0.001
Acetic:propionic ratio	3.60	3.31	0.188	<0.001
H_2 generated (µmol/mL)	114	145	6.8	<0.001
H_2 consumed (µmol/mL)	35.2	49.3	3.03	<0.001
Recovery (%)	30.9	33.9	1.33	<0.001
Fermentation efficiency (%)	72.9	73.8	0.45	<0.001

DM, dry matter; SEM, standard error of the mean.

Faba hulls promoted lower methane production (17.8 mL/g DM) than faba seeds (28.6 mL/g DM; Table 2, $P < 0.001$). The lowest methane production was observed with Cekin (19.8 mL/g DM) and the highest with Lielplatones, Aguadulce and Skrapari cultivars (25.3, 25.4 and 26.3 mL/g DM, respectively; Table 3, $P = 0.004$).

Table 3. Effects of broad and horse faba beans cultivars on *in vitro* fermentation parameters from 24-h batch incubations

	Broad faba					Horse faba			
	Cekin	Belshi	Aguadulce	Jögeva	Skrapari	Bauska	Lielplatones	SEM	P
Gas (mL/g DM)	194a	203abc	214bc	197a	215c	201abc	200ab	30.4	0.042
Methane (mL/g DM)	19.8a	20.4ab	25.4c	21.1ab	26.3c	24.2bc	25.3c	6.26	0.004
pH	6.11	6.11	6.02	6.16	6.10	6.18	6.12	0.081	0.590
Ammonia-N (mg/L)	67.3	71.6	67.4	71.7	73.3	68.1	70.4	4.15	0.705
Ammonia-N (mg/g DM)	11.1	11.8	11.1	11.8	12.0	11.2	11.6	0.68	0.712
Total volatile fatty acids (mmol/g DM)	10.3	10.5	10.7	10.0	10.6	10.5	10.1	0.63	0.157
Acetic acid (%)	62.5	62.7	63.1	62.4	62.2	62.4	62.4	1.54	0.479
Propionic acid (%)	18.4	18.3	18.1	18.3	18.3	18.3	18.2	0.69	0.927
iso-Butyric acid (%)	1.16b	1.16b	1.10a	1.19b	1.18b	1.15b	1.20b	0.099	0.002
Butyric acid (%)	13.3	13.1	13.2	13.4	13.5	13.4	13.4	0.95	0.822
iso-Valeric acid (%)	2.37b	2.35ab	2.27a	2.43b	2.41b	2.35a	2.44b	0.303	0.017
Valeric acid (%)	1.72a	1.77abc	1.72ab	1.80c	1.76abc	1.78bc	1.82c	0.137	0.009
iso-Caproic acid (%)	0.0119	0.0118	0.0099	0.0133	0.0128	0.0132	0.0121	0.001159	0.377
Caproic acid (%)	0.523	0.529	0.548	0.501	0.530	0.543	0.521	0.0728	0.502
Enanthic acid (%)	0.0339	0.0351	0.0359	0.0307	0.0327	0.0352	0.0347	0.01051	0.497

	Broad faba					Horse faba			
	Cekin	Belshi	Aguadulce	Jögeva	Skrapari	Bauska	Lielplatones	SEM	P
Acetic: propionic ratio	3.43	3.45	3.53	3.45	3.43	3.44	3.46	0.193	0.795
H_2 generated (µmol/mL)	128	131	134	125	132	131	126	7.1	0.201
H_2 consumed (µmol/mL)	41.7	42.4	43.2	40.6	43.4	43.2	41.3	3.16	0.412
Recovery (%)	32.4	32.2	32.1	32.4	32.7	32.5	32.5	1.35	0.699
Fermentation efficiency (%)	73.4	73.3	73.2	73.4	73.4	73.4	73.3	0.46	0.858

DM, dry matter; SEM, standard error of the mean.

pH was higher (6.21) for the incubation of faba hulls when compared to faba seeds (6.02; Table 2, $P < 0.001$). Fermentation pH was similar among faba cultivars after 24-h *in vitro* incubations (Table 3, $P = 0.590$).

Ammonia-N content and total volatile fatty acids production were respectively 32% and 24% higher with faba seeds than with faba hulls ($P < 0.001$, Table 2). Cultivars of faba did not affect these parameters ($P > 0.05$, Table 3).

Faba fraction greatly affected the proportion of individual volatile fatty acids ($P < 0.001$), except that of propionic acid ($P = 0.331$, Table 2). The acetic acid and *iso*-caproic acid proportions and the acetic acid to propionic acid ratio were higher with faba hulls than with faba seeds ($P < 0.001$), whereas the proportions of *iso*-butyric acid, butyric acid, *iso*-valeric acid, valeric acid, caproic acid, and enanthic acid followed the opposite trend ($P < 0.05$, Table 2). The faba cultivar only affected the proportions of the minor volatile fatty acids: *iso*-butyric acid, *iso*-valeric acid, and valeric acid ($P < 0.05$, Table 3). An interaction between the faba fraction and faba cultivar was also observed for *iso*-butyric acid ($P = 0.030$, Figure 1). The lowest *iso*-butyric acid proportion was recorded for the incubation of Aguadulce hull and the highest proportion for Jögeva and Lielplatones seeds.

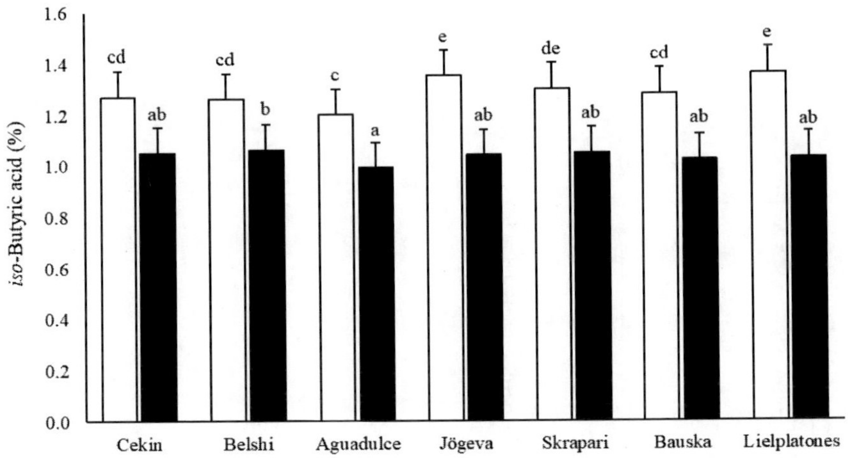

Figure 1. Effects of faba fraction and faba cultivars interaction on *iso*-butyric acid (%). Hull (black bar), seed (white bar). Mean values with different superscript letters (a, b, c, d, e) were significantly different ($P < 0.05$).

Hydrogen generated and hydrogen consumed during incubation, the fermentation efficiency, and the recovery were only affected by faba fraction. Seeds promoted the highest production and consumption of hydrogen, and hulls were the lowest ($P < 0.001$, Table 2). The fermentation efficiency and recovery reached their maximum with seeds (73.8% and 33.9%) and the minimum with hulls (72.9% and 30.9%, $P < 0.001$, Table 2).

Case Study Significance and Main Findings

In the last decades, there has been a revived interest in faba beans as an animal feed. This interest is motivated by the possibility of having a potential replacement for soybeans that can be grown in Europe, thus promoting the sustainability of local farming systems. The great genetic variability with faba beans unveiled by European breeding studies has been used to promote both seed composition and yield (Duc, 1997). Faba beans, like most legumes, are mainly consumed after removing the hulls through a simple industrial process. Hulls represent around 75% of the total faba bean yield production in the field, independently of plant density and genotype (Renna et al., 2020), thus creating an economic and environmental problem due to the large amount produced and elimination costs (Duc, 1997). At the same time, the disposal of hulls represents a waste of their nutritive value and agri-food waste. To promote a circular economy, this case study assessed the potential of hulls and seeds from seven faba bean cultivars as alternative dietary resources for ruminants by the evaluation of their composition and impact on *in vitro* rumen fermentation parameters.

Faba hulls chemical composition greatly differed from that of faba seeds. Starch was the main component of faba seeds (average 402 g/kg DM), followed by protein (average 324 g CP/kg DM), which agrees with the values found in the literature (Barratt, 1982; Crépon et al., 2010). Despite the protein content being lower than soybean meal, European faba seeds cultivars herein evaluated are valuable alternatives to imported protein sources used in ruminant diets or to balance other feed sources low in protein. Indeed, proteins from raw faba seeds are very soluble and highly degradable in the rumen (Crépon et al., 2010). However, heat treatment can reduce the degradability of faba proteins in the rumen, as described in soybean meal (Infascelli et al., 1995), thus shifting the effective protein degradation to the intestine and reducing the ruminal N losses (Crépon et al., 2010; Yu et al., 1998). An inverse relation between CP and carbohydrates was observed by Barratt (1982) in 30

cultivars or lines of *V. faba* and by Duc (1997) in 74 faba bean genotypes. In our study, no relation was found ($P > 0.05$; data not shown).

The gross energy of faba seeds averaged 18.5 MJ/kg DM, which is aligned with other reports (Yáñez-Ruiz et al., 2009) and similar to the gross energy of grain cereals, including maize, wheat, and barley (Bolarinwa & Adeola, 2012). Due to their low-fat content, the energy value of faba seeds depends on starch and protein contents and digestibility (Wiseman & Cole, 1988). Starch from faba seeds is known to be highly digestible (although slowly digested) in all animal species (FEDNA, 2010).

Due to their low content of lignin, fiber from faba seeds is considered to be easily digestible by the rumen microbial population (Gdala, 1998). However, the average ADL content (22.1 g/kg DM) of the studied seeds was greater than that reported in the literature (9.8-15 g/kg, DM basis; FEDNA, 2010; INRA, 2010), particularly of the horse bean Bauska and Lielplatones. Conversely, the average NDF content of seeds was similar to that earlier reported (FEDNA, 2010; INRA, 2010).

Raw faba seeds contain several secondary metabolites such as phytic acid, saponins, lectins, anti-trypsin factors, vicine, convicine, and condensed tannins that can exert anti-nutritional effects (Multari et al., 2015; Valente et al., 2019). These compounds are associated with low digestibility and may promote some pathological conditions (FEDNA, 2010; Gupta, 1987). Favism, an acute hemolytic disease caused by the consumption of faba beans by humans with glucose-6-phosphate-dehydrogenase deficiency, is the most common pathology (Luzzatto & Arese, 2018). Tannins are the most studied of these anti-nutritional compounds. Hydrolyzable tannins are mainly found in fruit pods and rarely in forages, while condensed tannins are common in trees, shrubs, and forage grain legumes (Min et al., 2003). The tannin content of faba bean seeds has been reported to be similar to that of soybean seeds (El-Shemy et al., 2000). In our case study, condensed tannins of faba seeds ranged from 0.275 g/kg DM in Cekin to 1.36 g/kg DM in Lielplatones, within the known broad range of condensed tannins in faba seeds, from 0.1-10.4 g/kg DM (Duc et al., 1999). The condensed tannins content of faba seed hulls from colored flower cultivars (average 6 g/kg) is higher than those of white flower cultivars (average 0.6 g/kg; FEDNA, 2010). This difference relies on two independent genes, zt1 (V. faba L. var. major origin) and zt2 (V. faba L. var. minor origin), that reduce the tannin content in the seeds (Picard, 1976) and the proportion of seed hulls (Duc et al., 1999). Tannins can be markedly reduced or even eliminated from seeds by soaking, germination, and extrusion (the cooking process under high pressure), among other processes (Alonso et al., 2000; Liu

et al., 2020). In addition, dehulling can also eliminate tannins as these compounds are concentrated in the seed hulls (FEDNA, 2010).

Information on faba hulls in the literature is scarce, but arabinans, galactans, and/or arabinogalactans have been reported to replace cell walls methylated rhamnogalacturonan regions (Mateos-Aparicio et al., 2010a). Faba hulls are undervalued and underexploited agro-food residues rich in dietary fiber and extractable polyphenols (greater than in other legumes such as chickpeas, yellow and black soybeans, yellow and green peas, black beans, and lentils) with free-radical scavenging activity and reducing power (Mateos-Aparicio et al., 2012; Renna et al., 2020; Valente et al., 2018). Therefore, faba hulls can be considered an abundant, inexpensive, and sustainable source of fiber and functional compounds (Mateos-Aparicio et al., 2010b; Renna et al., 2020).

After 24-h incubation, fermentation pH varied from 6.02 to 6.21, being in the range considered suitable for fibrinolytic enzyme activity (Yang et al., 2002). Compared with hulls, faba seeds were more extensively fermented, most probably due to the higher starch and lower fiber contents than those of hulls.

The higher fiber content of faba hulls compared to seeds promoted the acetic acid proportion and acetic acid to propionic acid ratio. However, the greater starch content of faba seeds failed to promote the propionic acid proportion compared to hulls. The similar propionic acid proportion observed may be due to the differences in the activity and/or content of the amylolytic bacteria of the inocula used. Indeed, faba beans were incubated with maize silage-based inocula expected to have high amylolytic activity, and with haylage-based inocula, where this amylolytic activity was expected to be low. This result suggests an effect of the inocula on the fermentation extension and also on the end-products profile of faba hulls and seeds, which should be further evaluated. Increased propionic acid production is of interest to the ruminant as it may increase glucose synthesis and acetate use efficiency, thus improving the nitrogen and energy balance (Blaxter, 1989).

The higher protein content of faba seeds promoted the proportions of *iso*-butyric acid, *iso*-valeric acid, and valeric acid, which probably resulted from protein degradation (El-Shazly, 1952). This is a key aspect for optimal fibrolytic activity (Wallace & Cotta, 1988), as ruminal cellulolytic bacteria require these branched- and odd-chain volatile fatty acids for growth and cellulose fermentation (Dehority et al., 1967; Muller, 1987). When compared to seeds, faba hulls presented reduced ammonia-N content, yet the value found was higher than the minimum proposed for optimal microbial activity in the

rumen (20-50 mg/L) (20-50 mg/L; Dewhurst & Newbold, 2022; Slyter et al., 1979).

Seeds were more extensively fermented than hulls, yet faba hulls were more efficient in decreasing enteric methane production. Therefore, faba hulls are suggested to be an affordable alternative dietary fiber source with the potential to tackle enteric methane emissions and, consequently, to reduce the environmental burden of ruminant production systems. In face of its lower fermentability, faba hulls may be of particular interest in the dietary regimens of animals with lower nutrient requirements and higher methane emissions. Additionally, faba hulls can constitute a source of functional compounds due to their contents of insoluble fiber and total phenolic compounds, the latter being shown to be correlated with antioxidant activity (Valente et al., 2018).

Broad and horse bean cultivars differently affected the gas and methane productions, as well as the proportions of *iso*-butyric acid, *iso*-valeric acid, and valeric acid (reflecting the different protein content). No effects were observed in pH, total volatile fatty acids production, and ammonia-N concentration, major volatile fatty acids (acetic, propionic, and butyric) proportions, acetic acid to propionic acid ratio, hydrogen generated and produced, fermentation efficiency, and recovery percentage between cultivars. Methane production was the lowest with Cekin, being similar to Belshi and Jogëva cultivars. The differences in methane production suggest that some European cultivars could be used as alternative feed sources with methane mitigation potential but without impairing the rumen fermentation, which is crucial for the ruminant welfare and health.

Condensed tannin content differed between hulls and seeds and also among cultivars. The lowest condensed tannins value found in Cekin and Belshi failed to promote the degradability of these cultivars, probably because the difference in condensed tannins content was insufficient to affect the rumen microbial activity. The absence of the tannins' effect is also supported by the effects on methane production. Although some tannins (depending on structure and concentration) can have beneficial effects in reducing enteric methane production (Beauchemin et al., 2008), the cultivar with the lowest condensed tannins (Cekin) led to the lowest methane production. Conversely, the cultivar with the highest tannin content (Lielplatones) promoted the highest methane production, not differing from that of Aguadulce and Skrapari with intermediate tannin content. Although not evaluated in the present study, the reviews from McMahon et al. (2000) and Min et al. (2003) state that condensed tannins of 20-45 g/kg DM (higher than the values herein reported) reduce ruminal protein degradation. This effect is due to a reversible binding

of tannins to proteins and a reduction of proteolytic rumen bacteria, but not the subsequent post-ruminal digestion of protein.

Conclusion

The quest for alternative protein sources has renewed the interest in faba bean seeds, with their agro-food by-product, faba bean hulls, emerging as a value-added alternative and underexploited dietary resource. The seven cultivars of European broad and horse faba beans differed in their chemical composition, yet the main constituent of hulls was fiber, and those of seeds were starch and protein. All faba hulls have been shown to be alternative sources of dietary fiber, and faba seeds may partially replace or complement the current protein sources available for ruminants. Faba hulls were less extensively fermented than seeds yet efficiently decreased methane production. Differences in the chemical composition of faba beans were only reflected in gas and methane productions and *iso*-butyric acid, *iso*-valeric acid, and valeric acid proportions. Cekin cultivar, along with Belshi and Jogëva promoted the lowest methane production, suggesting to be effective feed sources for methanogenesis mitigation without impairing rumen fermentation. Overall, hulls and seeds of the European cultivars of faba beans are alternative dietary resources that may play a key role in more sustainable and green animal production systems.

Funding

This case study was financially supported by Portuguese national funds (FCT/MCTES, Fundação para a Ciência e Tecnologia and Ministério da Ciência, Tecnologia e Ensino Superior) through the project UIDB/50006/2020 | UIDP/50006/2020. M.R.G.M. and I.M.V. thank FCT for funding through program DL 57/2016 – Norma transitória (SFRH/BPD/70176/2010 and SFRH/BPD/111181/2015). N.M. PhD grant was supported by the European Commission under the Erasmus Mundus project Green-Tech-WB:Smart and Green technologies for innovative and sustainable societies in Western Balkans (551984-EM-1-2014-1-ES-ERA MUNDUS-EMA2).

Acknowledgements

Authors greatly acknowledge Dr. Astrit Balliu, from the Agricultural University of Tirana, for offering the faba samples, and to Sílvia Azevedo, from School of Medicine and Biomedical Sciences, University of Porto, for the valuable technical assistance.

References

Aguilera, J. F., Bustos, M., & Molina, E. (1992). The degradability of legume seed meals in the rumen: effect of heat treatment. *Animal Feed Science and Technology, 36*(1–2), 101-112.

Alonso, R., Aguirre, A., & Marzo, F. (2000). Effects of extrusion and traditional processing methods on antinutrients and *in vitro* digestibility of protein and starch in faba and kidney beans. *Food Chemistry, 68*(2), 159-165.

Barratt, D. H. P. (1982). Chemical composition of mature seeds from different cultivars and lines of *Vicia faba* L. *Journal of the Science of Food and Agriculture, 33*(7), 603-608.

Beauchemin, K. A., Kreuzer, M., O'Mara, F., & McAllister, T. A. (2008). Nutritional management for enteric methane abatement: a review. *Australian Journal of Experimental Agriculture, 48*(2), 21-27.

Bjerg, B., Kbmeyer, E., Eggum, B. O., Larsen, T., Röbbelen, G., & Sørensen, H. (1988). The nutritive value of ten inbred lines of faba beans (*Vicia faba* L.) in relation to their content of antinutritional constituents and protein quality. *Plant Breeding, 101*(4), 277-291.

Blaxter, K. L. (1989). *Energy metabolism in animals and man*. Cambridge: Cambridge University Press.

Bolarinwa, O. A., & Adeola, O. (2012). Energy value of wheat, barley, and wheat dried distillers grains with solubles for broiler chickens determined using the regression method. *Poultry Science, 91*(8), 1928-1935.

Chalupa, W. (1977). Manipulating rumen fermentation. *Journal of Animal Science, 45*(3), 585-599.

European Commission (2018). *Report from the commission to the council and the European Parliament on the development of plant proteins in the European Union*. Brussels, Belgium.

Crépon, K., Marget, P., Peyronnet, C., Carrouée, B., Arese, P., & Duc, G. (2010). Nutritional value of faba bean (*Vicia faba* L.) seeds for feed and food. *Field Crops Research, 115*(3), 329-339.

Dehority, B. A., Scott, H. W., & Kowaluk, P. (1967). Volatile fatty acid requirements of cellulolytic rumen bacteria. *Journal of Bacteriology, 94*(3), 537-543.

Demeyer, D. I. (1991). Quantitative aspects of microbial metabolism in the rumen and hindgut. In J. P. Jouany (Ed.), *Rumen Microbial Metabolism and Ruminant Digestion* (pp. 213-237). Paris, France: INRA Editions.

Dewhurst, R. J., & Newbold, J. R. (2022). Effect of ammonia concentration on rumen microbial protein production *in vitro*. *British Journal of Nutrition, 127*(6), 847-849.

Duc, G. (1997). Faba bean (*Vicia faba* L.). *Field Crops Research, 53*(1–3), 99-109.

Duc, G., Moussy, F., Zong, X., & Ding, G. (1999). Single gene mutation for green cotyledons as a marker for the embryonic genotype in faba bean, *Vicia faba*. *Plant Breeding, 118*(6), 577-578.

Dvořák, R., Pechová, A., Pavlata, L., Klejdus, B., Kovařčík, K., Dostálová, J., Culková, J., Filípek, J., Švajdlenka, E., & Čapková, V. (2006). Reduction in the content of antinutritional substances in faba beans (*Vicia faba*) by different treatments. *Slovenian Veterinary Research, 43*(Suppl), 174-179.

El-Shazly, K. (1952). Degradation of Protein in the Rumen of the Sheep. 1. Some volatile fatty acids, including branched-chain isomers, were found *in vivo*. *Biochemical Journal, 51*, 640-641.

El-Shemy, H., Abdel-Rahim, E., Shaban, O., Ragab, A., Carnovale, E., & Fujita, K. (2000). Comparison of nutritional and antinutritional factors in soybean and faba bean seeds with or without cortex. *Soil Science and Plant Nutrition, 46*(2), 515-524.

FAOSTAT. (2021). Food and Agriculture Organization of the United States, Statistics Division.

FEDNA. (2010). *Tablas FEDNA de composición y valor nutritivo de alimentos para la fabricación de piensos compuestos* (C. Blas, G. G. Mateos, & P. García-Rebollar Eds. 3ª edición ed.). Madrid: Fundación Española para el Desarrollo de la Nutrición Animal.

Gdala, J. (1998). Composition, properties, and nutritive value of dietary fibre of legume seeds. A review. *Journal of Animal and Feed Sciences, 7*(2), 131-149.

Goelema, J. O., Spreeuwenberg, M. A. M., Hof, G., van der Poel, A. F. B., & Tamminga, S. (1998). Effect of pressure toasting on the rumen degradability and intestinal digestibility of whole and broken peas, lupins and faba beans and a mixture of these feedstuffs. *Animal Feed Science and Technology, 76*(1–2), 35-50.

Grela, E. R., & Günter, K. D. (1995). Fatty acid composition and tocopherol content of some legume seeds. *Animal Feed Science and Technology, 52*(3), 325-331.

Gupta, Y. P. (1987). Anti-nutritional and toxic factors in food legumes: A review. *Plant Foods for Human Nutrition, 37*(3), 201-228.

Infascelli, F., Di Lella, T., & Piccolo, V. (1995). Dry matter, organic matter and crude protein degradability of high protein feed in buffaloes and sheep. *Zootecnica e Nutrizone Animale, 21*, 89-94.

INRA. (2010). *Alimentation des bovins, ovins et caprins. Besoins des animaux - Valeurs des aliments*. Versailles, France.

Jansman, A. J. M. (1993). Tannins in feedstuffs for simple-stomached animals. *Nutrition Research Reviews, 6*(1), 209-236.

Köpke, U., & Nemecek, T. (2010). Ecological services of faba bean. *Field Crops Research, 115*(3), 217-233.

Liponi, G. B., Casini, L., De Vincenzi, S., & Gatta, D. (2009). Digestibility and nitrogen balance of diets based on faba bean, pea seeds and soybean meal in sheep. *Italian Journal of Animal Science, 8*(Suppl. 2), 353.

Liu, Y., Ragaee, S., Marcone, M. F., & Abdel-Aal, E.-S. M. (2020). Composition of phenolic acids and antioxidant properties of selected pulses cooked with different heating conditions. *Foods, 9*(7), 908.

Lopez, S., & Newbold, C. J. (2007). Analysis of methane. In H. P. S. Makkar & P. E. Vercoe (Eds.), *Measuring methane production from ruminants* (pp. 1-13): Springer Netherlands.

Luzzatto, L., & Arese, P. (2018). Favism and glucose-6-phosphate dehydrogenase deficiency. *New England Journal of Medicine, 378*(1), 60-71.

Maia, M. R. G., Fonseca, A. J. M., Cortez, P. P., & Cabrita, A. R. J. (2019). *In vitro* evaluation of macroalgae as unconventional ingredients in ruminant animal feeds. *Algal Research, 40*, 101481.

Maia, M. R. G., Fonseca, A. J. M., Oliveira, H. M., Mendonça, C., & Cabrita, A. R. J. (2016). The potential role of seaweeds in the natural manipulation of rumen fermentation and methane production. *Scientific Reports, 6*, 32321.

Makkar, H. P. S. (2000). *Quantification of tannins in tree foliage. A laboratory manual*. Vienna, Austria.

Mateos-Aparicio, I., Redondo-Cuenca, A., & Villanueva-Suárez, M.-J. (2010a). Isolation and characterisation of cell wall polysaccharides from legume by-products: Okara (soymilk residue), pea pod and broad bean pod. *Food Chemistry, 122*(1), 339-345.

Mateos-Aparicio, I., Redondo-Cuenca, A., & Villanueva-Suárez, M.-J. (2012). Broad bean and pea by-products as sources of fibre-rich ingredients: potential antioxidant activity measured *in vitro*. *Journal of the Science of Food and Agriculture, 92*(3), 697-703.

Mateos-Aparicio, I., Redondo-Cuenca, A., Villanueva-Suárez, M.-J., Zapata-Revilla, M.-A., & Tenorio-Sanz, M.-D. (2010b). Pea pod, broad bean pod and okara are potential sources of functional compounds. *LWT - Food Science and Technology, 43*(9), 1467-1470.

McMahon, L. R., McAllister, T. A., Berg, B. P., Majak, W., Acharya, S. N., Popp, J. D., Coulman, B. E., Wang, Y., & Cheng, K.-J. (2000). A review of the effects of forage condensed tannins on ruminal fermentation and bloat in grazing cattle. *Canadian Journal of Plant Science, 80* 469-485.

Melicharová, V., Pechová, A., Dvořák, R., & Pavlata, L. (2009). Performance and metabolism of dairy cows fed bean seeds (*Vicia faba*) with different levels of anti-nutritional substances. *Acta Veterinaria Brno, 78*(1), 57-66.

Min, B. R., Barry, T. N., Attwood, G. T., & McNabb, W. C. (2003). The effect of condensed tannins on the nutrition and health of ruminants fed fresh temperate forages: a review. *Animal Feed Science and Technology, 106*(1–4), 3-19.

Mordenti, A. L., Merendi, F., Fustini, M., & Formigoni, A. (2007). Effects of different protein plants in cows diet on milk for Parmigiano Reggiano production. *Italian Journal of Animal Science, 6*(Suppl. 1), 463-465. doi:10.4081/ijas.2007.1s.463

Mould, F. L., Kliem, K. E., Morgan, R., & Mauricio, R. M. (2005). *In vitro* microbial inoculum: A review of its function and properties. *Animal Feed Science and Technology, 123–124, Part 1*, 31-50.

Muller, L. D. (1987). Branched chain fatty acids (isoacids) and valeric acid for ruminants. *The Professional Animal Scientist, 3*(1), 9-12.

Multari, S., Stewart, D., & Russell, W. R. (2015). Potential of fava bean as future protein supply to partially replace meat intake in the human diet. *Comprehensive Reviews in Food Science and Food Safety, 14*(5), 511-522.

Palander, S., Laurinen, P., Perttilä, S., Valaja, J., & Partanen, K. (2006). Protein and amino acid digestibility and metabolizable energy value of pea (*Pisum sativum*), faba bean (*Vicia faba*) and lupin (*Lupinus angustifolius*) seeds for turkeys of different age. *Animal Feed Science and Technology, 127*(1), 89-100.

Picard, J. (1976). Aperçu sur l'hérédité du caractère absence de tanins dans les graines de féverole (*Vicia faba* L.). *Annales de l'Amelioration des Plantes, 26*, 101-106.

Porter, L. J., Hrstich, L. N., & Chan, B. G. (1985). The conversion of procyanidins and prodelphinidins to cyanidin and delphinidin. *Phytochemistry, 25*(1), 223-230.

Rawal, V., & Navarro, D. K. (2019). *The global economy of pulses*. Rome: FAO.

Renna, M., De Cillis, F., Leoni, B., Acciardi, E., & Santamaria, P. (2020). From by-product to unconventional vegetable: Preliminary evaluation of fresh fava hulls highlights richness in L-dopa and low content of anti-nutritional factor. *Foods, 9*(2), 159.

Slyter, L. L., Satter, L. D., & Dinius, D. A. (1979). Effect of ruminal ammonia concentration on nitrogen utilization by steers. *Journal of Animal Science, 48*(4), 906-912.

Taheri, F., Azadi, H., & D'Haese, M. (2017). A world without hunger: Organic or GM crops? *Sustainability, 9*(4), 580.

Theodorou, M. K., Williams, B. A., Dhanoa, M. S., McAllan, A. B., & France, J. (1994). A simple gas production method using a pressure transducer to determine the fermentation kinetics of ruminant feeds. *Animal Feed Science and Technology, 48*(3–4), 185-197.

Valente, I. M., Cabrita, A. R. J., Malushi, N., Oliveira, H. M., Papa, L., Rodrigues, J. A., Fonseca, A. J. M., & Maia, M. R. G. (2019). Unravelling the phytonutrients and antioxidant properties of European *Vicia faba* L. seeds. *Food Research International, 116*, 888-896.

Valente, I. M., Maia, M. R. G., Malushi, N., Oliveira, H. M., Papa, L., Rodrigues, J. A., Fonseca, A. J. M., & Cabrita, A. R. J. (2018). Profiling of phenolic compounds and antioxidant properties of European varieties and cultivars of *Vicia faba* L. pods. *Phytochemistry, 152*, 223-229.

Wallace, J., & Cotta, M. A. (1988). Metabolism of nitrogen containing compounds. In P. N. Hobson (Ed.), *The rumen microbial ecosystem* (pp. 217-249). London: Elsevier Applied Science.

Wiseman, J., & Cole, D. J. A. (1988). European legumes in diets for non-ruminants. In W. Haresign & D. J. A. Cole (Eds.), *Recent Advances in Animal Nutrition* (pp. 13–37). London: Butterworths.

Yáñez-Ruiz, D. R., Martín-García, A. I., Weisbjerg, M. R., Hvelplund, T., & Molina-Alcaide, E. (2009). A comparison of different legume seeds as protein supplement to optimise the use of low quality forages by ruminants. *Archives of Animal Nutrition, 63*(1), 39-55.

Yang, W. Z., Beauchemin, K. A., & Vedres, D. D. (2002). Effects of pH and fibrolytic enzymes on digestibility, bacterial protein synthesis, and fermentation in continuous culture. *Animal Feed Science and Technology, 102*(1), 137-150.

Yu, P., Holmes, J. H. G., Leury, B. J., & Egan, A. R. (1998). Influence of dry roasting on rumen protein degradation characteristics of whole faba bean (*Vicia faba*) in dairy cows. *Asian-Australasian Journal of Animal Sciences, 11*(1), 35-42.

About the Editors

Aamir Raina, PhD is an Assistant Professor at the Faculty of Life Sciences, Department of Botany, Aligarh Muslim University, Aligarh, UP, India. His current research interests focus on novel mutations induced by radiations in pulses, elucidation of physiological and molecular mechanisms in response to radiations and suitable germplasm donors for breeding purposes. Researching the mutagenesis of plants, Dr. Raina has found that radiation could help in developing elite mutants with traits such as high yield and resistance to stresses.

Dr. Raina has published over 30 journal articles, 4 books and 20 book chapters. He has received various research fellowships, including UGC Non-NET, CSIR JRF, DBT-SRF, GATE, and Nuffic OKP (The Netherlands), and awards, including the NESA Young Scientist Award and the Bharat Vikas Award. In 2018, he was awarded Young Scientist of the Year by the National Environmental Science Academy, New Delhi, India.

Samiullah Khan, PhD is a Professor and Head of Plant Breeding and Crop Genetics in the Department of Botany at Aligarh Muslim University, Aligarh, India. Dr. Khan obtained his PhD in 1990 from Aligarh Muslim University, specializing in Plant Breeding and Crop Genetics. He has 25 years of teaching experience in Embryology of Angiosperms, Genetics and Plant Breeding, and he has supervised several PhD, MPhil and MSc projects.

Dr. Khan has edited several books and published more than 100 research papers in journals of national and international repute. He has attended many conferences in India and abroad, including The XVI International Congress of Genetics in Toronto, Canada. Furthermore, he is a member of the editorial board of many scientific journals, and he is a fellow of the Indian Society of Genetics and Plant Breeding.

Index

D

E

F

G

H

I

L

M

N

O

P

R

S

T

U

V

W